T0302317

Microelectromechanical Systems — Materials and Devices

MATERIALS RESEARCH SOCIETY
SYMPOSIUM PROCEEDINGS VOLUME 1052

Microelectromechanical Systems — Materials and Devices

Symposium held November 26–28, 2007, Boston, Massachusetts, U.S.A.

EDITORS:

David A. LaVan
Yale University
New Haven, Connecticut, U.S.A.

Mark G. da Silva
Exponent Inc.
Natick, Massachusetts, U.S.A.

S. Mark Spearing
University of Southampton
Highfield, Southampton, United Kingdom

Srikar Vengallatore
McGill University
Montreal, Québec, Canada

Materials Research Society
Warrendale, Pennsylvania

CAMBRIDGE
UNIVERSITY PRESS

University Printing House, Cambridge CB2 8BS, United Kingdom

One Liberty Plaza, 20th Floor, New York, NY 10006, USA

477 Williamstown Road, Port Melbourne, VIC 3207, Australia

314-321, 3rd Floor, Plot 3, Splendor Forum, Jasola District Centre, New Delhi - 110025, India

79 Anson Road, #06-04/06, Singapore 079906

Cambridge University Press is part of the University of Cambridge.

It furthers the University's mission by disseminating knowledge in the pursuit of education, learning and research at the highest international levels of excellence.

www.cambridge.org
Information on this title: www.cambridge.org/9781558999909

Materials Research Society
506 Keystone Drive, Warrendale, PA 15086
http://www.mrs.org

First published 2009
First paperback edition 2012

Single article reprints from this publication are available through University Microfilms Inc., 300 North Zeeb Road, Ann Arbor, MI 48106

CODEN: MRSPDH

A catalogue record for this publication is available from the British Library

ISBN 978-1-558-99990-9 Hardback
ISBN 978-1-107-40858-6 Paperback

CONTENTS

*Invited Paper

POSTER SESSION

POSTER SESSION:
MEMS

MEMS MATERIALS AND PROCESSES I

MEMS MATERIALS AND PROCESSES II

SELECT PAPER FROM SYMPOSIUM N

PREFACE

This proceedings reports on the research presented in Symposium DD, "Microelectromechanical Systems—Materials and Devices," held November 26–28 at the 2007 MRS Fall Meeting in Boston, Massachusetts. This symposium was devoted to research involving MEMS materials and MEMS devices, including RF-MEMS, optical MEMS, MEMS metrology, tribology, materials characterization, and mechanical behavior, MEMS surfaces, surface modifications, and interfaces, MEMS reliability, packaging, and life assessment, MEMS modeling and software tools for materials integration, biocompatibility of MEMS materials and devices, new materials and fabrication methodologies for MEMS (including integration of nanostructured, nanocomposite, and biomimic materials with MEMS), microfluidics and nanofluidics, *in vivo* drug/gene/protein delivery, novel actuators, MEMS cell-based systems, MEMS neural interfaces, MEMS sensors and MEMS microengines and microfuel cells.

This symposium was organized as a continuation of the Materials Science of Microelectromechanical Systems (MEMS) Devices symposium first held during a 1998 Materials Research Society Meeting. In the past nine years, many sophisticated devices have been designed and studied. Many aspects of MEMS materials behaviors have also been characterized. However, there remain many basic questions about the relationship between process, properties and function for MEMS materials. Many experimental methods have been developed, but there is a lack of standardization that would allow comparison between laboratories and commercial vendors or the creation of materials specifications that would enable greater commercialization of MEMS.

This proceedings volume also includes one paper from Symposium N, "Materials, Integration and Technology for Monolithic Instruments II," which was organized by J. Theil, R. Thewes, D.S. Gardner, S. Miller, and P. Catrysse. Symposium N was a one-day invitation-only session also held at the 2007 MRS Fall Meeting. This paper, from the "Architecture and Integration" session, is included in the last section of this volume entitled "Select Paper from Symposium N."

The organizers would especially like to thank the invited speakers that brought depth and context to the symposium: Joerg Bagdahn, Ioannis Chasiotis, Maarten de Boer, Erik Deutsch, Hal Kahn, Scott Manalis, John Santini, William N. Sharpe Jr., and Brian Wardle. On behalf of all the participants, the organizers would also like to thank MEMS Industry Group, Silex Microsystems AB, Springer and MRS for providing financial support for the symposium.

David A. LaVan
Mark G. da Silva
S. Mark Spearing
Srikar Vengallatore

February 2008

MATERIALS RESEARCH SOCIETY SYMPOSIUM PROCEEDINGS

Volume 1024E —Combinatorial Methods for High-Throughput Materials Science, D.S. Ginley, M.J. Fasolka, A. Ludwig, M. Lippmaa, 2008, ISBN 978-1-60511-000-4
Volume 1025E —Nanoscale Phenomena in Functional Materials by Scanning Probe Microscopy, L. Degertekin, 2008, ISBN 978-1-60511-001-1
Volume 1026E —Quantitative Electron Microscopy for Materials Science, E. Snoeck, R. Dunin-Borkowski, J. Verbeeck, U. Dahmen, 2008, ISBN 978-1-60511-002-8
Volume 1027E —Materials in Transition—Insights from Synchrotron and Neutron Sources, C. Thompson, H.A. Dürr, M.F. Toney, D.Y. Noh, 2008, ISBN 978-1-60511-003-5
Volume 1029E —Interfaces in Organic and Molecular Electronics III, K.L. Kavanagh, 2008, ISBN 978-1-60511-005-9
Volume 1030E —Large-Area Processing and Patterning for Active Optical and Electronic Devices, V. Bulović, S. Coe-Sullivan, I.J. Kymissis, J. Rogers, M. Shtein, T. Someya, 2008, ISBN 978-1-60511-006-6
Volume 1031E —Nanostructured Solar Cells, A. Luque, A. Marti, 2008, ISBN 978-1-60511-007-3
Volume 1032E —Nanoscale Magnetic Materials and Applications, J-P. Wang, 2008, ISBN 978-1-60511-008-0
Volume 1033E —Spin-Injection and Spin-Transfer Devices, R. Allenspach, C.H. Back, B. Heinrich, 2008, ISBN 978-1-60511-009-7
Volume 1034E —Ferroelectrics, Multiferroics, and Magnetoelectrics, J.F. Scott, V. Gopalan, M. Okuyama, M. Bibes, 2008, ISBN 978-1-60511-010-3
Volume 1035E —Zinc Oxide and Related Materials—2007, D.P. Norton, C. Jagadish, I. Buyanova, G-C. Yi, 2008, ISBN 978-1-60511-011-0
Volume 1036E —Materials and Hyperintegration Challenges in Next-Generation Interconnect Technology, R. Geer, J.D. Meindl, R. Baskaran, P.M. Ajayan, E. Zschech, 2008, ISBN 978-1-60511-012-7
Volume 1037E —Materials, Integration, and Technology for Monolithic Instruments II, D. LaVan, 2008, ISBN 978-1-60511-013-4
Volume 1038— Nuclear Radiation Detection Materials, D.L. Perry, A. Burger, L. Franks, M. Schieber, 2008, ISBN 978-1-55899-985-5
Volume 1039— Diamond Electronics—Fundamentals to Applications II, R.B. Jackman, C. Nebel, R.J. Nemanich, M. Nesladek, 2008, ISBN 978-1-55899-986-2
Volume 1040E —Nitrides and Related Bulk Materials, R. Kniep, F.J. DiSalvo, R. Riedel, Z. Fisk, Y. Sugahara, 2008, ISBN 978-1-60511-014-1
Volume 1041E —Life-Cycle Analysis for New Energy Conversion and Storage Systems, V.M. Fthenakis, A.C. Dillon, N. Savage, 2008, ISBN 978-1-60511-015-8
Volume 1042E —Materials and Technology for Hydrogen Storage, G-A. Nazri, C. Ping, A. Rougier, A. Hosseinmardi, 2008, ISBN 978-1-60511-016-5
Volume 1043E —Materials Innovations for Next-Generation Nuclear Energy, R. Devanathan, R.W. Grimes, K. Yasuda, B.P. Uberuaga, C. Meis, 2008, ISBN 978-1-60511-017-2
Volume 1044 — Thermoelectric Power Generation, T.P. Hogan, J. Yang, R. Funahashi, T. Tritt, 2008, ISBN 978-1-55899-987-9
Volume 1045E —Materials Science of Water Purification—2007, J. Georgiadis, R.T. Cygan, M.M. Fidalgo de Cortalezzi, T.M. Mayer, 2008, ISBN 978-1-60511-018-9

MATERIALS RESEARCH SOCIETY SYMPOSIUM PROCEEDINGS

Volume 1046E —Forum on Materials Science and Engineering Education for 2020, L.M. Bartolo, K.C. Chen, M. Grant Norton, G.M. Zenner, 2008, ISBN 978-1-60511-019-6
Volume 1047 — Materials Issues in Art and Archaeology VIII, P. Vandiver, F. Casadio, B. McCarthy, R.H. Tykot, J.L. Ruvalcaba Sil, 2008, ISBN 978-1-55899-988-6
Volume 1048E—Bulk Metallic Glasses—2007, J. Schroers, R. Busch, N. Nishiyama, M. Li, 2008, ISBN 978-1-60511-020-2
Volume 1049 — Fundamentals of Nanoindentation and Nanotribology IV, E. Le Bourhis, D.J. Morris, M.L. Oyen, R. Schwaiger, T. Staedler, 2008, ISBN 978-1-55899-989-3
Volume 1050E —Magnetic Shape Memory Alloys, E. Quandt, L. Schultz, M. Wuttig, T. Kakeshita, 2008, ISBN 978-1-60511-021-9
Volume 1051E —Materials for New Security and Defense Applications, J.L. Lenhart, Y.A. Elabd, M. VanLandingham, N. Godfrey, 2008, ISBN 978-1-60511-022-6
Volume 1052 — Microelectromechanical Systems—Materials and Devices, D. LaVan, M.G. da Silva, S.M. Spearing, S. Vengallatore, 2008, ISBN 978-1-55899-990-9
Volume 1053E —Phonon Engineering—Theory and Applications, S.L. Shinde, Y.J. Ding, J. Khurgin, G.P. Srivastava, 2008, ISBN 978-1-60511-023-3
Volume 1054E —Synthesis and Surface Engineering of Three-Dimensional Nanostructures, O. Hayden, K. Nielsch, N. Kovtyukhova, F. Caruso, T. Veres, 2008, ISBN 978-1-60511-024-0
Volume 1055E —Excitons and Plasmon Resonances in Nanostructures, A.O. Govorov, Z.M. Wang, A.L. Rogach, H. Ruda, M. Brongersma, 2008, ISBN 978-1-60511-025-7
Volume 1056E —Nanophase and Nanocomposite Materials V, S. Komarneni, K. Kaneko, J.C. Parker, P. O'Brien, 2008, ISBN 978-1-60511-026-4
Volume 1057E —Nanotubes and Related Nanostructures, Y.K. Yap, 2008, ISBN 978-1-60511-027-1
Volume 1058E —Nanowires—Novel Assembly Concepts and Device Integration, T.S. Mayer, 2008, ISBN 978-1-60511-028-8
Volume 1059E —Nanoscale Pattern Formation, W.J. MoberlyChan, 2008, ISBN 978-1-60511-029-5
Volume 1060E —Bioinspired Polymer Gels and Networks, F. Horkay, N.A. Langrana, A.J. Ryan, J.D. Londono, 2008, ISBN 978-1-60511-030-1
Volume 1061E —Biomolecular and Biologically Inspired Interfaces and Assemblies, J.B.-H. Tok, 2008, ISBN 978-1-60511-031-8
Volume 1062E —Protein and Peptide Engineering for Therapeutic and Functional Materials, M. Yu, S-W. Lee, D. Woolfson, I. Yamashita, B. Simmons, 2008, ISBN 978-1-60511-032-5
Volume 1063E —Solids at the Biological Interface, V.L. Ferguson, J.X-J. Zhang, C. Stoldt, C.P. Frick, 2008, ISBN 978-1-60511-033-2
Volume 1064E —Quantum-Dot and Nanoparticle Bioconjugates—Tools for Sensing and Biomedical Imaging, J. Cheon, H. Mattoussi, C.M. Niemeyer, G. Strouse, 2008, ISBN 978-1-60511-034-9
Volume 1065E —Electroactive and Conductive Polymers and Carbon Nanotubes for Biomedical Applications, X.T. Cui, D. Hoffman-Kim, S. Luebben, C.E. Schmidt, 2008, ISBN 978-1-60511-035-6

Prior Materials Research Society Symposium Proceedings available by contacting Materials Research Society

Micromechanics I

Mater. Res. Soc. Symp. Proc. Vol. 1052 © 2008 Materials Research Society 1052-DD01-01

A Review of Tension Test Methods for Thin Films

William N. Sharpe, Jr.
Mechanical Engineering, Johns Hopkins University, 3400 North Charles Street, Baltimore, MD, 21218

ABSTRACT

Test methods for mechanical property measurement of the thin films used in MEMS have been developed and refined over the last decade. This brief review considers only tensile testing since that is the preferred method for measuring Young's modulus, strength, etc. for macroscale structural materials. There are basically two kinds of tensile specimens – framed specimens whose support strips are cut after mounting and semidetached specimens fastened to the substrate at one end. The loading and force measurement systems are similar and use commercial transducers. Strain is measured either by overall grip displacement or by digital imaging in most cases. Initial works are described and followed by descriptions of some recent applications. The paper concludes with a suggested test method that could be amenable to standardization.

INTRODUCTION

This paper is to a great extent an update of a presentation at the 2001 MRS Symposium [1]. It is not a comprehensive review; in fact, most of the references are from the past two years. They are chosen to present a 'state-of-the-art' view of the topic. Some readers may want a more thorough overview of the various approaches (not limited to tensile testing or to thin films) over the years. This author published such a review in 2001 [2], but it is now somewhat out of date. Hemker and Sharpe [3] have a recent overview that summarizes test methods and discusses results from a materials science viewpoint. Review articles by Haque and Saif [4] and Srikar and Spearing [5] will give the reader an appreciation for the innovation and sophistication of this area of experimental research.

The first issues in mechanical testing are how to manufacture a specimen and mount it in a testing machine; these become more challenging as the dimensions get smaller. A clear pattern of approaches has emerged for thin films. The tensile film specimen remains fastened at either one or both ends to the substrate upon which it has been deposited. In the first case, the substrate is fastened to one grip of the test machine and the free end was somehow fastened to the other grip. In the second case, the substrate ends are fastened to the test machine and temporary substrate supports are removed before testing. These will be referred to as 'semidetached' and 'framed' in this review.

The next and equally challenging issue is strain measurement. The first approach was obviously grip displacement measurement. Macroscale tensile testing avoids this because it is easy to apply resistance strain gages or clip gages in most cases – high temperature or other hostile environments being exceptions. There is a need to measure strain directly on the specimen, and interferometry was used early on. However, digital imaging has emerged as the predominant and preferred method; the images can be optical or otherwise such as from an AFM.

This paper is therefore organized around these two main themes – specimen mounting in test machines and strain measurement. Subheadings describe the various approaches in each case. This area of mechanical testing at the micron size scale is reaching a state of maturity that can and should lead to standardization. The paper concludes with a description of a preferred test method that may be appropriate.

SPECIMEN MOUNTING AND TEST MACHINES

One cannot simply pick up a thin-film specimen and put it in a test machine as is done with traditional macrospecimens. Such thin (not necessarily small) specimens are best handled when the substrates upon which they are deposited are attached to the test machine. These approaches fall into two general categories:

- framed specimens where both ends remain attached to the substrate. The intermediate portions of the substrate are removed to leave the free-standing tensile specimen attached to the test machine through its connection to the substrate.
- semidetached specimens where one end remains attached to the mounted substrate. A connection must be made to the free end of the specimen, but this is not very difficult .

This section presents these two approaches in turn. In each case, the original concept is referenced and then some recent applications are described to present the latest developments.

Framed specimens

Neugebauer was the first to use the framed specimen in a very clever approach back in 1960 [6]. He deposited gold films onto single crystals of rocksalt that had been cleaved to expose 100 planes and then highly polished. The films, which ranged in thickness from 50 nm to 1.5 μm, tended to contain large crystals with the same epitaxial orientation. This film/substrate was then cleaved to form a piece approximately 1 cm long and 1-2 mm wide. The ends of that piece were glued – film side down – to a special tensile apparatus. Then, the rocksalt between the gripped ends was dissolved with water to leave a completely freestanding tensile specimen. Force was measured from the current through a solenoid and strain from overall grip displacement. Numerous experiments gave a Young's modulus of ~50 GPa and an ultimate strength of ~260 MPa; these results are consistent with recent results. It is interesting to see the cleverness and the thoroughness of the presentation from researchers of that era.

It was another 30+ years before Read and Dally introduced "A New Method for Measuring the Constitutive Properties of Thin Films", which is the forerunner of the modern approach [7]. They deposited 2.2 μm thick Ti-Al-Ti films onto single crystal silicon wafers and then patterned and etched a set of four tensile specimens – each 0.25 mm wide by 1 mm long – into the center of a 6 mm by 8 mm rectangular frame. This produced 36 specimen sets on a 51 mm diameter wafer. The film side of the wafer was protected and matching smaller rectangles etched into the silicon wafer behind each frame. This left the four specimens suspended across a frame consisting of larger grip ends supported by side strips which were broken to leave the specimens completely free.

The frame was glued onto a loading system consisting of a micrometer translation stage, a spring for force measurement, and an eddy current transducer for overall displacement measurement. All four specimens were tested at the same time. This is an indeterminate system and requires that one assume the force is equally divided even though some of the specimens started yielding earlier than others.

The work of Read-Dally gave the author and colleagues the ideas for a similar system, but one testing only one specimen at a time and employing a linear air bearing to reduce friction. An innovation was the use of interferometry between reflective markers on the gage section to measure axial and lateral strain. This "New Technique for Measuring the Mechanical Properties of Thin Films" was published in 1997 [8]. Figure 1 shows a polysilicon specimen after the silicon substrate has been removed, and figure 2 shows a silicon nitride specimen that is ready for testing.

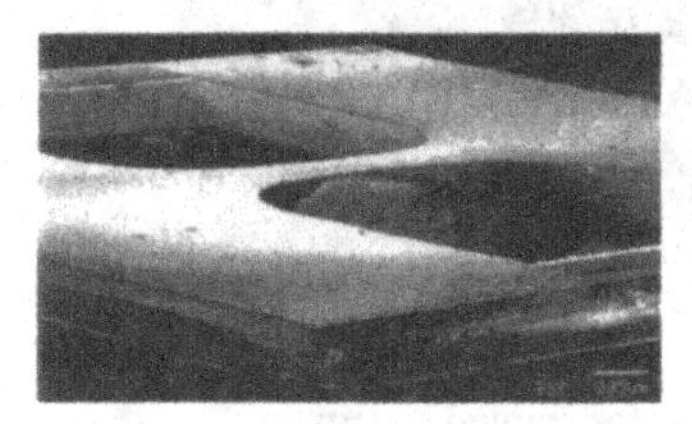

Figure 1. A framed polysilicon specimen.

Figure 2. A mounted silicon nitride specimen.

The silicon die in figure 1 is one centimeter square and 0.5 mm thick. Polysilicon is deposited onto it in the shape shown; its thickness is 3.5 μm. The grip ends are 3 mm by 10 mm rectangles, and the tensile specimen fairs into a 600 μm width at the center to avoid stress concentrations in this brittle material. As in Read-Dally, the front surface is protected and a rectangular window etched in the wafer from the back. It is easy to handle the die and mount it into a test machine where it is glued first with UV glue to allow final alignment and then with contact cement for strength. The support strips are then cut as shown in figure 2.

Figure 3 shows the current version of the test setup [9]. Gianola et al. used a similar system placed to make in situ X-ray measurements of diffraction peak broadening during deformation [10]. The linear air bearing supports the connection between the movable grip and the load cell. The five-axis stage enables alignment of the fixed grip to the movable one, and the capacitance gage measures the overall displacement of the grips to permit monitoring of the shape of the stress-strain curve during the test. The camera is for digital image processing – to be discussed later. Elongation is accomplished by one axis of the picomotor stage, and the test is of course run under computer control.

Isono et al. [11] use a variation on this theme with framed specimens of diamond-like-carbon (DLC) films as shown in figure 4. The 300 μm wide tensile sections ranged in thickness from 110 nm to 580 nm, and the frame was attached to the loading mechanism via hooks protruding up from the grips. Note that the side support strips are not cut, but three support strips around the free end are. This framed specimen is mounted on a compact tensile tester (see figure 5) consisting of a PZT actuator, linear variable differential transformer, and load cell. It is small enough to be positioned under an AFM head, which is used for full-field strain measurements to be discussed later.

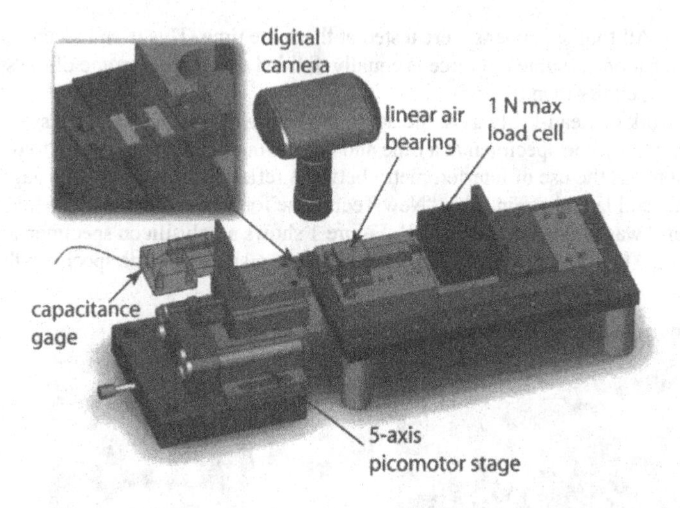

Figure 3. Schematic of a test system for framed specimens.

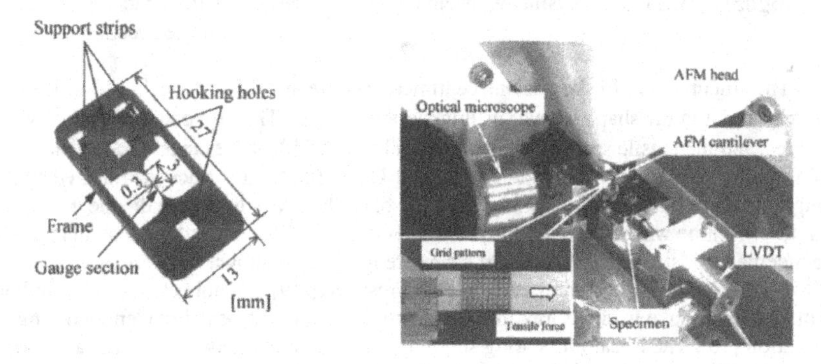

Figure 4. Framed DLC specimen.

Figure 5. Compact tensile tester under AFM.

Namazu (a co-author on the Isono paper) et al. used basically the same specimen shape and test setup to measure out-of-plane strain of gold-tin solder by X-ray diffraction [12]. The compact tensile tester was put under an X-ray machine instead of an AFM. This is an interesting innovation. Conceivably, one could test the same specimen in the elastic region by moving the compact tensile tester between two measuring systems and thereby gain an understanding of the anisotropy of films containing columnar grains.

The gold and gold-vanadium specimens tested by Ming-Tzer et al. [13] were basically the same configuration as those in figure 1 except that the grip ends were much larger and the specimens were straight-sided – 600 μm long, 100 μm wide, and 1 μm thick. The main difference here is that the load train was vertical, which not only eliminates the friction problem, but permits the test machine to be inserted in a resistance furnace for high temperature tests.

Though not thin films in the usual sense, Guo et al. strained 100 nm by 100 nm by 250 nm tensile specimens of metallic glass inside a transmission electron microscope [14]. First, small trenches approximately 150 nm wide were cut into both sides of a slice of the material to leave a 100 nm thick region between the thicker grip ends. Then, the little tensile specimens were formed by cutting rectangles on each side; all 'machining' was done in a focused ion beam microscope (FIB). The objective of this study was to observe the deformation inside a TEM, not to measure a stress-strain curve, so response of the frame material is not important. Metallic glasses are generally brittle, but these tests showed considerable ductility at this small size scale. As FIBs become more widely avalable, similar approaches may be used to prepare and test thin films.

Semidetached specimens

The concept of thin narrow tensile specimens fastened to the substrate at one end was introduced by introduced by Tsuchiya et al. [15] and Greek et al. [16]– both in 1997 publications that reported strength measurements of polysilicon and other thin films. The general concept has become widely used since then.

The shape of semidetached specimens is shown in figures 6 and 7 from the author's lab. The polysilicon structure is 3.5 μm thick and sits 2 μm above the silicon substrate except at the left end where it is fastened and at the four support strips around the large grip end at the right. These support strips are broken after the grip end is attached to the test machine or they can be broken by pushing on the grip end with the test machine while the gage section buckles. These specimens are released by liquid etching of the underlying sacrificial layer. A variation is to deposit the film directly on the silicon wafer and pattern the shape of the specimen as shown in figure 7. The silicon below the specimen is then removed with a gaseous etch. In both cases, etch holes must be included in the grip end and at the fairing to the fixed end to permit access of the etchant. A considerable advantage of this type of specimen is that many of them can be produced on a single die.

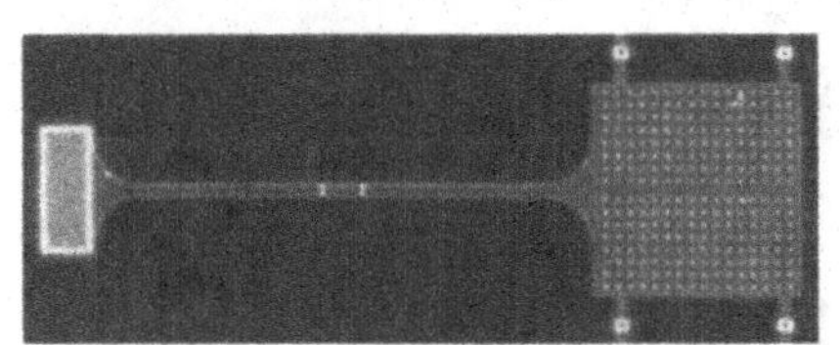

Figure 6. A polysilicon semidetached specimen. released by liquid etch.

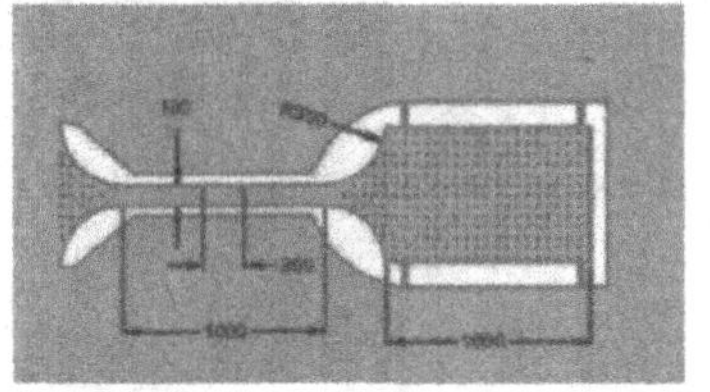

Figure 7. Drawing of a semidetached specimen for gaseous etch.

Gripping techniques are shown in figure 8. The left one is from Cho and Chasiotis who glue the glass grip to the grip end of the specimen with UV adhesive [17]. The advantage here is that one can align the system before curing the adhesive through the glass. Tsuchiya et al. use a similar system, but the probe has an insulating layer on the bottom [15]. The fixed end of the specimen is one end of the electrical circuit, and the voltage across the insulating layer generates enough friction force to pull the specimen. This permits rapid testing of specimens, but is limited to small cross-sections because higher voltages to enable larger friction cause arcing through the insulating layer.

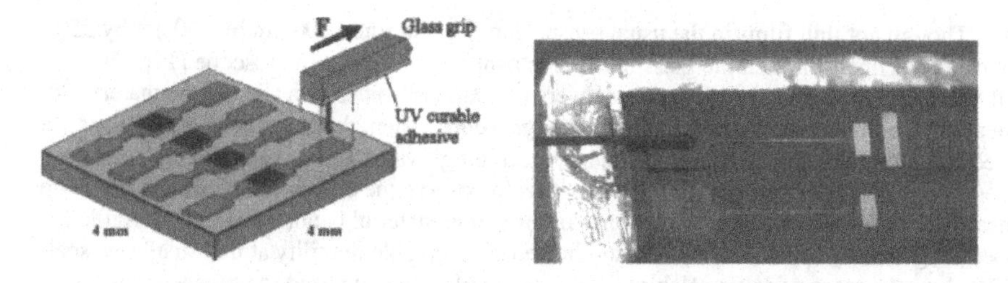

Figure 8. Gripping techniques for semidetached specimens.

The approach on the right glues a long 140 μm diameter silicon carbide fiber to the specimen with UV adhesive [18]. If a stronger bond is needed for larger cross-sections, a contact cement can be added after alignment is achieved.

A different and even easier approach is to fabricate the grip end as a ring which can be grabbed by a probe; see figure 9 from Boyce et al. [19]. The 'fixed' end actually allows the specimen to rotate for automatic alignment. The tip custom-made tungsten probe is held 0.5 μm above the substrate to avoid sliding friction as the specimen is pulled. It also permits current to pass through the gage section to heat it to as much as 800°C. This method has been used for strength measurements of polysilicon by Boyce et al. as well as by Boroch et al. who tested polysilicon specimens with holes and notches in the gage section. At least 38 specimens of each configuration were tested, which is a large enough sample for Weibull analysis.

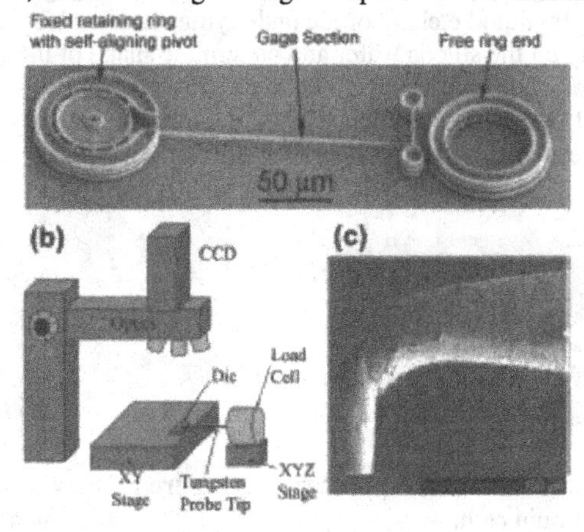

Figure 9. The ring grip approach showing the specimen, grip, and test setup.

The test system for these semidetached specimens is straightforward and also illustrated in figure 9. The die with multiple specimens is mounted on a stage to align them with the gripping device. This setup can be mounted horizontally or vertically depending upon whether the load cell is sensitive to off-axis forces.

STRAIN MEASUREMENT

It is easiest of course to determine strain from the overall displacement of the specimen grips, and this can work well if the elasticity of the loading mechanism is considered. Two markers on the specimen gage section remove this uncertainty. Full-field digital imaging has become widely used in just the past few years and appears to be the current preferred method.

Grip displacement

Neugebauer [6] measured the grip movement with a differential transformer, but this did not account for any deformation or slippage where the film was glued to the grips. Read and Dally [7] used the same approach, but sensed the displacement between an eddy current transducer and a vertical flag on the movable grip. Lin et al. [13] corrected the displacement of the piezoelectric actuator by accounting for the compliance of the load train.

Namazu et al. [12] used an elastic finite element analysis of their framed specimen to show that the uniform strain in the gage section was 68% of that determined from the overall displacement of the large grip ends. This does not account for elasticity of the load train. Gianola et al. [10] compared the strain from grip displacement to the full-field strain determined by digital image correlation to verify that it was giving accurate results. This direct comparison is preferred and allows one to use the easier real-time displacement signals instead of waiting for post-processing.

Greek and Johansson [16] tested semidetached specimens of different lengths and extracted Young's modulus from the difference of the slopes. This method is subject to errors arising from subtraction of similar numbers, but can give accurate results as demonstrated in Gianola and Sharpe [20].

Interferometry

It is always preferable to measure strain directly on the specimen, and Sharpe et al. showed that was possible in 1997 [8]. Figure 10 shows two thin (0.5 μm thick) gold lines on a 3.5 μm thick polysilicon tensile specimen; they serve as gage points for strain measurement. (Two markers are visible on the semidetached specimens in figures 6 and 8.) Four lines can be placed in a square array to permit measurement of axial and transverse strain.

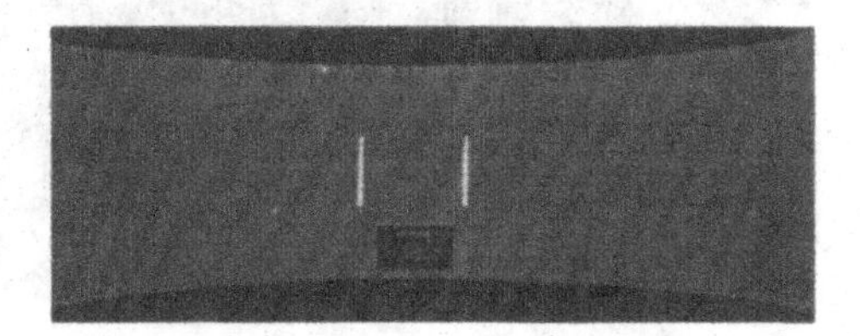

Figure 10. Two gold lines on a polysilicon specimen.

When they are illuminated with a laser (10 mw He-Ne lasers work well), interference fringes form at about 45° from diffraction at the individual markers and interference between the overlapping radiation. This is basically Young's optical interference phenomenon, but in reflection, not transmission. As the markers move due to strain, the fringes move. This motion can be monitored with linear diode arrays and converted to strain at the rate of 1-2 samples per second to enable real-time stress-strain curves to be displayed on the computer screen.

This 'strain gage' has been used to measure a variety of materials at room and elevated temperature over the years. However, it was recently discovered to be unusable on transparent silicon dioxide semidetached specimens. The incident laser beam goes through the specimen, and reflections from the diffuse silicon surface underneath overwhelm the fringe patterns.

Digital imaging

Similar gage points can be tracked by digital imaging as Tsuchiya et al. show [21]. Their markers on titanium films were almost identical to those of figure 10 and were 100 or 500 μm apart. Images were captured with a CCD camera and post-processed to generate the strain. This has the advantage of recording pictures of the specimen as it fails.

Strain in the transparent silicon dioxide specimens mentioned above were successfully measured by digital imaging as described in Sharpe et al. [18] where details are given. The markers were exactly the same as those in figure 10, and a strain resolution of ~ 100 microstrain was achieved over the 200 μm gage length. Comparisons between the interferometric method and digital imaging are discussed in that paper, but briefly:

- Interferometry has higher resolution (~ 10 microstrain), is real-time, measures strain at a single point, and can measure high-frequency strains.
- Digital imaging has lower resolution (~ 100 microstrain), requires post-processing, and enables full-field strain measurements as well as images of failure.

Most thin films are so smooth and reflective that there is not enough contrast for digital imaging. This can be remedied by depositing tiny particles on the specimen. Those in figure 11 are a ceramic adhesive powder and are simply sprinkled onto the specimen; some users blow them through a paper coffee filter. There is enough electrostatic attraction to hold the particles in place as the specimen is deformed. The image is from the 2007 thesis of Gianola [9], who used full-field digital image correlation (DIC) to study the deformation behavior of nanocrystalline aluminum films as thin as 100 nm. A schematic of the test system is shown in figure 3.

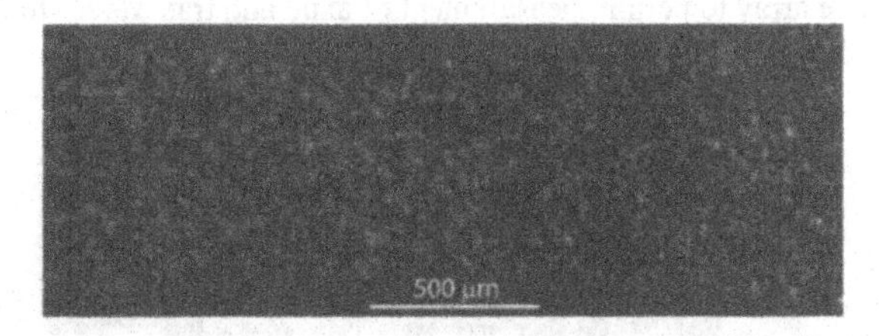

Figure 11. An aluminum film specimen with ceramic particles on it for digital imaging.

Read et al. were the first to use DIC of optical images on aluminum films in 2001 [22]. They concentrated on a 180 μm long region of the gage section and achieved a strain resolution of 55 microstrain over that length. The images do not have to be optical; AFM images often have sufficient contrast. Chasiotis introduced DIC of such images in 2002, and a more recent paper [17] presents a complete description of both the test system and the image processing. Localized strains on the order of 200 microstrain can be resolved. Isono et al. deposited a photoresist grid on the center of diamond-like carbon films to improve the resolution from AFM images [11]. The rectangular grid consisted of 20 μm wide lines spaced on 25 μm centers, which enabled strain resolutions of ~ 100 microstrain over a 130 μm gage length.

There are commercial DIC packages available, but Gianola, Eberl, and Thompson have developed a program that is posted at www.mathworks.com/matlabcentral/fileexchange (search for Eberl). Once the images (optical or otherwise) have been obtained, the user can elect to track two particles or markers, to compute full-field strain by DIC, or compute full-field strain by tracking many points, which is faster. It is currently being used at Johns Hopkins for a variety of thin-film tests.

CONCLUDING REMARKS

Da Silva and Bouwstra discuss the metrology of MEMS in a 2007 paper [23]; their paper includes an extensive list of references. They focus on mechanical material properties measurement, but the theme throughout the paper is a call for standardization of test methods for all properties – modulus, strength, CTE, creep, etc. The MEMS industry has grown to a point where accepted standards are needed. NIST has some activity in this direction [24], and SEMI has added MEMS to its coverage [25].

This paper has focused on only a very limited aspect of MEMS metrology – destructive tensile tests – and there is evidence that the methods are have matured. Most of the authors referenced here have considerable experience and have refined their techniques and procedures . Those that have tested polysilicon report Young's modulus values in the 160-165 GPa range.

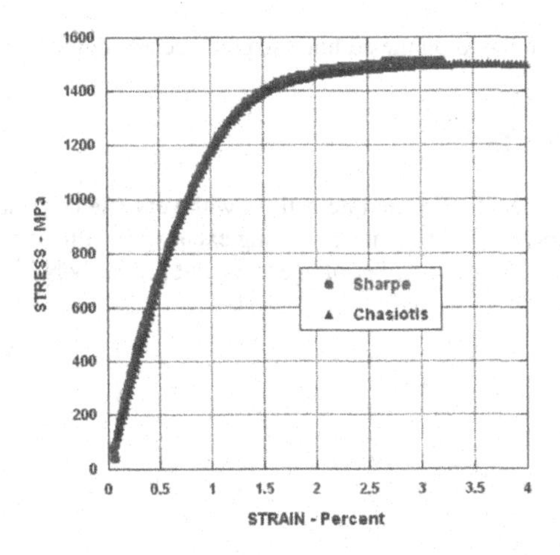

Figure 12. Stress-strain curves of platinum film tested in two labs.

An example of this maturity is shown in figure 12. Semidetached platinum film specimens 0.5 μm thick were provided by the Army Research Lab and tested at Johns Hopkins and in Chasiotis's lab at the University of Illinois. The results are nearly identical although the tests were in two different facilities.

Adoption of standards acceptable to an entire industry usually involves a 'round robin' test(s) to involve the entire community. The same material may be distributed to the participants who fabricate their own specimens or a single facility produces specimens of the participant's design with the same set of processes. The latter is the appropriate approach for MEMS and has been followed in two exercises. An AFOSR-sponsored round robin in 1998 [26] showed considerable variation in the polysilicon results from tensile and bending tests, but this was in the early days. Sandia sponsored a cross-comparison of strengths of polysilicon [27]in 2001 with much closer agreement except for one test method. A useful round robin would have to be sponsored by an organization with the facilities or financial support to provide specimens according to the designs of the participants.

The following is a preferred method for determining the mechanical properties of thins films at the MEMS size scale.

- The specimen shape would be semidetached. These are cheaper and easier to release than the framed specimens. Suitable force and displacement transducers are available commercially and can be easily calibrated or traced to standards. Further, the transducers can be separated from the specimen and grips to enable testing in hostile environments.

- Strain would be measured directly on the specimen by digital imaging of the gage section – either between two markers or over the full field. A displacement transducer or perhaps the output of the translation stage would be useful to monitor the overall shape of the stress-strain curve in real time.

This approach is chosen based on the author's experience and could be developed into an industry standard.

ACKNOWLEDGMENTS

The author has benefited from interactions with very talented undergraduate and graduate students over the years as well as from stimulating exchanges with other researchers in the field. Support has been provided by a number of agencies, but primarily NSF, DARPA, and the Army Research Laboratory.

REFERENCES

1. W. N. Sharpe Jr. in *Materials Science of Microelectromechanical Systems (MEMS) Devices IV,* (Mater. Res. Soc. **687**, Boston, MA, 2002) pp. 293-304.

2. Sharpe, William N. Jr., in *Mechanical Properties of MEMS Materials* (CRC Press, Boca Raton, FL, 2001), pp. 3-1 to 3-33.

3. K. J. Hemker and W. N. Sharpe Jr., *Annual Review of Materials Research* **37**, 92 (2007).

4. M. A. Haque and M. T. A. Saif, *Exp. Mech.* **43**, 248 (2003).

5. V. T. Srikar and S. M. Spearing, *Exp. Mech.* **43**, 238 (2003).

6. C. A. Neugebauer, *J. Appl. Phys.* **31**, 1096 (1960).

7. Read, D.T. and Dally, J.W., *Journal of Materials Research* **8**, 1542 (1992).

8. W. N. J. Sharpe, B. Yuan, and R. L. Edwards, *J. Microelectromech. Syst.* **6**, 193 (1997).

9. D. S. Gianola., thesis Johns Hopkins University (2007).

10. D. S. Gianola *et al.*, *Acta Materialia* **54**, 2253 (2006).

11. Y. Isono, T. Namazu, and N. Terayama, *J. Microelectromech. Syst.* **15**, 169 (2006).

12. T. Namazu, H. Takemoto, H. Fujita, and S. Inoue, *Science and Technology of Advanced Materials* **8**, 146 (2007).

13. Ming-Tzer Lin *et al.*, *Microsystem Technologies* **12**, 1045 (2006).

14. H. Guo *et al.*, *Nature Materials* **6**, 735 (2007).

15. T. Tsuchiya, O. Tabata, J. Sakata, and Y. Taga, in *Proceedings IEEE. The Tenth Annual International Workshop on Micro Electro Mechanical Systems* (IEEE Robotics & Autom. Soc., New York, NY, 1997) pp. 529-524.

16. S. Greek and S. Johansson, *Proceedings of the SPIE - the International Society for Optical Engineering* **3224**, 344 (1997).

17. S. W. Cho and I. Chasiotis, *Exp. Mech.* **47**, 37 (2007).

18. W. N. Sharpe Jr. *et al.*, *Exp. Mech.* **47**, 649 (2007).

19. B. L. Boyce, J. M. Grazier, T. E. Buchheit and M. J. Shaw, *J. Microelectromech. Syst.* **16**, 179 (2007).

20. D. S. Gianola and W. N. Sharpe Jr., *Exp. Tech.* **28**, 23 (2004).

21. T. Tsuchiya, M. Hirata and N. Chiba, *Thin Solid Films* **484**, 245 (2005).

22. D. T. Read, Y. Cheng, R. R. Keller and J. David McColskey, *Scr. Mater.* **45**, 583 (2001).

23. M. G. Da Silva and S. Bouwstra, in *Reliability, Packaging, Testing, and Characterization of MEMS/MOEMS VI* (SPIE -International Society for Optical Engineering, Bellingham WA, 2007) pg. 64360.

24. http://www.eeel.nist.gov/812/MNT/index.html

25. http://www.semi.org/

26. W. N. Sharpe Jr. *et al.*, *Microelectromechanical Structures for Materials Research* (Mater. Res. Soc, Warrendale, PA, 1998).

27. D. A. LaVan *et al.* in *Mechanical Properties of Structural Films* (American Society for Testing and Materials, West Conshohocken, PA, 2001) pp. 16-27.

Mater. Res. Soc. Symp. Proc. Vol. 1052 © 2008 Materials Research Society 1052-DD01-02

Reliability of MEMS Materials: Mechanical Characterization of Thin-Films using the Wafer Scale Bulge Test and Improved Microtensile Techniques

Joao Gaspar, Marek Schmidt, Jochen Held, and Oliver Paul

Department of Microsystems Engineering - IMTEK, Microsystems Materials Laboratory, University of Freiburg, Georges-Koehler-Allee 103, Freiburg, 79110, Germany

ABSTRACT

This paper reports on recent improvements of the bulge and microtensile techniques for the reliable extraction of material parameters such as the Young's modulus E, Poisson's ratio ν, plane strain modulus $E_{ps} = E/(1-\nu^2)$, prestress σ_0, fracture strength μ, Weibull modulus m and strain hardening coefficients n, and on the direct comparison between the two methods. The bulge technique is extended to full wafer measurements enabling throughputs of data with statistical relevance whereas key improvements of a previous fabrication process of microtensile specimens lead now to much higher yields, approaching 100%. Both techniques are applied to an extensive set of materials, brittle and ductile, typically used in MEMS applications. These include thin films of silicon nitride, silicon oxide, polycrystalline silicon and aluminum deposited by techniques such as thermal oxidation, LPCVD, PECVD and PVD.

INTRODUCTION

The mechanical characterization of thin films is of great importance for their use in integrated circuits and as structural materials in microelectromechanical systems since device functionality and reliability depend strongly on mechanical parameters. Deviations of materials mechanical behavior from the macro to the microscale demand novel characterization methods or adjustment of current ones for the reliable extraction of both elastic and fracture properties. Established techniques that have been used to characterize a wide variety of thin film materials include the wafer curvature method, beam buckling analysis, nanoindentation, pull-in voltage experiments, measurements of resonance frequency, and bulge and microtensile tests [1].

The bulge test consists on the measurement of the load-deflection response of membranes under applied pressure P [1]. A multilayer mechanical model for the plane strain deformation of long membranes, where the compliance of the supporting edges is included through a spring constant K, has been recently developed [2], and is here used to characterize homogeneous and composite diaphragms made of different materials. A bulge setup, automated for the measurement of up to 80 membranes on a silicon wafer, enables the extraction of mechanical data with statistical significance, as never reported before for this technique. In addition, the stress distribution within the diaphragms at the moment of failure is computed and used to analyze the brittle material strength via Weibull distributions [2-4], extending this technique to the consistent fracture analysis of poly-Si thin films.

The tensile method is one of the most fundamental, simplest and best-standardized mechanical tests [5]. Its main advantage resides in the fact that uniaxially loaded specimens experience a uniform stress distribution allowing a simple analysis and straightforward extraction of E, ν, σ_0 and μ [5]. Major difficulties in performing tensile tests with microscale specimens concern sample handling and attachment to actuators. In order to avoid an improper load application occurring from misaligned sample clutching, Kamiya *et al.* developed a test structure consisting of fixed and movable crystalline silicon (c-Si) frames, connected through springs, whose configuration restricts the deformation of bridging polycrystalline (poly-Si) specimens to uniaxial strains [6]. The microtensile method presented here relies on the same concept and is extended to other materials. However, in contrast to the previous measurement procedure, an optical interference technique is adopted for localized strain measurements [7]. In addition, whereas previous structures were separated by rather rough dicing, they are now defined by the same deep reactive ion etching (DRIE) step employed to define the springs and gaps. The test structures can be broken off the wafer by gentle twisting with highly improved yield.

BULGE MEASUREMENTS

Figure 1 (a) schematically shows the deflection profile $w(x,y)$ of a membrane with width a and length b, bulged with a pressure P. Its aspect ratio $a{:}b$ being smaller than 1:4, an extended middle section of the structure responds by a plane strain deformation $w(x)$, Fig. 1 (b), with no dependence on y [1,2]. From the model in [2], which takes into account the films prestress, structural stretching and bending stiffness, and compliance of the supporting edges, the relation between the center deflection w_0 and pressure P is given by

$$w_0 = \frac{Pa^2}{8S} + \left[\frac{P}{S}\left(D_2 + \frac{Ka}{2}\right) + S_1\right] \frac{1 - \cosh\left(\sqrt{S/D_2}\, a/2\right)}{S\cosh\left(\sqrt{S/D_2}\, a/2\right) + K\sqrt{S/D_2}\,\sinh\left(\sqrt{S/D_2}\, a/2\right)}, \quad (1)$$

where K, S, S_1 and D_2 are the rigidity of the membrane supporting edges, effective line force, initial bending moment per unit length and bending stiffness, respectively, which depend on the thickness h and elastic constants of each individual layer in the stack composite [2]. E_{ps} and σ_0 are extracted from the fit of Eq. (1) to the data obtained from membranes with widths and thicknesses ranging from 400 to 800 μm and from of 0.1 and 2 μm, respectively.

The membranes are produced by bulk micromachining of the backside of silicon substrates with KOH/TMAH solutions. Silicon nitride (Si_3N_4) deposited by low-pressure chemical vapor deposition (LPCVD) is used as an etch mask layer, whereas oxide films or LPCVD Si3N4 are used as etch stop layers depending on the mechanical layers to be processed and characterized. Moreover, strongly compressive films are stacked with highly tensile LPCVD Si_3N_4 layers in order to obtain stress-compensated diaphragms and consequent unbuckled, plane-

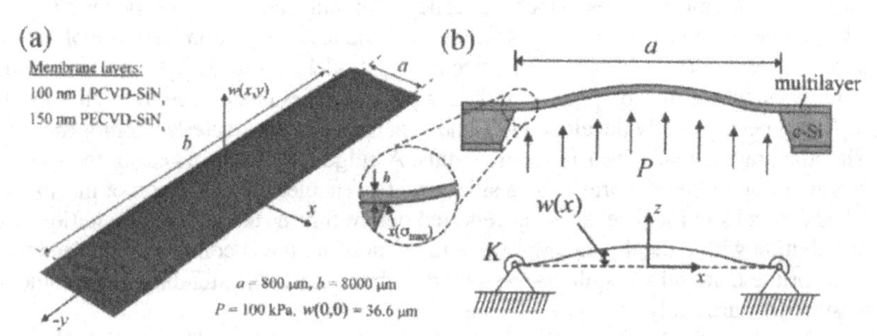

Figure 1. (a) Measured plane-strain deflection profile of a long bilayer membrane, bulged with a pressure of 100 kPa. (b) Schematics of the extended middle section of a diaphragm with compliant supports defined by a c-Si frame.

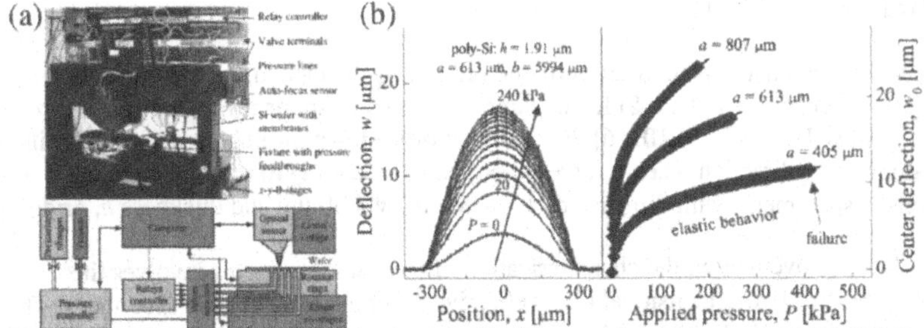

Figure 2. (a) Automated bulge test setup. (b) Measured membrane profiles and w_0 vs. P responses of 1.91-μm-thick poly-Si diaphragms with different widths a.

strain response [1,2]. The films here characterized using the bulge technique include silicon nitride (Si_3N_4 or SiN_x), silicon oxide (SiO_2 or SiN_x), poly-Si and aluminum (Al) grown by techniques such as LPCVD, plasma-enhanced chemical vapor deposition (PECVD), thermal oxidation and physical vapor deposition (PVD) with different conditions.

The high throughput bulge test is performed with a setup that is able to measure sequentially up to 80 membranes [2], Fig. 2 (a): a wafer is mounted on a fixture with feedthroughs whose pressure inlet is adjusted with a pressure controller and multiplexed to the membranes through solenoid valves. At the same time, the deflection profiles are monitored using an optical autofocus sensor and scanned via automated x-y-θ stages. Deflection profiles, measured at different pressures, of a 613-μm-wide poly-Si membrane and the variation of w_0 with P for diaphragms with different dimensions are shown in Fig. 2 (b). The lines are fits to Eq. (1), showing the predicted width scaling effects [1,2]. Figure 3 (a) shows the values of E_{ps} thus obtained at each membrane location on the wafer and histograms of the plane strain modulus and prestress values, in this case for poly-Si grown by LPCVD at 625°C and fully annealed at 1050°C [E_{ps} = 160±16 GPa, σ_0 = −8 ±16 MPa]. Both elastic constants and prestress values extracted from the bulge test of the several films here characterized are listed in Table I.

The membranes are bulged until fracture occurs and the stress distribution at the moment of fracture [2] is used to analyze the material strength using standardized pooled Weibull distributions [2-4], as depicted in Fig. 3 (b) for silicon, oxide and nitride films (see also Table I).

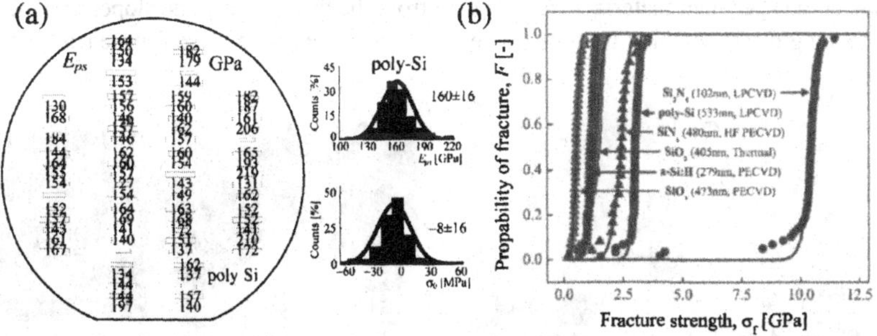

Figure 3. (a) Wafer distribution of extracted E_{ps}-values, and E_{ps}- and σ_0-histograms for poly-Si films. (b) Pooled Weibull data and fits obtained for silicon, silicon nitride and silicon oxide films grown using different techniques.

MICROTENSILE TESTING

The microtensile method relies on the previously developed test structure shown in Fig. 4 (a), which consists of a c-Si fixed frame connected to an inner moving frame through parallel springs defined by DRIE [6]. Its design allows a linear in-plane motion of the inner frame along the x-direction when a force is applied to it, leading to an axial elongation of four microtensile specimens, with dimensions $L = 400$ µm, $w = 50$ µm and thickness h, bridging the gap.

One improvement in the characterization of these test structures involves the implementation of two-dimensional (2D) reflective gratings on the tensile specimens. The variation of the periods Δx and Δy due to the local strains is measured from the interference pattern arising from monochromatic illumination of the grating [8] with charge-coupled device [CCD] arrays, as illustrated in Fig. 4 (a) and shown in Fig. 4 (b) [7]. Two CCDs make it possible to distinguish slight sample rotations from pure elongations. (marker)

Another enhancement concerns the fabrication of the test structures. After defining the springs and gaps by backside DRIE, the devices used to be separated by dicing, this step being relatively rough for the delicate tensile specimens [6]. The same DRIE is now used to define the test structures and, in addition, connectors that keep the devices on the wafer, as shown in Fig. 5 (a). These can now be broken by gentle twisting prior to the characterization of one test structure. Variants of the previous fabrication process for poly-Si specimens have been explored in order to obtain test structures with microtensile specimens made of other materials, such as silicon nitride or aluminum. Scanning electron microscope (SEM) graphs are shown in Fig. 5 (b).

Figure 6 (a) shows the measured and simulated diffraction patterns by one CCD for a given applied force. The position of diffraction peaks shifts towards $s_c \rightarrow 0$ with increasing force implying that Δx decreases [8]. From the detailed variation of the maxima locations, it is possible to obtain the strain variation with total applied force, i.e. $d\varepsilon/dF$, as explained in detail elsewhere [7,8]. Typical load-displacement curves of test structures with poly-Si and silicon nitride specimens are shown in Fig. 6 (b). After the initial contact point, the slope of the curve is the sum of the spring constants of the c-Si parallel springs and those of four microtensile specimens, according to

$$F = (k_1 + k_2)\delta = [4(hw/L)E + k_{c\text{-}Si}]\delta. \tag{2}$$

The abrupt transitions in the diagrams evidence the fracture of the four specimens. The Young's modulus of the specimen materials are extracted from the difference in the slopes and the fracture strength μ is obtained from the force drops, necessary to fracture four tensile specimens.

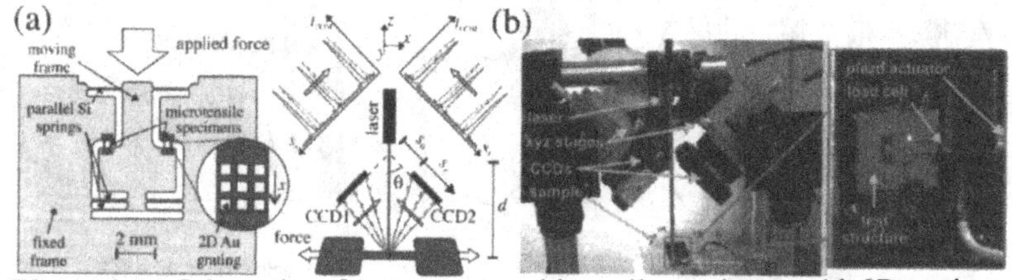

Figure 4. (a) Schematics of test structure with tensile specimens with 2D gratings and strain measurement principle. (b) Experimental setup used to measure strain components evaluated locally on the microspecimen via optical interference.

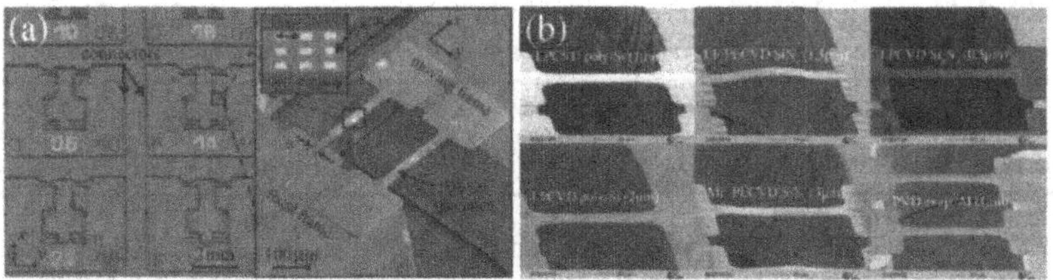

Figure 5. (a) Detachable test structures at the wafer-level and respective SEM of poly-Si microtensile specimens with 2D gratings. (b) SEM micrographs of specimens made of various thin films.

The mechanical properties thus extracted from the microtensile testing of the several films are listed in Table I as well, consistent with the bulge results.

DISCUSSION AND CONCLUSIONS

Wafer-level bulge measurements enable the reliable characterization of both elastic and fracture properties of thin films. Young's modulus and prestress values thus obtained show no dependence on film thickness and with values well within literature data [9], however with narrower uncertainty. Because of the relatively high throughputs of data achieved, almost free of human error, the bulge technique can now be extended to the consistent extraction of fracture parameters of brittle materials. For instance, this enables one to obtain Weibull moduli as high as 50 in the case of LPCVD Si_3N_4, making this material highly predictable from the fracture point of view. Also, among the different thin films tested, LPCVD Si_3N_4 is the one having the highest strength, larger than PECVD layers or silicon films. In the opposite side, silicon oxide films are the weakest and less predictable. The fracture properties extracted are within the ranges of literature data using different characterization tools [9]. Moreover, ductile parameters of materials such as Cu or Al can be extracted from bulge tests [10]. The wafer-scale version presented here will certainly be valuable for the more accurate, reliable and meaningful extraction of ductile properties. The mechanical data obtained from the optimized measurement

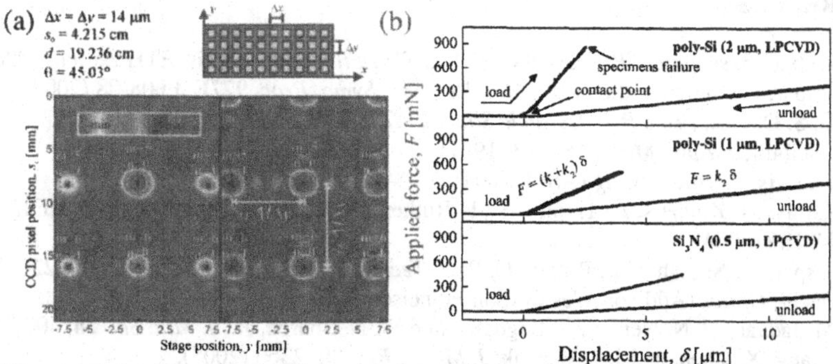

Figure 6. (a) Measured and modeled interference patterns resulting from the monochromatic illumination (λ = 633 nm) of a test structure with 2D gratings. (b) Load-displacement curves of structures with poly-Si and Si_3N_4 tensile samples.

Table I. Mechanical parameters extracted from bulge measurements of thin films. Data obtained from microtensile tests are also shown (printed in bold).

	Deposition		Thin film	Mechanical parameters			
Technique	Temp. [LC]	React. Freq.		E_{ps}, E [GPa]	σ_0 [MPa]	μ [GPa]	m
LPCVD	625(1)	-	Poly-Si (0.5-2μm, **1-2μm**)	160±16 **156±16**	−8±16	3.00-1.72 **1.9-1.39**	22.9 **16.3**
	425	-	SiO_x (0.4μm)	53±8	−321±15	0.26	5.1
	770	-	Si_3N_4 (0.1-0.3μm, **0.5μm**)	278±10 **251**	+1194±30	10.39-7.91 **3.76**	50.2
PECVD	220	13.56 MHz	a-Si:H (0.3μm)	126±13	−163±17	1.13	8.8
	330	187 kHz	SiN_x (0.1-0.4μm, **0.5μm**)	166±29 **129**	−1274±64	**0.97**	-
		13.56 MHz		170±23	+412±28	1.32	7.3 **6.1**
		mixed		143±12 **112±9**	−323±25	0.80-2.38 **2.10**	8.2
		187 kHz	SiO_x (0.4μm)	76±8	−395±13	0.49	10.1
		13.56 MHz		72±11	−312±16	0.54	5.8
PVD (Evap.)	-	-	Al (1μm, **1μm**)	48(2)	30	**0.36±0.04**(3)	-
Oxidation	950	-	SiO_2 (0.4μm)	60±5	−318±15	1.32	5.4

(1) Films subjected to a post-deposition annealing at 1050°C during 1 hour.
(2) Obtained from the unloading parts of cyclic pressure-deflection responses. A strain hardening coefficient of $n = 0.34$ is extracted from equivalent uniaxial stress-strain reconstructed curves.
(3) Ultimate strength.

of tensile specimens defined on an optimized test structure are in agreement with those from the bulge measurements. Progress is being made in developing a microtensile test setup at the wafer-level, analogous to the current bulge test setup, for the sake of a truly reliable tensile characterization of thin films [11].

REFERENCES

1. V. Ziebart, *Mechanical Properties of CMOS Thin Films*. PhD thesis, ETH Zurich, 1998.
2. J. Gaspar, P. Ruther, and O. Paul, *Mat. Res. Soc. Symp. Proc.* **977E**, FF08-08 (2007).
3. D. G. S. Davies, *Proc. Brit. Ceramic Soc.* **22**, 429 (1973).
4. W. Weibull, *J. Appl. Mech.* **18**, 293 (1951).
5. J. R. Davis, *Tensile Testing* (ASM, 2004).
6. S. Kamiya, J. Kuypers, A. Trautmann, P. Ruther, and O. Paul, *Tech. Dig. IEEE MEMS Conf. 2004*, 185 (2004).
7. J. Gaspar, Y. Nurcahyo, P. Ruther, O. Paul, *Tech. Dig. IEEE MEMS Conf. 2007*, 223 (2007).
8. E. Hecht, *Optics* (Addison Wesley, San Francisco, 2002)
9. O. M. Jadaan, N. N. Nemeth, J. Bagdahn, and W. N. Sharpe, Jr., *J. Mat. Sci.* **38**, 4087 (2003).
10. Y. Xiang, X. Chen, and J. J. Vlassak, *J. Mater. Res.* **20**, 2360 (2005).
11. J. Gaspar, M. Schmidt, J. Held, and O. Paul, *Tech. Dig. IEEE MEMS Conf. 2008*, in press.

Mater. Res. Soc. Symp. Proc. Vol. 1052 © 2008 Materials Research Society 1052-DD01-06

Fast Characterization of Silicon Membrane Structures by Laser-Doppler Vibrometry

Ronny Gerbach, Matthias Ebert, and Joerg Bagdahn
Components in Microelectronics, Microsystems and Photovoltaics, Fraunhofer Institute for Mechanics of Materials, Walter-Huelse-Str. 1, Halle, 06120, Germany

ABSTRACT

Micromechanical structures were investigated nondestructively via laser-Doppler-vibrometry to determine defect structures. Silicon membrane structures were characterized by their measured resonant frequencies and mode shapes. The influence of defects on the micromechanical structures on the measured dynamic properties is shown. Defective samples were indentified on the basis of the ratios of measured resonant frequencies and the quantified comparison of mode shapes without an identification of unknown parameters. The investigations showed that a fast determination of defect structures is possible by measuring dynamic properties.

INTRODUCTION

The nondestructive characterization of microelectromechanical systems (MEMS) is a key element for the monitoring of manufacturing processes of MEMS. It is necessary to describe all components of these systems including electronic and non-electronic components. There are a lot of well established measuring techniques for the characterizing of the electrical components. In contrast, there are few nondestructive methods the testing of MEMS.

Optical measurement techniques have gained great importance. They allow a fast and nondestructive characterization of mechanical components of MEMS. They should follow general requirements such as being a reliable measurement without a changing of the behavior of the structure [1]. Some measuring techniques have already been developed for these tasks [1-3]. Laser-Doppler vibrometry is a well established method for the characterization of micromechanical structures on basis of their dynamic properties. Different investigations were already presented results for the identification of unknown properties [4-7]. Validation experiments were performed to verify identified values with an error less than 2% [7]. In the following paper, different approaches were presented for the determination of defect structures without an identification of parameters. The approaches introduced here are based on the use of measured resonant frequencies and mode shapes.

PRINCIPLES OF THE METHOD

Mechanical structures can be analyzed on basis of their dynamic properties. Dynamic properties can be characterized in the time and frequency domain. Investigations in the time domain comprise the behavior of the mechanical structure under a pulsative or homogeneous actuation. A characterization of micromechanical structures in the frequency domain can occur by examining the frequency spectrum generating to a wide-band excitation and also by their resonant frequencies and mode shapes. Micromechanical structures vibrate at high frequencies with small amplitudes because of the small size of the mechanical structures which leads to a high signal to noise ratio in the time domain. Micromechanical structures have a relatively high quality factor. This leads to large amplitude vibrations at the resonant frequencies and small amplitude vibrations at frequencies which are not resonant frequencies of the structure. It is possible to measure the frequency spectrum of the structure and extract the resonant frequencies and mode shapes. These properties can be used to characterize mechanical structures; these two possibilities are presented in the following paragraphs.

System Characterization Based on Measured Resonant Frequencies

The first possibility for the characterization of mechanical systems is the use of their frequency spectrum and the extracted resonant frequencies. The frequency spectrum is the measured response of the system by a wide-band excitation e.g. by sweep or noise signals. Figure 1 shows the frequency spectrum of a square shaped silicon membrane structure measured via laser-Doppler vibrometry. Five peaks can be clearly identified. These peaks may be assigned to resonant frequencies of the structure by analytic or numerical calculations of the value of the natural frequencies or by analyzing the mode shapes of the structure in comparison to the expected ones.

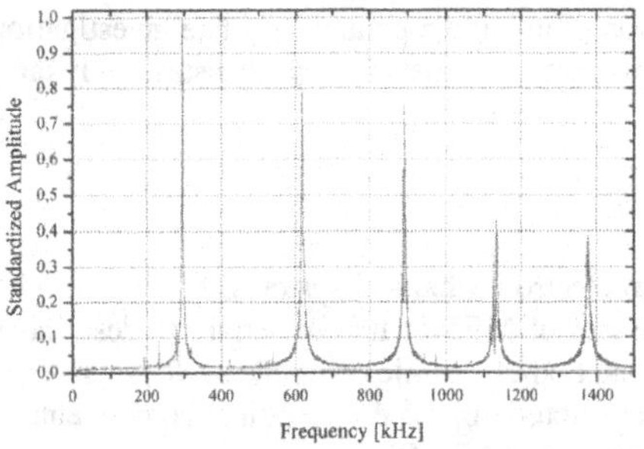

Figure 1 Measured frequency spectrum of a square shaped silicon membrane structure

Mechanical structures can be directly characterized by their resonant frequencies for the determination of defective samples. The characterization can be realized based on:

- the number and value of measured resonant frequencies in comparison to analytical or numerical calculations or to reference structures and
- the ration between measured resonant frequencies in comparison to calculated or measured ratios from reference structures.

An example for the characterization of mechanical structures by their resonant frequencies is shown in Figure 2. The diagram shows the measured frequency spectra of quadratic silicon membrane structure with and without a defect. The defect results in a significant change of the frequency spectrum by increasing the number of the resonant frequencies and enables a classification.

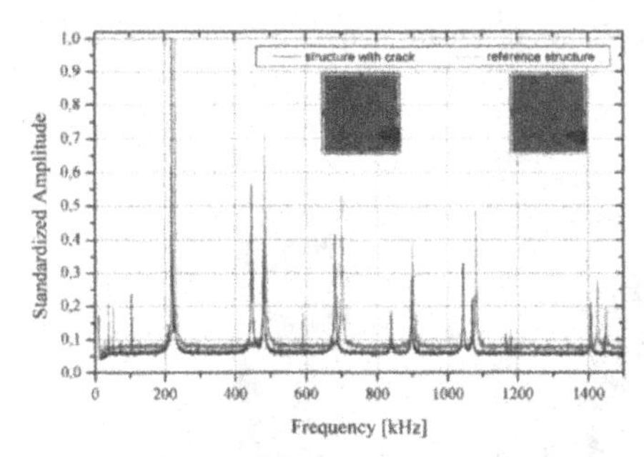

Figure 2 Measured frequency spectra of square shaped silicon membrane structures with a defect (black) and defect less reference structure (grey)

Furthermore, the resonant frequencies can be used for the identification of unknown mechanical or geometric parameters. The determination is based on the combination of dynamic measurements and analytical or numerical calculations. The mechanical system has to be described with all relevant parameters to use this method. Possible errors occur if a significant parameter is not taken into account. An optimization process is required to identify more then one parameter. The optimization process can be performed directly in the developed numerical or analytical model or with an interpolated function, which has to be calculated ahead of time. Different results and methods to identify parameters on silicon membrane structures [4, 7], MEMS optical scanners [5] and cantilever beams [6] have been presented.

System Characterization Based on Measured Mode Shapes

Mechanical systems can also be characterized by the mode shapes of their resonant frequencies. Each resonant frequency has typical mode shapes. These mode shapes depend on the construction and clamping of the structure. Large defects (e.g. cracks or residue from the etching processes) lead to a significant modification of the frequency spectrum (shown in Figure 2) and mode shapes. If a small defect is present, the change in the frequency spectrum is too small for the identification of a faulty structure but it does result in a change in the mode shapes in comparison to a reference structure [7]. Table I shows a summary of measured mode shapes of defective and faultless square shaped silicon membrane. The maximal displacements of the mode shape of the resonant frequency f_1 and f_3 show a small observable change. The mode shape of f_2 for a square shaped membrane structure is a superposition of mode shapes of two resonant frequencies at the same frequency and is influenced significantly by small and large defects. Large defects can lead to a split of the frequencies into two separate mode shapes. This effect can be used for the determination of defective elements.

Table I Maximal displacement of measured mode shapes of square shaped silicon membrane structures

	f_1	f_2	f_3
faultless sample			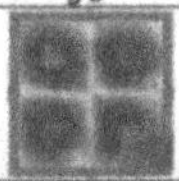
sample with large defect		 	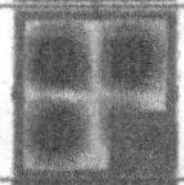
sample with small defect		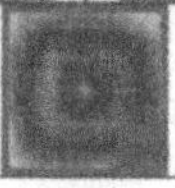	

A quantified correlation of mode shapes is necessary for the identification of mechanical structures with defects. The quantified comparison can occur with MAC-value *(Modal Assurance Criterion)* defined below.

$$MAC = \frac{\left(v_S^T * v_R\right)^2}{\left(v_S^T * v_S\right)\left(v_R^T * v_R\right)} \cdot 100\% \qquad (1)$$

The MAC-value compares the two eigenvectors with same dimensions. Here, v_S is an eigenvector of a mode shape of the sample structure and v_R of a reference structure without defects. v^T means the transposed eigenvector. The MAC-value can adopt values between 0% and 100% where 0% means a complete independence of the two vectors and 100% a complete conformity. Statistical investigations have to be performed to determine a boundary value for the MAC for the characterization of defective structures. First results applying this method have been presented on membrane structures with generated defects [7].

EXPERIMENTS

Investigated Structures

Bonded silicon membrane structures were used for the investigations. The dynamic properties can be described with the analytic model of an all-side clamped plate. A cross section of the investigated structure is shown in Figure 3. A cavity with a quadratic shape is etched in a silicon bulk wafer by wet chemical etching with KOH. The edge length of the cavity is 900µm. A membrane wafer is directly bonded to the bulk wafer and grinded and polished to the desired, generating a membrane above the etched cavity. A passivation layer of silicon dioxide is deposited by the following process. Fluctuations of thickness are possible inside the area of one membrane about the average thickness of a wafer.

Figure 3 Cross section of the investigated structures in SEM (2000x)

Structures with three different nominal thicknesses (18µm, 20µm and 22µm, ten samples for each case) were used for the investigations described in this paper. The structures with a nominal thickness of 18µm do not have a passivation deposit. Structures 6 and 10 in this row have visible cracks at one edge of the membrane and are shown in Figure 4.

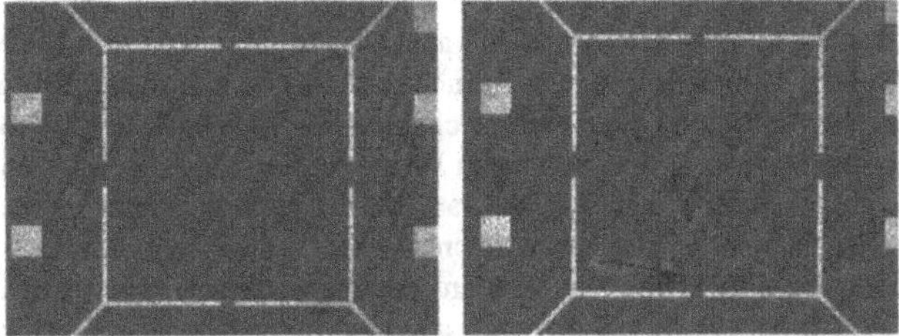

Figure 4 Microscopic image (50x) of the defective samples with a nominal thickness of 18µm (left: sample 6, right: sample 10)

Measuring Setup

The measuring setup shown in Figure 5 was used to characterize the dynamic properties of the membrane structures in out-of-plane direction. A probe tip was positioned above the membrane by a probe head. The mechanical system was stimulated to vibrate by an applied voltage at different frequencies from a function generator. The voltage of the signal was amplified 400V peak-to-peak. Typical mode shapes of the membrane were generated when the excitation was at a resonant frequency of the mechanical structure. The commercial laser-Doppler-vibrometer (MSA400 from Polytec) was used to measure the dynamic properties [3]. The frequency spectrum was measured with only a few measurement points. Mode shapes were measured by scanning the laser beam in a defined reticule which results in longer measuring times. To perform investigations on the wafer level the measuring setup was integrated on a manual probe station. This enables an exact and repeatable realization of the measurements. The first three resonant frequencies (f_1, f_2 and f_3) and mode shapes were measured with the described setup.

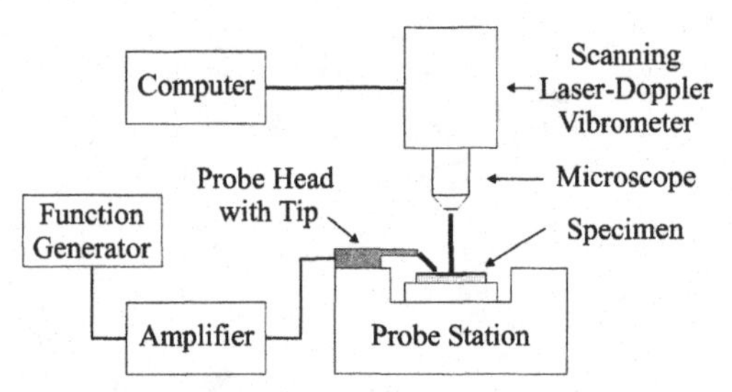

Figure 5 Schematic description of the used measuring setup

RESULTS & DISCUSSION

The measured resonant frequencies and mode shapes were used for the characterization of the investigated structures. Because of differences of the membrane thickness caused by the manufacturing process the value of the resonant frequency can't be used directly for analysis. The ratio of resonant frequencies is useful property for the characterization. This ratio should be constant for an all-side clamped plate. The real mechanical structure shows a small dependence of the thickness to the frequency ratios caused by the non-ideal clamping of the membrane due to the elastic behavior of the membrane material. The diagrams in Figure 6 show the ratio between f_2 and f_1 and f_3 and f_1, respectively. It can be seen that the ratios are almost constant for the structures with the same nominal thickness. The ratios decrease with increasing thickness without consideration for the defective structures. The standard deviations for all investigated thicknesses are smaller than 0.2% of the mean. In the row with a nominal thickness of 18µm the structure with number 10 has a significantly higher ratio between f_2 and f_1 (increase of 7%) than between f_3 and f_1. This implies that sample is defective. Structure 6 has less on an increase in values for both ratios. Because of the thickness dependence of the frequency ratios it could be assumed that this structure is thinner than the other structures of this row and can not be directly characterized as a defective structure by the ratio of the investigated resonant frequencies. Structure 7 shows an increase in the ratio between f_3 and f_1 but does not have a visible crack.

The samples in the rows with a nominal thickness of 20µm and 22µm· do not show conspicuous or visible defects. The investigated ratios are nearly constant. Small changes in the ratios are caused by small variations of the membrane thickness.

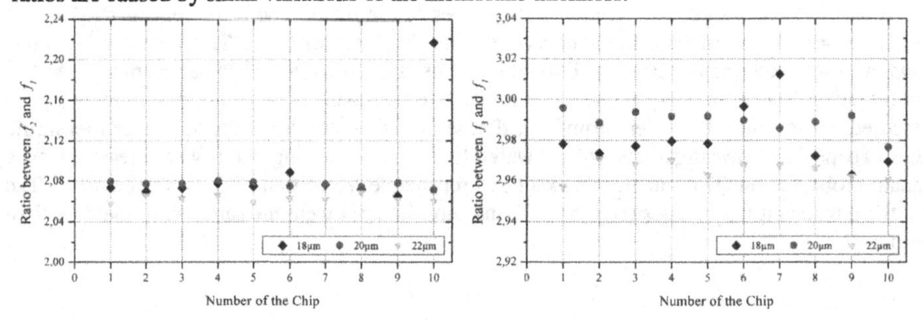

Figure 6 Ratio between measured resonant frequencies (left: f_2/f_1, right f_3/f_1)

The MAC-value was calculated from the maximal values of the displacement of each point of the mode shapes. The results from the mode shapes of f_1 and f_2 are shown in Figure 7. For each row, structure 1 was used as reference structure. This results in MAC-values of 100%. The mode shapes of faultless structures show a good agreement for the mode shape of f_1 with values of about 99% with a standard deviation of less than 0.5%. The agreement of the mode shapes of f_2 is with MAC-values larger than 93% and standard deviations up to 2.5%. Structure 10, with a nominal thickness of 18µm, show, significant deviations for the investigated mode shapes (6% to the average MAC-value for f_1 and 18% for f_2) and can be identified as a defective structure. The defect of sample 6 does not lead to a variation of the mode shape of f_1 (MAC=98.9%) of the reference structure. So, this structure can not be characterized with this mode shape. A distinct classification of sample 6 can be performed by the usage of the mode shape of f_2. The error of this structure leads to reduce of the MAC-value to 65%.

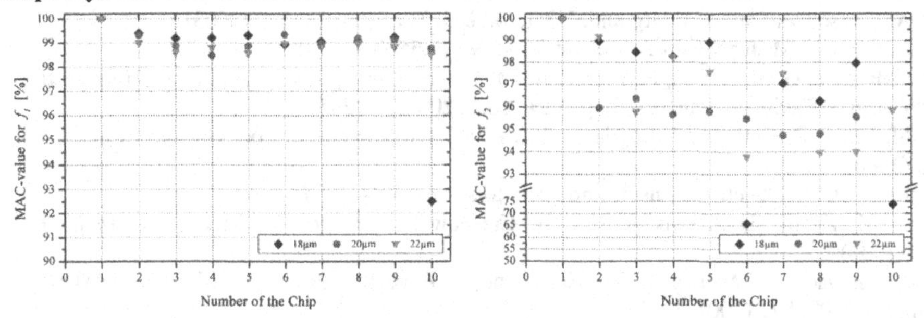

Figure 7 MAC-value for the measured resonant frequencies f_1 (left) and f_2 (right)

CONCLUSIONS

Methodical and experimental investigations were performed for the characterization of defective structures. Different possibilities were described to characterize micromechanical structures via their dynamic properties. Experimental investigations were performed on silicon membrane structures with different thicknesses to characterize the structures and determine defective ones. The first three resonant frequencies and mode shapes were measured with a laser-Doppler vibrometer and electrostatic stimulation. The values of the measured resonant frequencies can not be used directly for the characterization because of differences of the membrane thicknesses caused by the manufacturing process. Classification was performed via the ratios of resonant frequencies. These ratios depend on the membrane thickness and are almost constant. A determination of all defective structures was not achieved.

The mode shapes of the resonant frequencies are further properties for the characterization of mechanical structures. The MAC-value was established for a quantitative comparison of two mode shapes. This value was calculated from the maximal value of the displacement. It was shown that MAC-values for faultless structures are larger then 99% for f_1 and 93% for f_2 which implies a very good agreement of the measured mode shapes. A defect in a mechanical structure leads to a modification of the form of the vibration which results in lower MAC-values in comparison to defect-less structures; which was shown on two samples. Because of the significant reduction of the MAC-value for f_1 and f_2 for sample 10 the defect structure was determined. The error of sample 6 leads to small decrease of the MAC-value for f_2 but to significant decrease for f_2. A classification for this structure as defective is only possible by considering the mode shape of the second investigated resonant frequency.

ACKNOWLEDGMENTS

The financial support of the project "PAR-TEST" by the German Federal Ministry of Education and research (BMBF) (contract 16SV1940) is gratefully acknowledged. Furthermore the authors like to thank Siegfried Hering from X-Fab Semiconductor Foundries AG for the preparation of test structures.

REFERENCES

1. A. Bossoboeuf and S. Petitgrand, J. Micromech. Microeng 13, S23 (2003)
2. C. Gorecki, M. Jozwik and L. Salbut, Proc. SPIE 5543, 63 (2004)
3. C. Rembe, G. Siegmund, H. Steger and M. Woertge, in *Optical Inspection of MEMS*, edited by W. Osten (Taylor & Francis Group, New York, 2007), p. 245
4. M. Ebert, R. Gerbach, J. Bagdahn, S. Michael and S. Hering, Proc. 7th EuroSimE, Como, 208 (2006)
5. Kurth and W. Doetzel, Sensors and Actuators A 62, 760 (1997)
6. L. M. Zhang, D. Uttamchandani, B. Culshaw and P. Dobson, Meas. Sci. Technol. 1, 1343(1990)
7. R. Gerbach, F. Naumann, M. Ebert and J. Bagdahn, Proc. of MikroSystemTechnik Kongress 2007, 877 (2007)

Mater. Res. Soc. Symp. Proc. Vol. 1052 © 2008 Materials Research Society 1052-DD01-09

MEMS Lubrication: An Atomistic Perspective of a Bound + Mobile Lubricant

Douglas L Irving, and Donald W Brenner
Materials Science and Engineering, North Carolina State University, Box 7907, Raleigh, NC, 27695-7907

ABSTRACT

The adhesive pressure needed to separate two ocatdecyltrichlorosilane (ODTS) coated surfaces both with and without the addition of tricresyl phosphate (TCP) as a function of separation rate is characterized using molecular dynamics simulation. The simulations predict that when TCP is added between surfaces the adhesive pressure needed for separation is reduced compared to the system containing ODTS only. Both the adhesive pressure and the break up of the TCP layers exhibit a separation rate dependence that appears unrelated to the rate of diffusion of TCP on the ODTS. The ability of the TCP to remain localized to defected areas of the ODTS layer upon normal separation of the contact is also characterized. It is found that the TCP remains localized to defect sites and, thus, effectively coats the damaged area.

INTRODUCTION

Monolayer lubricants, such as self-assembled monolayers (SAMs), have proven to be very successful in the protection of microelectricalmechanical systems (MEMS) from stiction related failure that often occurs during processing of the device. This success is related to the ability of the SAM to transform the surface chemistry from hydrophilic to hydrophobic while at the same time introducing a low self-adhesion. Unfortunately, protection against failure due to in-use friction is marginal, which limits many devices to "single shot" type applications. Further complicating the issue of lubrication is the limited frictional protection only occurs within a narrow range of temperatures and environments. It is, therefore, of interest to explore new lubrication strategies that extend the lifetime of devices in which reciprocating contact is needed and that broaden the environmental conditions under which the lubricant can be used.

To overcome the shortcomings of the above lubricant a new "bound + mobile" scheme for MEMS, similar to that used in hard drives, has been studied [1-4]. It was first shown by Eapen et al. that a combination of chemisorbed and physisorbed Fomblin Zdol extends the lifetimes of a MEMS device [1]. This extension was attributed to the ability of the physisorbed layers to diffuse to defected areas and protect the uncovered surface. Although this scheme extends the lifetime, it does not necessarily extend the conditions under which the device can be used. Environmental insensitivity of the lubricant could possibly be achieved by incorporation of one or more mobile species with the traditional bound lubricant. In this direction, a combination of a bound octadecyltrichlorosilane (ODTS) with a mobile tricresyl phosphate (TCP) has been studied both experimentally [4] and computationally [2, 3]. Neeyakorn et al. showed that 3-5 monolayers of TCP readily absorb onto an ODTS covered surface, while at the same time maintaining a reasonable slippage as measured by a quartz crystal microbalance [4]. Irving and Brenner used molecular dynamics simulations to quantify single molecule diffusion of TCP on an ODTS substrate [2]. These simulations predicted an anisotropic energy barrier for inclusion of TCP into a cylindrical defect in the ODTS overlayer that depended on chain tilt. Brenner et al. used a multi-scale analysis technique together with the simulated diffusion coefficients to predict conditions under which TCP could successfully lubricate reciprocating contacts [3]. Although much has been done to characterize the transport, incorporation, and absorption, it is still unclear whether addition of a mobile layer changes the low self-adhesion of the bound layer, and whether upon release the mobile layer will remain in defected areas.

This paper presents results of atomistic simulations that suggest that adhesion is not adversely affected by the addition of the mobile overlayer. Break up of the trapped mobile layer is rate dependent on the timescale of the simulations, which appears to be related to a correlated motion in the creation of defects that evolve slower than the diffusion process. Finally, initial results for TCP between a defected and ideal SAM system suggest that the mobile layer will remain in the defected areas of the SAM after the contact is separated.

EXPERIMENT

Atomistic molecular dynamics simulations were used to address the issues discussed above. Finite difference solutions to Newton's equations of motion were solved within the DL_POLY 2.14 software package [5] with a constant time step of 0.25 femtoseconds. The inter- and intra- molecular interactions were described by the AMBER force field [6]. Further details regarding the parameters used in the force field are given in [2].

The SAMs were modeled by fixing the Si head groups in a close packed hexagonal array with the experimental area per head group of 22 Å^2. Periodic boundary conditions were used in all three directions creating an infinite flat plane of the SAM. Sufficient room was added in the direction normal to the surface to eliminate interaction between periodic images. A single SAM was initially equilibrated at 300 K for times greater than one nanosecond [2]. The tilt angle of the initially equilibrated SAM was roughly 32° and the in-plane unit cell parameters were 70.56 Å and 69.84 Å, which includes 224 ODTS molecules. A second surface was created by copying, translating, and rotating the SAM such that the surface normals were opposite in direction and point toward the opposing surface. These two SAMs make up the first interfacial system, which is referred to below as the SAM/SAM interface. To study the effects of the mobile layer on adhesion 144 TCP molecules were inserted between the SAMs. This corresponds to approximately 3-4 monolayers as calculated using the liquid density of TCP. This system is referred to as the SAM/TCP/SAM interface. The final system of interest was created by replacing one of the SAMs with a structure containing a cylindrical defect as described in [2]. All molecules were terminated with a CH_3 end group and are referred to as SAM(d)/TCP/SAM. After construction the SAM/SAM and SAM/TCP/SAM systems were compressed to an initial pressure of ~1.5 GPa and then equilibrated for ~75 ps. The SAM/TCP/SAM system was then unloaded at constant velocities of 20, 2, and 0.2 m/s by simultaneously separating both layers by giving the Si head groups of each layer half the unloading velocity. Similarly, the SAM/SAM interface was unloaded at 2 m/s. The SAM(d)/TCP/SAM system was lightly loaded, equilibrated, then unloaded at 20 m/s. During the unloading no additional constraints, such as a thermostat, were placed on the systems.

DISCUSSION

Plotted in Figure 1a is the pressure versus separation as calculated in the molecular dynamics simulations of the SAM/TCP/SAM and SAM/SAM systems. In this figure a positive pressure corresponds to the system under compression while a negative pressure corresponds to tension. The curves are shifted such that a separation equal to 0 Å corresponds to the first instance of zero pressure on unloading. The pressure is calculated by taking the negative derivative of the total energy vs. separation, which gives the external force needed to maintain the imposed constant velocity of the Si head groups, divided by the in-plane area as defined by the periodic boundaries. These results are compared to the measured forces on the Si atoms and, on average, equivalent results are obtained. All values presented are from individual molecular dynamics trajectories. The highest adhesive pressure is seen for the SAM/SAM system with a value of ~210 MPa, which is consistent with other measured [7] and calculated [8] values of this adhesive pressure. For the SAM/TCP/SAM systems the adhesive pressures are calculated to be 140, 105, and 95 MPa for rates of 20, 2 and 0.2 m/s. All values are reduced in comparison to the SAM/SAM system suggesting that addition of the mobile layer is not detrimental to the low self-adhesion found in the SAM/SAM system. Although the adhesive forces are smaller for the SAM/TCP/SAM system they extend to larger separations. In certain instances the extended separation regions with lower adhesive pressures may come at the expense of higher energy consumption.

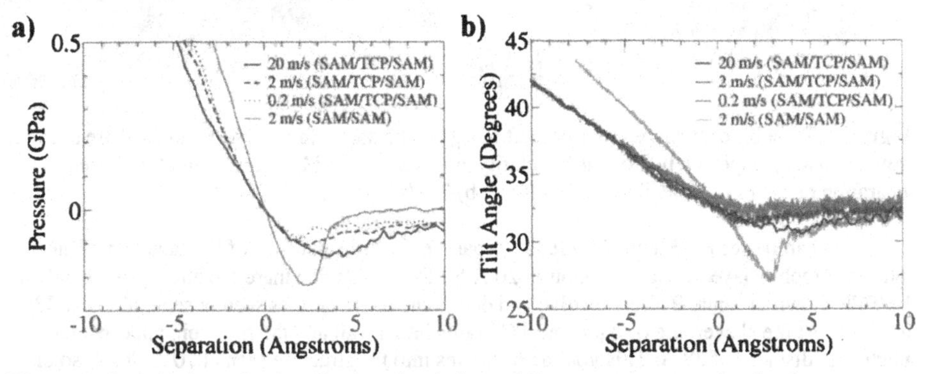

Figure 1. Results from the de-adhesion simulations. a) Pressure vs. separation for the SAM/TCP/SAM system (black) and the SAM/SAM system (grey). b) Average tilt angle with respect to the surface normal of the ODTS molecules vs. separation.

The mechanical instability of the SAM, as measured by the change in average tilt angle of the molecules relative to the surface normal during unloading, is also of interest [9]. Plotted in Figure 1b is the tilt angle as a function of separation, with separation defined as is in Figure 1a. In both the SAM/SAM system and the SAM/TCP/SAM system the initial tilt angle of the SAM starts at ~42° when under ~1.5 GPa of pressure. This differs from the 32° equilibrium value of a free SAM at 300 K. As the system is unloaded by increasing the separation at the above rates the tilt angle gradually approaches its equilibrium value of 32° at 0 Å separation. Beyond this point the two systems diverge in behavior. In the SAM/SAM system a mechanical instability is seen with the tilt angle continuing to decrease until it reaches a minimum value of 27° at a 3 Å separation. Upon further separation this system snaps back to the equilibrium angle and the surfaces are separated. Unlike the SAM/SAM system there is only a subtle change to the tilt angle in the SAM/TCP/SAM system past zero separation, which, like pressure, also appears to be rate dependent. It appears that slower rates allow the trapped mobile species to take on more of the applied tension, lessening the response of the SAM beyond a separation of 0 Å.

Illustrated in Figures 2 a) and b) are time ordered snap shots, from left to right, of the unloading of the SAM/TCP/SAM systems at rates of 20 and 2 m/s, respectively. The top of the figures shows the view down the surface normal with the atoms in the SAM and TCP shaded grey and black, respectively. For further clarity the atoms in the SAM are transparent. The bottom portion of each figure shows a side view of the separation. The four time ordered frames correspond to total separations of -10, 20, 38, and 70 Å.

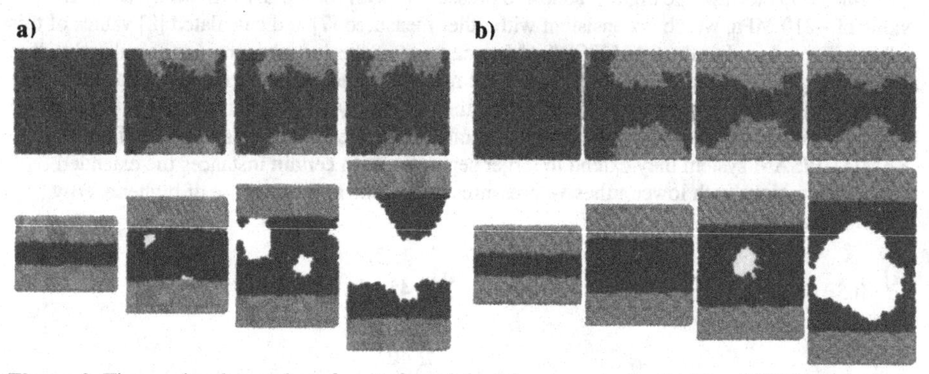

Figure 2. Time ordered snapshots from left to right. The atoms in the SAM and TCP are shaded grey and black, respectively. For further clarity atoms in the SAM are transparent. Different separation rates are illustrated in a) 20 m/s and b) 2 m/s.

As can be seen in Figure 2 there is a separation rate dependence of the break up of the entrapped mobile layer. At a separation of 20 Å for the 20 m/s rate there are already two voids in the TCP shown in figure 2 a) that evolve and determine the structures seen at separations of 38 and 70 Å. At the slower rate of 2 m/s the TCP pulls into a wall and the first void appears at a much later distance of 38 Å. This void then evolves into the structure seen at 70 Å. It is also of note that the molecular contact between surfaces has not been broken at this point, as it has in the 20 m/s case. Final breaking of contact of the molecular layer is also rate dependent, with the TCP breaking off at 52.0 and 70.3 Å for rates of 20 and 2 m/s, respectively. The single molecule diffusion length for one simulation time step of TCP on ODTS at 300 K is 3.0×10^{-2} Å. This is much faster then displacement per time step for the constant velocity separations, which are 5×10^{-5} and 5×10^{-6} Å for 20 and 2 m/s, respectively. This suggests that there is another process, such as correlated motion in the formation of voids or TCP inter-diffusion, that is on the on the order of the separation rates used in the simulation. Exceeding this rate drives the system into a higher energy configuration as is illustrated by comparing to two plots in figure 3a. Plotted in figure 3b is the energy difference between the 20 and 2 m/s simulations with both systems starting from an identical configuration. Divergence begins almost immediately and persists to a 10 Å separation covering the full range plotted in figure 1.

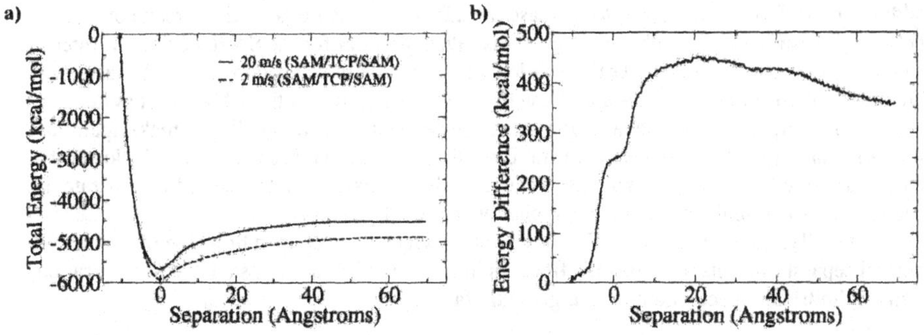

Figure 3. Results from de-adhesion simulations. a) Total energies as a function of separation for 20 m/s and 2 m/s. b) Energy difference as a function of separation for 20 m/s and 2 m/s

An energetic preference for the TCP to remain in a cylindrical defect after overcoming a small anisotropic barrier for inclusion was shown in [2]. It is of interest to explore whether the mobile TCP will remain in this defected area, effectively coating the surface, when contact is broken via normal separation. The evolution of the TCP layers is illustrated in Figure 4.

Figure 4. Time ordered snapshots of the SAM(d)/TCP/SAM system at a separation rate of 20 m/s. The frames correspond to separations of 0, 15, 30, 45, and 60 Å. The SAM and TCP molecules are shaded grey and black, respectively. Molecules in the SAM are transparent for clarity. The separations cannot be directly compared to those above because of different initial pressure states.

As can be seen in Figure 4 the break up of the TCP layers is similar to that seen in Figure 2. Voids form near the defected surface that then grow and evolve as the surfaces are separated. Although most of the TCP is removed from the defected surface the majority of molecules that remain are localized in the defect, effectively coating the exposed area.

CONCLUSIONS

Molecular dynamics simulations have been used to study the adhesion properties of a SAM/SAM and SAM/TCP/SAM interface. SAMS are known to have a low self-adhesion and addition of TCP does not appear to increase the adhesive pressure needed to break contact between the surfaces. The adhesive pressure is rate dependent for the SAM/TCP/SAM interface and separation rates varying over three orders of magnitude have been explored. Adhesive pressures for all separation rates are below the calculated values of the SAM/SAM system.

Break up of the mobile layer is also rate dependent. The mobile layer breaks apart via the formation and growth of voids in the film. Even though the separation rate is much slower than the diffusion length of TCP it appears to exceed another unknown rate. Exceeding this rate has been shown to drive the system into a higher energy configuration.

Finally, it is found that TCP will remain in defected areas upon breaking of contact via normal separation. This reaffirms the findings in reference 2 that suggested that incorporation into a cylindrical defect would be energetically favorable.

ACKNOWLEDGMENTS

This work has been supported by the Air Force Office of Scientific Research, MURI grant # FA9550-04-1-0381. DLI would also like to acknowledge many valuable discussions with J. Krim, A. Hook, B. Miller, L. Sun, Y. Hu, Z. Hu, and C. W. Padgett.

REFERENCES

1. K. C. Eapen, S. T. Patton, and J. S. Zabinski, *Tribol. Lett.* **12**, 35 (2002).
2. D. L. Irving and D. W. Brenner, *J. Phys. Chem. B*, **110**, 15426 (2006).
3. D. W. Brenner, D. L. Irving, A. I. Kingon, J. Krim, and C. W. Padgett, *Langmuir*, **23**, 9253 (2007).
4. W. Neeyakorn, M. Varma, C. Jaye, J. E. Burnette, S. M. Lee, R. J. Nemanich, C. S. Grant, and J. Krim, *Tribol. Lett.*, **27**, 269 (2007).
5. W. Smith and T. R. Forester, *J. Mol. Graphics*, **14**, 136 (1996).
6. W. D. Cornell, P. Cieplak, C. I. Bayly, I. R. Gould, K. M. Merz, D. M. Ferguson, D. C. Spellmeyer, T. Fox, J. W. Caldwell, P. A. Kollman, *J. Am. Chem. Soc.*, **117**, 5179 (1995).
7. A. R. Burns, J. E. Houston, R. W. Carpick, T. A. Michalske, *Langmuir*, **15**, 2922 (1999).
8. M. Chandross, G. S. Grest, and M. J. Stevens, *Langmuir*, **18**, 8392 (2002).
9. K. J. Tupper, R. J. Colton, and D. W. Brenner, *Langmuir*, **10**, 2041 (1994).

Micromechanics II

Mater. Res. Soc. Symp. Proc. Vol. 1052 © 2008 Materials Research Society 1052-DD02-01

Microstructural and Geometrical Factors Influencing the Mechanical Failure of Polysilicon for MEMS

Krishna Jonnalagadda, and Ioannis Chasiotis
Aerospace Engineering, University of Illinois at Urbana-Champaign, 104 S. Wright Street, Urbana, IL, 61801

ABSTRACT

The mechanical strength of polycrystalline silicon is discussed in terms of activation of critical flaws, as well as material microstructure and inhomogeneity. The Weibull probability density function parameters were obtained to deduce the scaling of material and component strength and to identify critical flaw populations, especially when two or more flaw types are concurrently active. It was shown that scaling of strength can change for self-similar micron-sized features, which limits the applicability of strength data from large MEMS components to small MEMS components. On the other hand, the probability of failure for small components is described by a larger Weibull material stress parameter, which makes uniaxial strength data appropriate for conservative design. Furthermore, according to mode I and mixed mode I/II fracture studies for polysilicon, it is concluded that variation in the local critical energy release rate, owed to microstructural inhomogeneity, accounts for up to 50% scatter in strength (with reference to the minimum recorded value.) Thus, the conditions for the applicability of the Weibull probability density function in polycrystalline silicon are rather weak, because flaws of the same length that are subjected to the same macroscopic stresses are not always critical.

Introduction

The mechanical strength and toughness of brittle materials for microelectromechanical systems (MEMS) are rather orthogonal properties. Common materials, such as silicon and polysilicon, are not stronger than 3.5-7 GPa [1-3]. At the same time the toughness of polysilicon is on average about 1 MPa√m [4-6]. Other materials, such as ta-C have larger strengths (7-11 GPa) [7] and average mode I toughness of about 4.5 MPa√m [8]. Similarly the strength of SiC is on the order of 3-5 GPa while its toughness reaches 3-4 MPa√m [9-10]. Thus, the strengths of these thin film materials are significant but their toughnesses are low.

In polysilicon, mechanical strength is defect-controlled and it is affected by the conditions of deposition, micromachining, and post-processing [11-13]. Recent experiments with single crystal silicon structures suitable for nanoelectromechanical systems (NEMS) pointed out that nanometer-scale smooth surfaces limit the size of detrimental flaws and allow for mechanical strengths that approach the theoretical values [14]. Brittle fracture, on the other hand, is controlled by interatomic bond strengths. Polysilicon falls in this category and its mechanical strength is influenced by local properties and stresses. Device design is conducted according to strength data, due to the lack of appreciable toughness. A design methodology presented in [15] incorporated strength data from uniform tension experiments applied to finite element models so that probabilities of failure could be estimated for more elaborate specimen geometries. Recently, it was shown that multiple flaw populations with comparable severity may be in competition and therefore higher order Weibull analyses are needed [16].

This paper aims at providing perspective on unpublished and previously reported data on strength and toughness of polysilicon fabricated by the Multi User MEMS Processes (MUMPs), so that the concepts of material and device failure are put in perspective while the influence of microstructure (grain) on mechanical strength is elucidated.

Experiments and Discussion

A methodology to quantify, and potentially predict, the probability of brittle failure involves uniaxial tension experiments where the ordinal probability of failure is plotted as a function of "material" strength [15,17]. This analysis is based on the premise that, under a certain local stress field, a flaw is always critical and catastrophic. Theoretically speaking, such an analysis can also account for self-similar geometries through scaling of the material volume or surface area containing the critical flaws. Scaling is employed to rationalize experimental results but very rarely the extracted parameters are used to make or to confirm predictions [15]. Hence, it is not possible to assess the fidelity and transferability of Weibull probability density function parameters reported in literature. Two examples are presented in this paper to discuss the effects of (a) modification of scaling of strength due to material processing, and (b) the competition between coexisting flaw populations in specimens with self-similar geometries. In the first example, microscale as-fabricated and Tungsten (W)-coated polycrystalline silicon uniform tension specimens were manufactured by the SUMMiT process at the Sandia National Labs. A 15-nm thick, conformal on all outer specimen surfaces, W-coating was applied by chemical vapor deposition [18].

Figure 1 shows the probability of failure as a function of strength for different specimen dimensions and surface finish. The analysis assumes that the Weibull conditions of the weakest link approximation do hold, i.e. a critical flaw always leads to catastrophic failure and that strength values scale with the size of the surface area or volume where the flaw population is located. This last point is applied in the present discussion, in other words, strong scaling of strength data is an indication of the location of the active flaw population.

Table I. Weibull parameter estimates for as-fabricated and W-coated polycrystalline silicon specimens grouped by length.

Weibull Parameter	Specimen Length			
	W-coated samples		Uncoated samples	
	250 μm	1000 μm	250 μm	1000 μm
m	7.86	7.75	8.45	9.71
σ_0 (GPa)	2.32	2.14	3.50	3.06

Tensile strength tests were conducted with the aforementioned materials and the Weibull parameters are listed in Table I. The strength data from the as-fabricated specimens were analyzed in a previous publication [1,17]. Specifically, it was shown that the failure inducing flaws resided on the specimen sidewalls since both the average and the characteristic strengths scaled with the specimen sidewall surface area. In comparison, the strength data from W-coated specimens did not abide to the same scaling law as the uncoated samples. W-coated polysilicon demonstrated strength values on average smaller by 1 GPa. Generally speaking, this may be considered an unexpected outcome as surface coatings even out surface markings and therefore alleviate stress concentrations. However, a previous literature report pointed out to W-clusters (particles) near the W-polysilicon interface that were as large as 20 nm [18], which may lead to the lower strength values are reported here. Additionally, one must consider the state of stress at the interface when evaluating strength data in composite films. The W layer in the particular composite is characterized by large tensile stress that was reported to be 1 GPa [18]. The latter value agrees very well with the shift in the two pairs of curves in Figure 1. This tensile stress acts as offset on the stress applied to surface flaws, therefore increasing their probability to become critical. Nevertheless, this stress mismatch alone cannot fully explain the change in scaling of mechanical strength that is discussed in the next paragraphs.

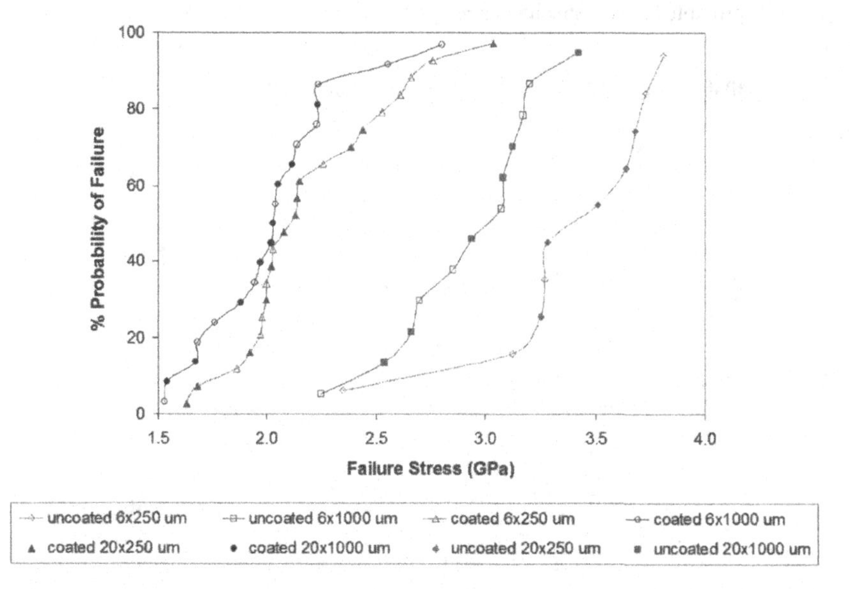

Figure 1. Probability of failure as a function of tensile strength for W-coated and as-fabricated polysilicon specimens grouped by specimen length. The two numbers in the legend are the specimen width and length respectively. All specimens were 2.5 μm long.

Specifically, it appears that the mechanical strength of W-coated specimens does not scale with the specimen length. In order to identify further correlations, the source of reduction in strength is taken into account by plotting all strength data according to the specimen surface area. Since the coating is conformal, any relevant flaws would be distributed on the outside surface of the specimens. Figure 2 presents a plot of the strength data following the aforementioned considerations. The plots of the W-coated samples are not smooth due to the small number of available specimens. When the strength values are grouped according to the surface area, the flaw population is shown to scale with the specimen length at low probabilities of failure, and with the surface area at large probabilities of failure.

Given the data groupings in Figure 2, the following interpretation is offered: At probabilities of failure smaller than 35%, the sidewall grooves generated by reactive ion etching (RIE) were the cause of failure as in the case of uncoated specimens. The severity of RIE flaws was exacerbated by W-clusters at the troughs of sidewall ridges. The addition of stress concentrations due to W-clusters led to smaller strength values for the W-coated specimens compared to the uncoated specimens with the same flaw scaling. At probabilities of failure larger than 55% the strength values were ordered according to the outside specimen surface as the newly introduced subsurface W clusters acted as stress raisers and initiated failure. In this case the W-clusters covering the top and bottom specimen surfaces were responsible for specimen failure. These surfaces are smoother compared to the sidewalls. Therefore, the compounding of stress concentrations (surface roughness and subsurface particles) was less severe and it promoted higher specimen strengths.

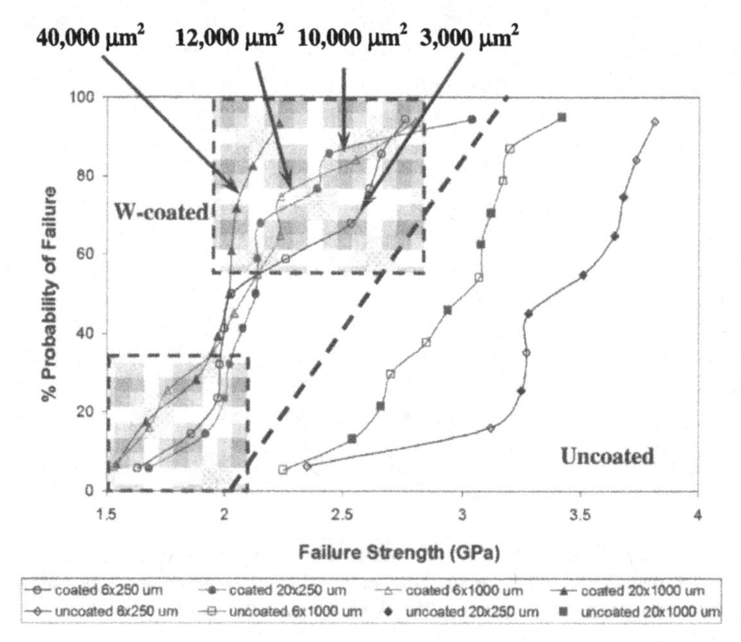

Figure 2. Probability of failure as a function of tensile strength for specimens grouped by length for as-fabricated polysilicon, and by top and bottom surface area for W-coated polysilicon. The dashed line separates the data from the two sets of specimens: (left) coated and (right) uncoated.

Hence, it is concluded that an improved approach in describing failure integrates the effect of multiple flaw populations and potentially any knowledge about local stress fields. In the case of uniform tension it is possible to derive analogies from strength values because the stress is constant. This may not be possible, however, for samples subjected to non-uniform stresses. Previous studies with MUMPs polysilicon [1,16] showed that strength values from uniaxial tension experiments provide scalable characteristic Weibull strengths only for some specimen geometries. MUMPs polysilicon has significant surface roughness compared to SUMMiT polysilicon and failure in the former originates in the specimen surface. Experiments conducted with notched specimens with various radii of curvature and stress concentration factors showed that there are two possible origins of failure depending on the local stress distribution and the absolute dimensions of the specimen (see references [8,9] for details):

1. Small radii of curvature or small stress concentration factors (one can be controlled independently of the other) allow for smaller sidewall surface under high stress compared to the top surface area, and consequently, active flaws that are mainly due to the top surface roughness. In other words, this case corresponds to sidewall flaws with a smaller weight factor (severity) than the top surface flaws that initiate fracture.

2. Large radii of curvature or large stress concentration factors (as before, one can be controlled independently of the other) result in larger weight factors for the sidewall surface flaws than for the top surface flaws. Consequently, the majority of failure inducing flaws reside at sidewalls, which are the result of the RIE process employed to pattern polysilicon films.

Figure 3 is instructive in this regard. Samples that failed due to flaws originating at the specimen top surface provided characteristic strength values between 1.7-2.0 GPa, while samples that failed due to flaws at the specimen sidewalls resulted in smaller characteristic strengths between 1.0-1.4 GPa. The lack of a consistent dominant flaw population implies that a single Weibull probability density function is not sufficient to predict failure.

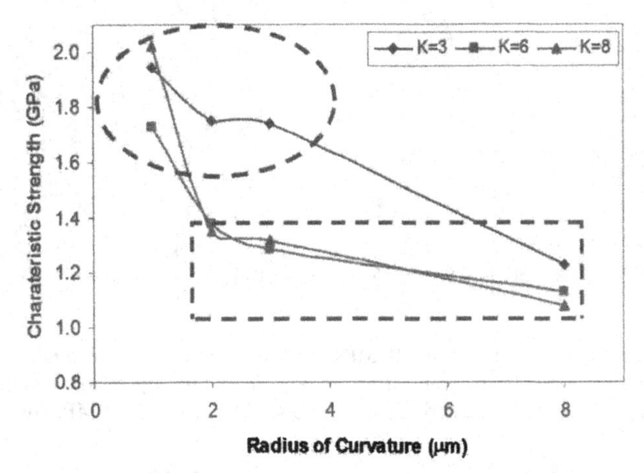

Figure 3. Characteristic strength for various specimen geometries with nominal stress concentration factors K= 3, 6, 8 (true stress concentration factors K= 3, 8, 11) and radii of curvatures 1, 2, 3, and 8 μm. The dashed lines include specimen geometries for which failure initiated at the same specimen surface.

Moreover, local anisotropy and inhomogeneity are material aspects that are not accounted for in a Weibull analysis. An experimental investigation of polysilicon fracture with pre-cracked specimens showed that fracture is controlled by local surface energies both in mode I and mixed mode I/II loading. Macroscopic mode I fracture toughness values obtained from edge pre-cracked polysilicon samples, measured with an accuracy of a few kPa√m, spanned the range between 0.85-1.25 MPa√m and averaged 1.0 MPa√m. The smaller values of critical stress intensity factor correlate well with the fracture toughness (0.82 MPa√m) for {111} single crystal silicon planes. The 50% variation in fracture toughness (with the smallest value used as reference) is the effect of local in-grain anisotropy and more importantly due to grain boundaries that provide enhanced resistance to crack initiation [4]. An example of two pre-cracks with the same length and orientation with respect to the applied far-field stress, i.e. nominally the same severity, is shown in Figure 4. The crack in Figure 4(a), whose tip was located in a grain, resulted in 30% smaller effective critical stress intensity factor, K_{Ic}, compared to the crack in Figure 4(b) that was of the same length but its tip was located at a triple junction point. The variation in K_{Ic}, due to local control in crack initiation, is directly proportional to the variation in the macroscopically applied stress. For a critical flaw of given size and orientation, the latter results in an equivalent variation in the measured tensile strength. In other words, 50% variation in material strength does not strictly reflect a large distribution of defect sizes as calculated by Griffith's criterion assuming homogeneity.

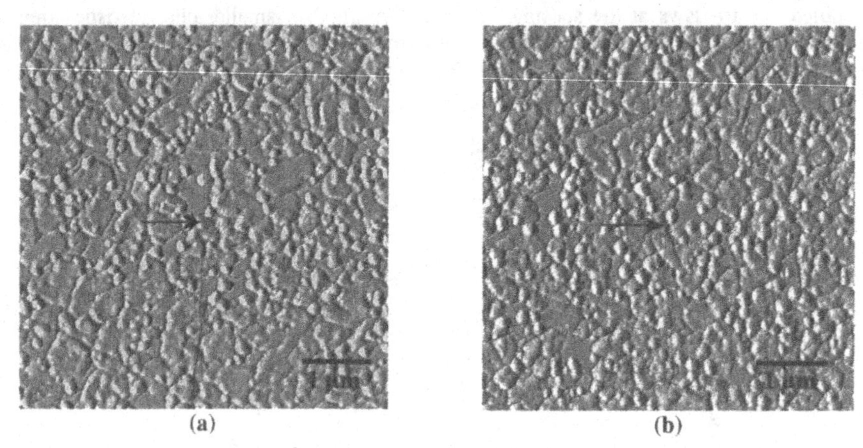

Figure 4. AFM images (5×5 μm^2) of MUMPs polysilicon specimens with edge pre-cracks. The crack tip location resulted in different effective K_{IC}: **(a)** Crack tip residing inside grain, K_{IC} = 0.85 MPa√m, vs. **(b)** crack tip located at the grain boundary, K_{IC} = 1.13 MPa√m.

Analogous conclusions are derived from mixed mode I/II fracture experiments. Figure 5(a) shows the locus of normalized mode I/II stress intensity factors for crack angles between 0-41°. An additional datum point from a specimen with pre-crack angle of 55° is included in the figure, which deviates significantly from the theoretical curves. The latter provide predictions for fracture initiation in homogeneous brittle materials according to the maximum tensile strength (MTS) and the maximum energy release rate (MERR) criteria [21]. The geometry of the MEMS fracture specimens was subject to fabrication and mechanical testing constraints and it did not allow for comparisons with predictions by MTS and MERR criteria for > 45°. This deviation is clear for the datum point that corresponded to the 55° crack angle, which is in agreement with reports for macroscopic tests where $K_{II}\backslash K_{Ic}$ decreased for inclination angles larger than a threshold angle [23]. It is easily deduced that the values of mode II stress intensity factor increase according to the pre-crack angle whose values are listed next to each point in Figure 5. On the contrary, the values of K_I present the same spread as K_{IC} that was discussed in the previous paragraph. The scatter in K_I values contributes to the overall spread in the $K_I\backslash K_{IC}$ vs. $K_{II}\backslash K_{IC}$ plot. AFM images showed that two oblique pre-cracks with small and large K_I (for example the two specimens with pre-crack angle 12.5° in Figure 5(a)) had their crack tips inside a grain and at a grain boundary, respectively [19].

Mixed mode I/II fracture experiments with tetrahedral *amorphous* diamond-like carbon (ta-C) helped to emphasize the defining role of inhomogeneity in polysilicon fracture [20]. Figure 5(b) shows a small data set of K_I/K_{IC} vs. K_{II}/K_{IC} measurements from edge cracks in ta-C films also fabricated for MEMS and hard coatings. In the absence of stress gradients [8], the fracture data agreed with predictions for fracture initiation by MTS and MERR criteria [21]. Similarly, experiments on mode I toughness of ta-C resulted in small standard deviations in K_{IC} as long as residual stresses and their gradients were accounted for [8]. As a conclusion, the distribution in polysilicon K_{IC} values is owed to the material microstructure that also contributes significantly to the stochastic nature of its strength.

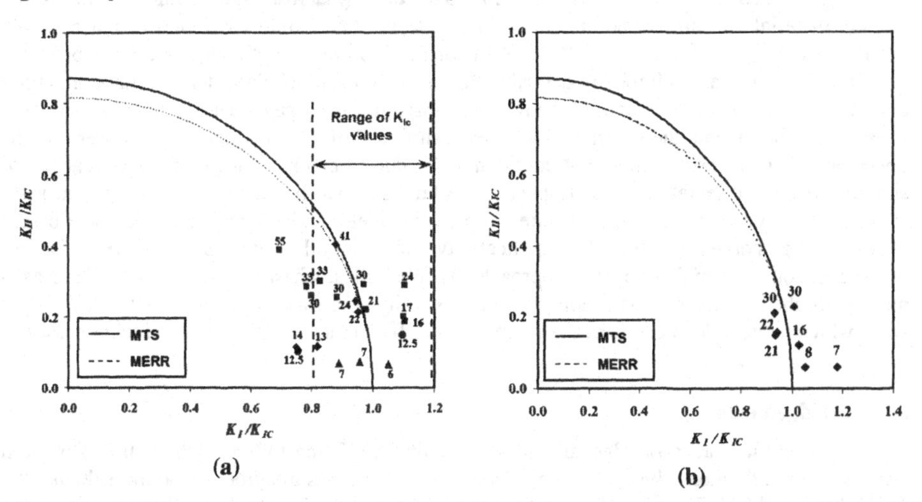

Figure 5. K_I/K_{IC} vs. K_{II}/K_{IC} computed from oblique edge pre-cracks for **(a)** polysilicon, **(b)** tetrahedral amorphous diamond-like carbon (ta-C). The experimental data are compared to predictions by MERR (dashed line) and MTS (solid line) criteria. The numbers next to the data points are the pre-crack angles.

There are additional factors of lesser importance that influence failure initiation. Local inhomogeneity affects not only the local energy release rate but also the transmission of load from the far field to the crack tip. Finite element modeling alluded to modest crack tip shielding [5] due to material inhomogeneity. Experiments with SUMMiT polysilicon concluded to a minimum representative volume element for polysilicon of 15×15 columnar grains [22]. For

MUMPs polysilicon (300 nm average grain size) that was used for the notched and pre-cracked samples presented in this paper, an ensemble of 15×15 columnar grains is equivalent to specimen domains that are approximately 5×5 μm^2. As a consequence, load transmission in small material volumes at acute stress concentrations may not be described appropriately by homogeneous fields, which imposes additional constraints to the application of Weibull probability functions to describe polysilicon failure.

Conclusions

The mechanical strength and fracture toughness of polycrystalline silicon was addressed by probabilistic and local approaches. It was shown that a comprehensive treatment of polysilicon failure with predictive potential requires accounting for multiple flaw populations that are simultaneously active and become dominant according to sample size and the local details of stress. In addition to device geometry driven activation of flaw populations, material treatments (e.g. conformal coatings) not only may shift the strength data but also change the nature and distribution of active flaws and finally the scaling law in mechanical strength.

It was also shown that the variation in the effective (macroscopic) fracture toughness due to local material anisotropy and inhomogeneity reflects directly into the values of macroscopic strength and accounts for a large portion of the distribution of strength data. In other words, the spread in the latter, as traditionally described by the Weibull probability density function, is not only due to a spread of flaw sizes in the material. Instead, to a large extent it is the result of local stress fields that dictate the severity of otherwise similar flaws. Consequently, the lower and the upper bounds of measured tensile strengths may be due to cracks or material flaws that are of similar sizes and orientations with respect to the far field load but are subject to different local energy release rates and stress fields due to crack tip shielding. This realization to some extent weakens the conditions for the applicability of Weibull statistics. However, for an inhomogeneous material, one may approach the Weibull stochastic treatment of failure as a method to obtain scaling rules according to the severity of a flaw population (not just the flaw size), which is the combined effect of distributions of flaw/crack lengths and local toughnesses.

Acknowledgements

The authors acknowledge the support by the Air Force Office of Scientific Research (AFOSR) through grant F49620-03-1-0080 with Dr. B.L. Lee as monitor. The authors thank Drs. T. Buchheit and T.A. Friedman from the Sandia National Labs for providing the polysilicon and ta-C specimens, respectively.

References

1. I. Chasiotis, W.G. Knauss, "The Mechanical Strength of Polysilicon Films: 2. Size Effects Associated with Elliptical and Circular Perforations", J. of the Mechanics and Physics of Solids 51, pp. 1551-1572, (2003).
2. B. L. Boyce, J.M. Grazier, T.E. Buchheit, M.J. Shaw, "Strength Distributions in Polycrystalline Silicon MEMS," Journal of Microelectromechanical Systems, 16 (2), pp. 179-190, (2007).

3. Y. Isono, T. Namazu, T. Tanaka, "AFM Bending Testing of Nanometric Single Crystal Wire at Intermediate Temperatures for MEMS," 14th IEEE Intl. Conf. on Microelectromechanical Systems, pp. 135-138, (2001).
4. I. Chasiotis, S.W. Cho, K. Jonnalagadda, "Fracture Toughness and Subcritical Crack Growth in Polycrystalline Silicon", Journal of Applied Mechanics 73 (5), pp. 714-722, (2006).
5. R. Ballarini, R. L. Mullen, A.H. Heuer, "The effects of heterogeneity and anisotropy on the size effect in cracked polycrystalline films," International Journal of Fracture 95, pp. 19-39, (1999).
6. R. Ballarini, R. L. Mullen, Y. Yin, H. Kahn, S. Stemmer, A.H. Heuer, "The fracture toughness of polysilicon microdevices: A first report," Journal of Materials Research 12(4), pp. 915-922. (1997).
7. S.W. Cho, I. Chasiotis, T.A. Friedman, J. Sullivan, "Young's modulus, Poisson's ratio and failure properties of tetrahedral amorphous diamond-like carbon for MEMS devices", Journal of Micromechanics and Microengineering 15 (4), pp. 728-735, (2005).
8. K. Jonnalagadda, S. Cho, I. Chasiotis, T.A. Friedman, J. Sullivan, "Effect of Intrinsic Stress Gradient on the Effective Mode-I Fracture Toughness of Amorphous Diamond-like Carbon Films for MEMS," (in press) Journal of the Mechanics and Physics of Solids, (2008).
9. W.N. Sharpe, O. Jadaan, G.M. Beheim, G.D. Quinn, N.N. Nemeth, "Fracture Strength of Silicon Carbide Microspecimens," Journal of Microelectromechanical Systems, 14(5), pp. 903-913, (2005).
10. S. Roy, R.G. DeAnna, C.A. Zorman, M. Mehregany, "Fabrication and Characterization of Poycrystalline SiC Resonators," IEEE Transactions on Electron Devices, 49(12), pp. 2323, (2002).
11. I. Chasiotis, W.G. Knauss, "The Mechanical Strength of Polysilicon Films: 1. The Influence of Fabrication Governed Surface Conditions", J. of the Mechanics and Physics of Solids 51, pp. 1533-1550, (2003).
12. H. Kahn, C. Deeb, I. Chasiotis, A.H. Heuer, "Anodic oxidation during MEMS processing of silicon and polysilicon: native oxides can be thicker than you think", Journal of Microelectromechanical Systems 14 (5), pp. 914-923, (2005).
13. D.C. Miller, B.L. Boyce, K. Gall, "Galvanic Corrosion Induced Degredation Of Tensile Properties In Micromachined Polycrystalline Silicon," Applied Physics Letters, 90,191902, pp. 1-3, (2007).
14. T. Alan, M. Hines, and A. Zehnder, "Effect of Surface Morphology on the Fracture Strength of Silicon Nanobeams," Applied Physics Letters, 89, pp. 091901, (2006).
15. J. Bagdahn, W.N. Sharpe, O. Jadaan, "Fracture strength of polysilicon at stress concentrations", Journal of Microelectromechanical Systems 12 (3) pp. 302-312, (2003).
16. A. McCarty, I. Chasiotis, "Description of Brittle Failure of Non-uniform MEMS Geometries", Thin Solid Films 515, pp. 3267-3276, (2007).
17. D.A. LaVan, T. Tsuchiya, G. Coles, W.G. Knauss, I. Chasiotis, D. Read, "Cross Comparison of Direct Tensile Techniques on SUMMiT Polysilicon Films", Mechanical Properties of Structural Films, ASTM STP 1413, pp. 1-12, ASTM, W. Conshohocken, PA (2001).
18. S.S. Mani, J.G. Fleming, J.A.Walraven, J.J. Sniegowski, M.P. de Boer, L.W. Irwin, D.M. Tanner, D.A. LaVan, M.T. Dugger, J. Jakubczak, W.M. Miller, "Effect of W Coating on Microengine Performance", 38th Annual International Reliability Physics Symposium, pp.146-151, (2000).
19. S.W. Cho K. Jonnalagadda, I. Chasiotis, "Mode I and Mixed Mode Fracture of Polysilicon for MEMS," Fatigue and Fracture of Engineering Materials and Structures 30 (1), pp. 21-31, (2007).
20. K. Jonnalagadda, I. Chasiotis, "Mixed Mode Fracture in Brittle Materials for MEMS," Proceedings of 2006 ASME International Mechanical Engineering Congress and Exposition, IMECE2006 - Microelectromechanical Systems, (2006).
21. T.L. Anderson, Fracture Mechanics, 2nd Edition, CRC Press, (1995).
22. S.W. Cho, I. Chasiotis, "Elastic Properties and Representative Volume Element of Polycrystalline Silicon for MEMS," Experimental Mechanics 47 (1), pp. 37-49, (2007).
23. P.S. Theocaris, G.A. Papadopoulos, "The influence of edge cracked plates on K_I and K_{II} components of the stress intensity factor studied by caustics", Journal of Physics D: Applied Physics 17, 2339-2349, (1984).

Mater. Res. Soc. Symp. Proc. Vol. 1052 © 2008 Materials Research Society 1052-DD02-02

Passive devices for determining fracture strength of MEMS structural materials

H Kahn[1], R Ballarini[2], and A H Heuer[1]
[1]Materials Science and Engineering, Case Western Reserve University, Cleveland, OH, 44106-7204
[2]Civil Engineering, University of Minnesota, Minneapolis, MN, 55455

ABSTRACT

In microelectromechanical systems (MEMS) device design and fabrication there exists a need for rapid determination of fracture strength. This report describes several passive devices that use the residual stresses contained within structural MEMS materials to determine fracture strength. Stress concentrations of varying degrees are generated at micromachined notch roots, and the critical stress required for failure indicates the fracture strength. A variety of devices have been fabricated from materials such as polysilicon, silicon nitride, and aluminum, with widely varying residual stresses, including devices with both tensile and compressive residual stresses.

INTRODUCTION

To accurately design microelectromechanical systems (MEMS) devices and predict their reliabilities, the mechanical strengths of the structural materials must be known. Since MEMS are typically made from brittle materials such as polysilicon, it is essential that the strength measurements be made on materials that have gone through the same fabrication steps, namely deposition and etching, that would be seen in device processing. For brittle materials, the strength is governed by the inherent flaws that are generated during fabrication. Rapid determination of fracture strength could also then be used to verify that any changes in a fabrication process had not caused detrimental effects to the device integrity.

It is highly desirable that the strength measurements be made using "passive" devices – devices that perform their function upon release with no further actuation. Then, visual inspection is all that is needed to obtain the results. Devices that fit this description (which have previously been reported) are clamped-clamped beams produced from materials that contain residual tensile stresses [1-3]. Upon release, the residual tension creates stresses within straight beams [1], notched beams [2], or beams with sharp pre-cracks [3]. If the stresses are high enough, the beams will fail catastrophically, and if not, no changes will be observed. Given accurate knowledge of the residual stress in the material, finite element analysis (FEA) can predict the stresses in the device. Then, upper and lower bounds can be established for the strength of the material.

However, for a given device geometry, the usable range of residual stresses can be small. For some materials, the residual stress can be tailored by varying the deposition conditions, but in other cases, this ability is limited. In this report, we describe several passive device designs that can be used to determine the strength of materials with a wide range of residual stresses, including devices with both tensile and compressive residual stresses.

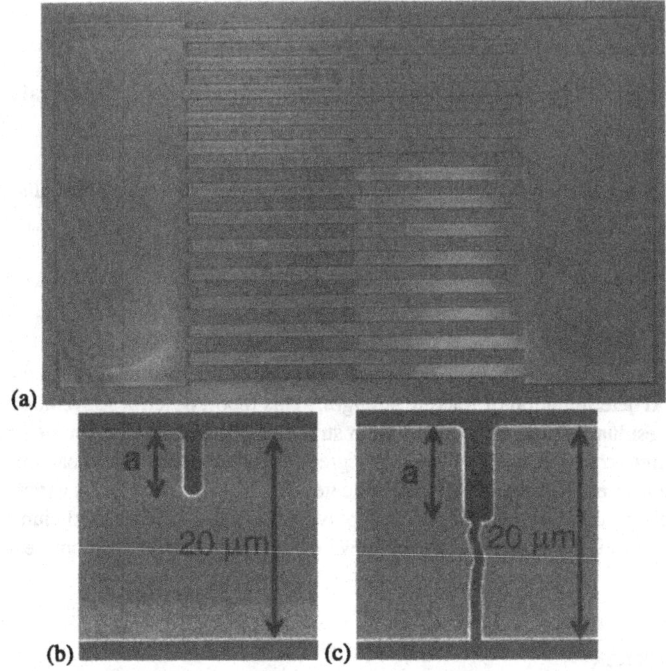

Figure 1. (a) SEM micrograph of a polysilicon device comprising a series of 500 µm long clamped-clamped beams containing notches with increasing depths. (b) and (c) Higher magnification images of the 4th and 7th beams from the top.

EXPERIMENTAL DETAILS

The devices reported here were all fabricated using standard surface micromachining technology. Device dimensions were determined using scanning electron microscopy (SEM). The residual stress of each material was quantified by laser wafer curvature measurements. For the polysilicon and silicon nitride films, the residual stresses were also determined using rotating microstrain gauges [4] fabricated on the same wafers, with good agreement between the two techniques. Two-dimensional FEA was used to predict the stress concentrations at the notch roots.

RESULTS AND DISCUSSION

Fig. 1 shows the device designed to investigate tensile strength for notched beams in materials with moderate tensile stresses. It consists of 500 µm long clamped-clamped beams with micromachined notches with root radii of 0.9 µm. The notches in the 13 beams increase in depth, *a*, from 3.2 µm to 15.2 µm, with 1 µm intervals. The device is fabricated from a 2 µm thick undoped polysilicon film deposited at 570°C by low-pressure chemical vapor deposition (LPCVD) and subsequently annealed at 615°C to produce a fine-grained microstructure [5] with a tensile residual stress of 318 MPa. The polysilicon is etched in a Cl_2 plasma, and the SiO_2 release layer is dissolved in liquid HF. The release is timed, to leave in place the sacrificial SiO_2 beneath the large anchors seen on both sides of the beams in Fig. 1a. This results in extremely stiff anchors.

As seen in Fig. 1a, the top 5 beams (with notch depths of 7.2 μm and smaller) survived after release, exemplified by Fig. 1b, and the lower 8 beams (with notch depths of 8.2 μm and greater) failed after release, as illustrated by Fig. 1c. Due to the stochastic nature of brittle failure and etching-induced flaws, several identical devices were fabricated and released. Out of 14 devices, the beam with $a = 8.2$ μm failed 11 times. Statistical analysis using the standard Weibull distribution for brittle fracture gave the fracture strength as 3.9 GPa.

Fig. 2 shows the same device fabricated from a 1 μm thick sputtered Al film, with a residual tensile stress of 44 MPa. In this case, the fracture strength was determined to be 1.0 GPa. While the intact beam (Fig. 2b) shows a small amount of bending, the broken beam (Fig. 2c) does not. This implies that no plastic deformation occurred before fracture. However, this could be investigated in greater detail with devices designed specifically for this purpose.

In Fig. 2, only the 2 beams with the deepest notches fractured upon release. If the residual stresses were slightly lower, none of the beams would have broken, and the fracture strength could not have been determined. For such low residual tensile stresses, the device shown in Fig. 3 has been designed. The beams are wider at the ends and narrower in the center. This effectively increases the stress in the central portion of the beams.

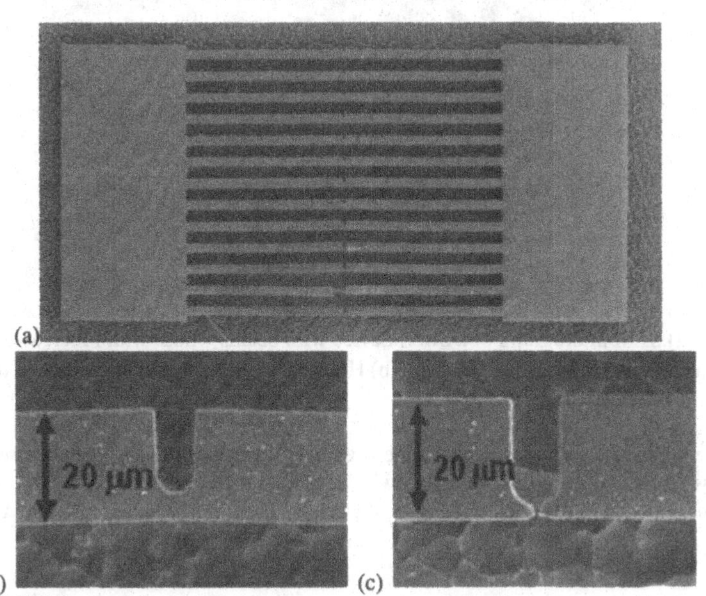

Figure 2. (a) SEM micrograph of the same device shown in Fig. 1, made from aluminum. (b) and (c) Higher magnification images of the 11th and 13th beams from the top.

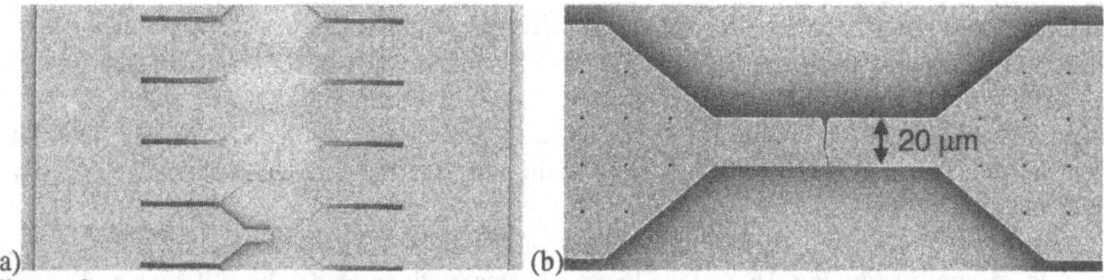

Figure 3. (a) SEM micrograph of a device with tapering beams. The total beam length is 500 µm. (b) Higher magnification image showing the center of the top beam. The release holes are visible.

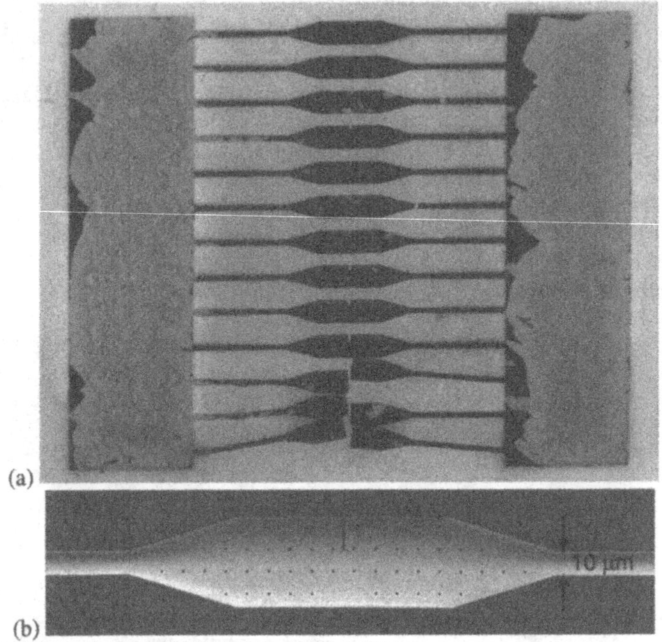

Figure 4. (a) SEM micrograph of a Si_3N_4 device with beams whose width increases in the center. The total beam length is 500 µm. (b) Higher magnification image showing the center of one beam. The release holes are visible.

For the opposite case – very high tensile stresses – the device in Fig. 4 can be used. This device was fabricated from LPCVD Si_3N_4 with a residual tensile stress of 800 MPa. In a similar manner but opposite to the device of Fig. 3, the central portion of the beams, where the notches are located, are wider than the ends. This reduces the stress concentrations at the notch roots. An additional complication for this device is that the stress at the notch roots can be greater after partial release. That is, after the HF has dissolved the release oxide from beneath the sides of the beams, but a thin strip of SiO_2 remains beneath the central axis of the beam, the stress at the notch root can actually be higher than after the beam is fully released. To prevent this situation, the release holes visible in Fig. 4b were designed so that the last release oxide to be removed is located directly opposite the notch. FEA analysis shows that this geometry effectively shields the notch root from high stresses during the release process. The Si_3N_4 devices reveal a fracture strength of 6 GPa.

Finally, for materials with compressive residual stresses, the devices in Fig. 5 have been designed. The devices shown up to this point can be considered MEMS equivalents of familiar uniaxial tensile specimens used for bulk materials. The devices in Fig. 5 are the MEMS equivalent of the "Theta" specimen, in which compressive loading and Poisson expansion of a brittle sample shaped like the Greek letter theta leads to tensile stresses in the cross beam sufficiently large to cause tensile failure [5]. Because the release holes are difficult to see in Fig. 5, the anchors have been labeled. Upon release, the material expands due to its residual

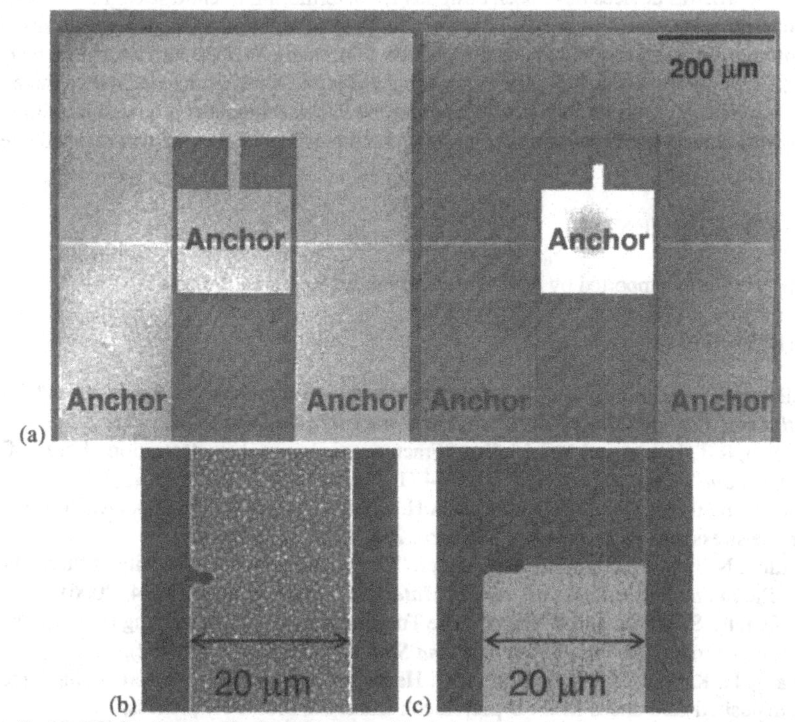

Figure 5. (a) SEM micrograph of two devices for determining fracture strength in compressive materials. (b) and (c) Higher magnification images showing the notched area of the two devices. The device with the greater notch depth has failed.

compression. The wide, long beams on the sides of the devices expand more than the thin, short beams in the center, leading to a tensile stress in the central beams. The notches are naturally located within the central beams. These devices were fabricated from LPCVD polysilicon deposited at 615°C, with a columnar microstructure [6] and a compressive residual stress of -300 MPa.

From the notch depths required for failure, the fracture strength was determined to be 3.0 GPa. This is smaller than that measured for tensile fine-grained polysilicon. It is possible that the columnar microstructure creates a larger surface roughness and larger surface flaws when etched with Cl_2 plasma. It is also possible that these devices were inadvertently over-etched, which resulted in higher surface roughness. This possible reduction in strength due to non-optimized fabrication procedures emphasizes the need for a straightforward and rapid measurement of the fracture strength of micromachined MEMS materials.

SUMMARY

Passive MEMS devices have been designed for determining fracture strength. These devices use the residual stresses contained within the structural MEMS materials to generate stress concentrations at micromachined notch roots. By fabricating devices with a range of notch depths, in conjunction with FEA, the critical stress required for fracture can be determined. Various devices have been fabricated to measure the fracture strengths of materials with widely varying residual stresses, including devices with both tensile and compressive residual stresses.

ACKNOWLEDGMENT

This work was supported by the Defense Advanced Research Projects Agency.

REFERENCES

1. M. Biebl, H. von Philipsborn, "Fracture strength of doped and undoped polysilicon" *Proc. Intl. Conf. On Solid-State Sensors and Actuators, Transducers 95*, 72-75 (1995).
2. L.S. Fan, R.T. Howe, and R.S. Muller, "Fracture toughness characterization of brittle films," *Sensors and Actuators A*, **21-23**, 872-874 (1990).
3. H. Kahn, R. Ballarini, J.J. Bellante, and A.H. Heuer, "Fatigue failure in polysilicon not due to simple stress corrosion cracking," *Science*, **298**, 1215-1218 (2002).
4. H. Kahn, N. Jing, M. Huh, and A.H. Heuer, "Growth stresses and viscosity of thermal oxides on silicon and polysilicon," *Journal of Materials Research*, **21**, 209-214 (2006).
5. A.J. Durelli, S. Morse, and V. Parks, "The Theta specimen for determining tensile strength of brittle materials," *Materials Research and Standards*, **2**, 114-117 (1962).
6. J. Yang, H. Kahn, S.M. Phillips, and A.H. Heuer, "A new technique for producing large-area as-deposited zero-stress LPCVD polysilicon films: the MultiPoly process," *J. Microelectromech. Syst.* **9**, 485-494 (2000).

Mater. Res. Soc. Symp. Proc. Vol. 1052 © 2008 Materials Research Society 1052-DD02-03

Effect of Surface Oxide Layer on Mechanical Properties of Single Crystalline Silicon

Kenji Miyamoto, Koji Sugano, Toshiyuki Tsuchiya, and Osamu Tabata
Microengineering, Kyoto University, Yoshidahonmachi, Sakyo-ku, Kyoto, 606-8501, Japan

ABSTRACT

This paper reports on the tensile testing of single crystal silicon (SCS), whose specimen surface was intentionally oxidized, and the effect of the oxide thickness on the mechanical properties in order to investigate the fatigue fracture mechanism under cyclic loading. SCS specimens were fabricated from silicon-on-insulator (SOI) wafer with 3-µm -thick device layer and oxide layer were grown to the specimens using thermal dry oxidation at 1100 °C. The specimen test part was 120 or 600 µm long and 4 µm wide. Quasi-static tensile testing of SCS specimen without oxide layer, with 100-nm-thick oxide, and with 200-nm-thick oxide was performed. As the results, the fracture origin location changed from the surface of the specimen of SCS without oxide to inside of silicon of oxidized specimen. This change may be caused by the smoothing of the surface and formation of oxide precipitation defects in silicon during oxidation. The estimated radius of the defects in specimen with 100 -nm-thick oxide and with 200-nm-thick oxide was 26 nm and 45 nm, respectively, which is well agreed with the fracture-initiating crack sizes calculated from the measured strengths.

INTRODUCTION

Single crystal silicon (SCS) is widely used for MEMS devices as structural material, due to its excellent mechanical properties and its compatibility to the microfabrication process. Since silicon is a brittle material, the mechanical reliability, especially the fatigue fractures under long term cyclic loading, is being concerned for and the fatigue properties and mechanisms are being investigated widely [1,2,3].

Three features were reported in the previous works on the fatigue of SCS. One is the relationship between the fatigue life and applied stress. Like metals, as the applied stress increases, the fatigue life decreases. Another is the necessity of cyclic loading for fatigue failure. Unlike glass, fatigue fracture doesn't occur under static loading. The other is the effect of the humidity of the test environment. In higher humidity environment, the fatigue life becomes shorter [3]. From these features Muhlstein *et al.* proposed reaction layer fatigue theory as a fatigue mechanism of silicon [1]. In this theory, fatigue process of silicon is the combination of the surface oxide layer growth and the crack propagation within it under cyclic loading, which is enhanced by the surrounding moisture. However, the role of the surface oxide layer is not understood.

The goal of our research is to reveal the effect of surface oxidation on the mechanical properties, such as tensile strength, Young's modulus, and fatigue fracture. For this purpose, SCS specimens covered with oxide layer of known-thickness were prepared. This paper reports on quasi static tensile testing of the SCS specimens to evaluate the fracture strength and fracture behavior prior to fatigue test.

EXPERIMENTAL

Specimen

Three types of specimens were prepared; SCS specimens without oxide layer, with 100-nm-thick oxide layer, and with 200-nm-thick oxide layer. To fabricate the specimen we used silicon on insulator (SOI) wafers. The fabrication process was divided into four steps. The first step was specimen patterning process that was photolithography and inductively coupled plasma-reactive ion etching (ICP-RIE). The second was backside silicon etching using ICP-RIE to remove the substrate under the specimens. The third was removal of the sacrificial oxide layer by etching using buffered hydrofluoric acid (BHF). The fourth was dry thermal oxidation to form thickness-controlled oxide layer. The oxygen flow rate was 0.7 l/min, and the temperature was 1100 °C. The oxidation time was 30 and 90 minutes, which correspond to oxide thicknesses of 100 nm and 200 nm, respectively.

The test part length, width, and thickness were 120 or 600 μm, 4 μm, and 3 μm, respectively. The specimen surface orientation was (100), and tensile axis was aligned to <110> direction. The fabricated SCS specimen of 120 μm in the test part length is shown in Figure 1.

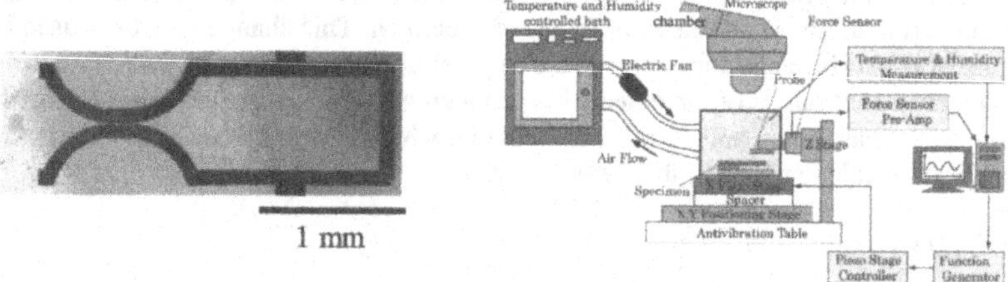

Figure 1. (left) Single crystal silicon specimen of 120-nm-long test part.
Figure 2. (right) Schematic diagram of environment-controlled tensile tester.

Tensile testing

Quasi static tensile testing was carried out using the environment-controlled tensile tester we have developed before [3]. In this tester, electrostatic grip system was used for chucking the specimen. Tensile load was applied using a piezoelectric-actuator-driven stage, and measured by a commercially available load-cell (KYOWA, LTS-50GA). Specimen was put in a small chamber which was connected to a temperature and humidity controlled chamber. The humidity and temperature were monitored by a sensor placed at the vicinity of the specimen. In this experiment, the test environment was controlled at 26 °C and 50 %RH. Force loading rate was controlled at the stage speed of 0.5 μm/sec.

Finite element analysis

Finite element (FE) analysis was performed in order to evaluate the stress distribution in the specimen cross section due to the compressive stress of thermal oxide layer. The FE model was a silicon bar of 100 μm long covered with a 100 nm or 200 nm-thick oxide layer. The compressive stress of 460 MPa was introduced to the oxide layer [4]. One end of the bar was fixed and the initial displacement of 1.7 μm was applied on the other end which is equivalent to the mean fracture strain of SCS specimens.

RESULTS

Tensile testing of SCS specimens (w/o oxide), SCS specimens with 100 nm-thick oxide (100-nm-oxide), and SCS specimens with 200 nm-thick oxide (200-nm-oxide) was carried out. The number of tested specimens of the w/o oxide, 100-nm-oxide, and 200-nm-oxide were 10, 10 and 15, respectively.

Tensile force-stage displacement curves

Figures 3 show the tensile force-stage displacement curves of the 100-nm-oxide and 200-nm-oxide, respectively. The tensile force increases linearly with the stage displacement and the specimens fractured as a brittle material.

Mechanical properties

The nominal Young's modulus of the specimen was calculated from the difference in the slopes of the tensile force-stage displacement curves between 120 and 600 μm-length in order to cancel the deformation except for the test part. The nominal Young's modulus of w/o oxide, 100-nm-oxide, and 200-nm-oxide were 163, 136, and 119 GPa, respectively. Figure 4 shows the fracture strain of each specimen. Each fracture strain was calculated from the measured fracture force and the nominal Young's modulus. The mean fracture strain of specimens w/o oxide, 100-nm-oxide, and 200-nm-oxide were 0.018, 0.020, and 0.016, respectively. In the specimens with oxide, the mean fracture strain was higher than specimens w/o oxide. As for the oxidized specimens, the mean fracture strain decreased with increasing the oxide thickness. The Weibull moduli of w/o oxide, 100-nm-oxide, and 200-nm-oxide were 2.5, 4.7, and 4.5, respectively. The distribution of fracture strain of the oxidized specimens was lower than that of SCS specimens.

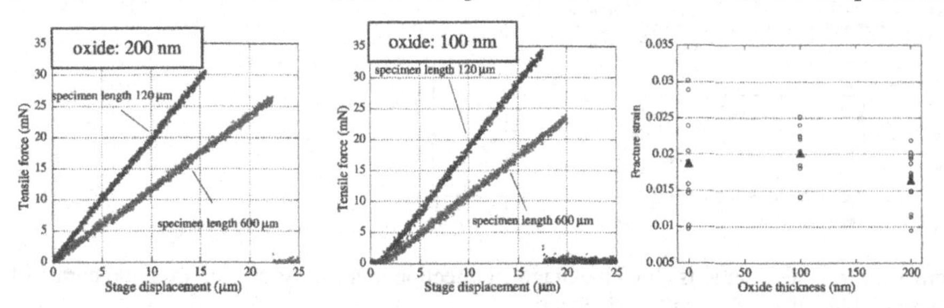

Figure 3. Tensile force-stage displacement curves of specimens with 200-nm-thick (left) and 100-nm-thick oxide layer (center).
Figure 4. (right) Fracture strain of specimens as a function of the oxide layer thickness. Triangles show the average.

Fractography

Figures 5 show the scanning electron microscope (SEM) images of the fracture surface of the specimens. In all specimens, the fractured surface was in (111) plane and river patterns were observed. We assume the center of the radial pattern as the fracture origin. The fracture origin of specimen w/o oxide was on the surface of the specimen, and those of specimens with oxide were not on the surface, but about 200 nm inside the interface of Si/SiO_2.

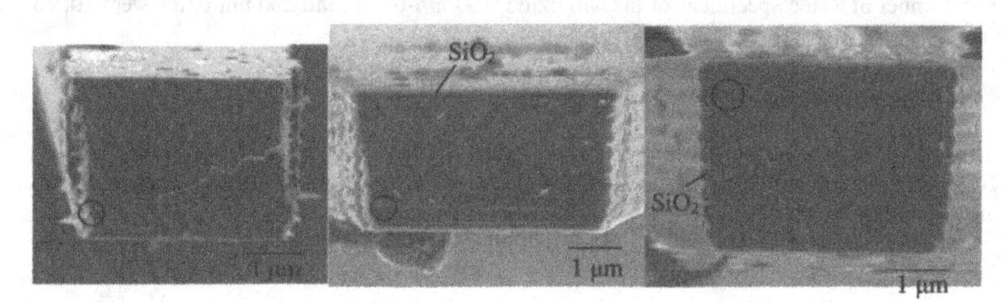

Figure 5. Fracture surfaces of SCS specimens without oxide (left), with 100-nm-thick (center) and 200-nm-thick (right) surface oxide. Circles indicate the estimated fracture origins.

Finite element analysis

Figures 6 show the axial stress profile near the corner of the cross section of the oxidized specimen when the mean fracture strain of silicon was applied as initial displacement. The horizontal axis shows the distance from the bottom corner along the diagonal direction. The maximum axial stress along length direction in the oxide was 846 MPa, and that in silicon was 3.3 GPa. The stress in silicon was slightly higher near the corner interface.

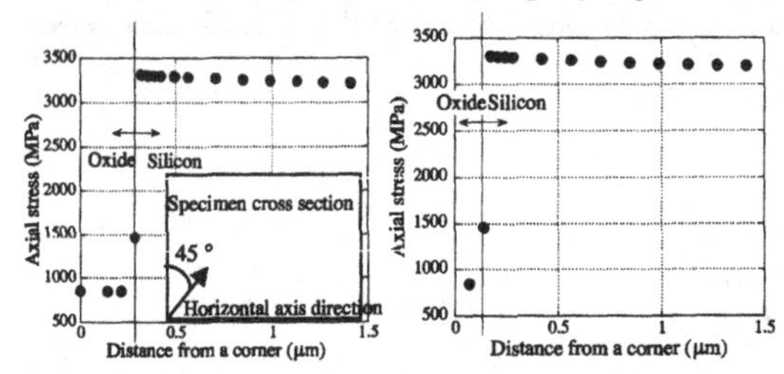

Figure 6. Stress distributions along the diagonal direction in the cross section from the corner of 200-nm-oxide (left) and 100-nm-oxide (right).

DISCUSSION

Young's modulus

The nominal Young's modulus decreased as the oxide thickness increased. The measured nominal Young's modulus was lower than the estimated values from the specimen model with uniform oxide thickness, which was 170, 159, and 148 GPa. The difference of nominal Young's modulus between measured and estimated was due to the large distribution of width and thickness of the specimens.

Fracture origin of oxidized SCS opecimen

SCS specimens with oxide showed very distinctive fracture behaviors; the fracture initiated from the silicon near the corner of the Si/SiO_2 interface. According to the FE analysis of the 200-nm-oxide, the maximum axial stresses in oxide and silicon were 846 MPa and 3.3 GPa, respectively. The fracture stress of silicon dioxide film of 2 μm thick was reported as 1.85 GPa [5]. The maximum stress of surface oxide layer was much smaller than the fracture stress. We thought the fracture does not initiate from surface or oxide layer. Figure 7 shows the maximum axial stress in the silicon part calculated from the fracture strain and the maximum axial stress from the FE analysis. The mean maximum axial stress in silicon of 100-nm-oxide and 200-nm-oxide were 3.8 GPa and 3.2 GPa. Comparing to the theoretical SCS tensile strength of 32 GPa [6], we could estimate the presence of fracture-initiating defects of 31 nm and 21 nm in radius inside silicon.

From the FE analysis, there was stress concentration in the cross section of the specimen due to the compressive stress in the oxide layer. The axial stress in silicon was slightly higher near the corner of Si/SiO_2 interface. Therefore, the fracture may be initiated at the defect near the corner of Si/SiO_2 interface.

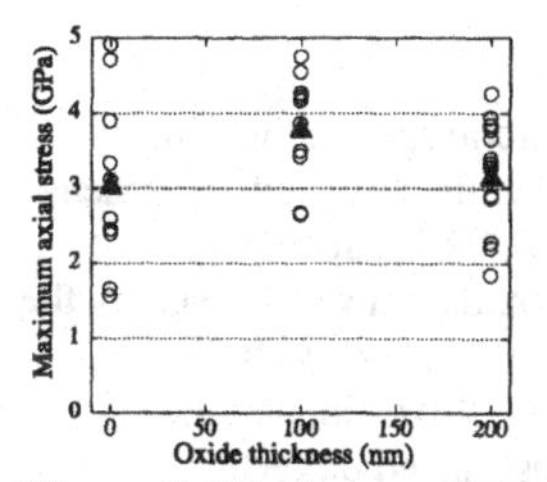

Figure 7. Maximum axial stress in silicon at fracture as a function of oxide thickness. Triangles show the average value.

Oxidation effect on fracture behavior

There are three features from these results. First, the mean fracture strain was increased by oxidation, but was decreased as the oxide thickness increased. Second, the fracture origin was on the surface in w/o oxide, but inside silicon in 100-nm-oxide and 200-nm-oxide as shown in Figures 5. Third, fracture doesn't initiate from oxide layer due to the compressive stress by oxidation according to the FE analysis results.

In w/o oxide, the fracture was estimated to be initiated from the surface of the specimens, especially at the scallops on the sidewall. The large distribution in the fracture strain of w/o oxide also indicates the fracture from surface of the specimens. In 100-nm-oxide and 200-nm-oxide, we considered that there may be defects inside silicon and they were grown during oxidation process. We assume the defects as the oxygen precipitation defects grown during thermal oxidation.

Defects in silicon

In this section, we verified the assumption that the oxygen precipitation defects are the fracture origins of oxidized SCS specimens from a view point of the defect size. The radius of the oxygen precipitation defects is described by the following equation [7],

$$r(t) = \sqrt{2D\left(\frac{C_0 - C'}{C_p - C'}\right)t} \tag{1}$$

Where r is the radius of the precipitate, t is the oxidation time, D is the diffusion constant of oxygen in silicon and C_0 is the interstitial oxygen density in silicon, C_p is the oxygen density of silicon dioxide, C' is the oxygen concentration in the silicon matrix immediately adjacent to the precipitates.

In our experiment the oxidation temperature was 1100 °C, and the time was 30 and 90 minutes. The D, C', C_p, and C_0 are determined to be 8×10^{-11} cm^2/s, 3×10^{17} atoms/cm^3, 4.6×10^{22} atoms/cm^3, and 7×10^{17} atoms/cm^3, respectively[7,8]. The radii of the precipitate grown in the 30 and 90 minutes oxidation were 26 nm and 45 nm, respectively, which were close to the calculated size of defects of 21nm and 31nm from the mean fracture strength. To confirm this evaluation, the presence of the defects needs to be investigated using defect etching or transmission electron microscope (TEM).

CONCLUSIONS

Quasi static tensile testing was carried out for single crystal silicon specimens with surface oxide layer of controlled-thickness. As the result, the mean fracture strains of specimens without oxide, with 100-nm-oxide, and with 200-nm-oxide were 0.018, 0,020, and 0.016, respectively. We think that the increase of mean fracture strain after oxidation was caused by the change of fracture origin due to the formation of defects in silicon. The decrease with the increase of the oxide thickness was caused by the growth of the defects. We estimated the defects as oxygen precipitation defects grown during oxidation. The estimated oxygen precipitation defect size of 26 nm and 45 nm were close to defect size of 21nm and 31nm calculated from the measured maximum axial stress.

REFERENCES

1. C.L. Muhlstein, S.B. Brown, R.O. Ritchie, Acta Mater. **50**, 3579 (2002).
2. H. Kahn, R. Ballarini, J. Bellante, A.H. Heuer, Science. 1215 (2002).
3. Y. Yamaji, *Proceedings of the MEMS'2007*, Kobe, Japan. 267 (2007).
4. J.T. Fitch, C.H. Bjorkman, and G. Lucovsky, J.Vac. Sci Technol. **B7**, 775 (1989).
5. T. Yoshioka, T Ando, M Shikida, K Sato, Sensors and Actuators. **82**, 291 (2000).
6. E. Orowan, Rep. Prog. Phys. **12**, 185 (1948).
7. A.Borghesi, BPivac, A.Sassella, and A. Stella, J. Appl. Phys.**77** (9), 4169 (1995).
8. F.M. Livingston, S Messoloras, R C Newman, B. C. Pike, R.J. Stewrt, W. P. Brown, and J.G.Wilkes, J. Phys. C **17**. 6253 (1984)

Mater. Res. Soc. Symp. Proc. Vol. 1052 © 2008 Materials Research Society 1052-DD02-06

Indentation Characterization of Fracture Toughness and Interfacial Strength of PECVD Nitrides after Rapid Thermal Annealing

H-Y Yan[1], K-S Ou[1], and K-S Chen[1,2]
[1]Mechanical Engineering, National Cheng-Kung University, 1 University Rd., Tainan, 70101, Taiwan
[2]Center for Micro/Nano Science and Technology, National Cheng-Kung University, 1 University Rd., Tainan, 70101, Taiwan

Abstract

This work presents the result of mechanical characterization of the fracture toughness and interfacial strength of PECVD silicon nitride films deposited on silicon subjected to rapid thermal annealing (RTA) processing between 200 and 800 °C. Both micro- and nano-indentation techniques are employed to perform the experiments. In conjunction with the model proposed by Marshall and Lawn for data reduction, the fracture toughness of as-deposited nitride is obtained as 2.2 MPa√m based on a series of Vickers micro-indentation tests and this value is essentially unchanged if the RTA temperatures are below 400°C. Further increase in RTA temperature would significantly enhance the fracture toughness. On the other hand, using nanoindentation testing in conjunction with the model proposed by Marshall and Evans, the interfacial strength between the nitride and silicon is determined as 17.2 J/m^2 for as-deposited nitrides and it could also be significantly enhanced by RTA processes with temperatures exceeding 400°C. These results should be useful to improve the structural integrity and performance of PECVD silicon nitride films for MEMS and IC fabrication.

Keywords: Nanoindentation, Nitride, Fracture Toughness, Rapid Thermal Annealing (RTA)

I. Introduction

Controlling of thin film fracture has its inherently importance in virtually all aspects. Plasma enhanced chemical vapour deposited (PECVD) silicon nitrides have been widely used in microelectromechanical systems (MEMS) and integrated circuits (IC) devices as mask or structural materials [1]. However, their performance is largely determined by the material properties such as residual stresses and fracture toughness of the nitride after deposition, which is processing parameter dependent. As a result, controlling of thin film fracture has its inherently importance in virtually all aspects. In particular, the rapid thermal annealing (RTA) process, which has been widely used to minimize thermal budget for the state-of-the art integrated circuit (IC) fabrication, would certainly alter the thermo-mechanical properties of materials. Previous investigation indicated that the RTA process parameters significantly influence the microstructure of polysilicon [2]. Consequently, the residual stress would also be varied significantly. On the other hand, in our previous study, we utilized indentation techniques to investigate the modulus and residual stress variation of PECVD silicon nitride subjected to RTA between 200 - 800°C and the results concluded that the RTA process brought significantly variation on residual stress of films [3]. In this work, we like to further investigate the influence of RTA process parameter on the effect of fracture toughness of silicon nitride as well as the interfacial strength between the nitride films and silicon substrates for the reference of MEMS and IC process engineers.

II. Experimental Setup and Mechanics

Fig.1 shows the experimental flowchart for characterizing the fracture toughness of nitride. 5000 Å PECVD nitrides were deposited on 4-inch silicon wafers using a Nano-Architect Research/BR-2000LL PECVD system located at the semiconductor research center of National Chiao-Tung University (NCTU) at temperatures between 250 and 400°C with a pressure of 5 Torrs based on the following reaction formula:

$$SiH_4(g)+NH_3(g)\xrightarrow[\Delta]{RF+N_2(g)}SiN_x:H(s)+3H_2(g) \tag{1}$$

After deposition, the wafers were die-sawed into square chips (10 mm × 10 mm). After measuring the residual stress using a KLA-Tencor profilemeter and Stoney's equation[4], specimens were then subjected to RTA processes using an Annealsys/AS-One 100 RTA system at temperatures of 200, 400, 600, and 800 °C with an annealing period of 60 seconds. After the RTA process, the curvatures of the corresponding specimens were re-measured for the evaluation of the residual stress generation during the RTA period. A MTS-XP nanoindenter with Berkovitch indentor as well as a Shimadzu HMV-2 microindenter with a Vickers diamond tip are then used to characterize the modulus, hardness, and toughness of the material [3].

It is known that once the strain energy release rate G exceeds the critical strain energy rate G_c, cracks will propagate. As a result, it is possible to estimate G_c using indentation experiments. The strain energy release rate for channelling cracking can be expressed as [5]:

$$G=1.976\frac{(1-\nu_f)\sigma^2 h_f}{E_f}, \tag{2}$$

where E_f, ν_f, and h_f are the elastic modulus, Poisson's ratio, and thickness of the associate thin film material. σ is the applied stress, which is essentially equal to the residual stress in this study.

On the other hand, the interfacial strength can also be obtained via indentation test. Based on the theory developed by Marshall and Evans [6], the interfacial toughness can be expressed as

$$G=\frac{(1-\nu_f)h_f}{E_f}[(1-\lambda)\sigma_r^2+0.5(1+\nu_f)\sigma_I^2-(1-\lambda)(\sigma_I-\sigma_B)^2], \tag{3}$$

where σ_r is the residual stress of thin film, σ_I and σ_B are the indentation and the buckling stress, respectively. λ is the slope of the buckling load. Their detail definition can be found elsewhere[6]. Physically, one can simultaneously measure the indentation depth and the size of delamination zone to estimate the magnitude of the required quantities and applying Eq.(2) to obtain the final result.

III. Experimental Results

■ Thin Fracture Toughness Estimation

In order to characterize the fracture toughness of the silicon nitride films using Eq.(2), residual stress variation must be firstly characterized and the results are shown in Fig. 2, which indicated that a significant residual stress generation occurring after a 400°C RTA. After residual stress characterization, the specimens were indented by the Vickers microindentor before and after RTA with a load of 245 mN. Fig. 3 shows a typical optical microscope image for a specimen. Some penny-shaped and stable channelling cracks were observed around the indentation zone. By Eq.(2), using the following data: $\nu_f = 0.27$, $h_f = 5000$ Å, $\sigma = 1.47$ GPa, and $E_f = 90$ GPa, the critical strain energy release rate G_c is estimated to be 17.3 J/m^2.

In parallel, using the method proposed by Marshall and Lawn based on measured crack length [7] and further modified by Zhang et al [8], the fracture toughness can be estimated via Vickers indentation test as

$$K_{IC} = 0.016\left(\frac{E_f}{H}\right)^{1/2}\left(\frac{P_{max}}{c^{3/2}}\right) + \frac{3.545}{c^{1/2}}\sigma_r h_f , \tag{4}$$

where E_f (90 GPa) and H (11 GPa) obtained in our previous study [3] are the modulus and hardness of the film, P_{max} (245 mN) is the indentation lading and c the average crack length of the four indented cracks generated by a Vickers indentation. By Eq.(4), the K_{IC} of the nitride film prior to RTA is estimated as 2.2 MPa$\sqrt{m}$. Based on the same procedure, the fracture toughness of nitride after various RTA processing are reported and shown in Fig. 4. It can be found that RTA processing at higher temperature would enhance the toughness. For example, after 800°C RTA, the fracture toughness of nitride would be enhanced approximately by a factor of 30%.

■ Interficial Fracture Toughness Variation w.r.t. RTA process

For RTA temperatures between 200 – 800°C with a load between 300 – 450 mN, the interfacial toughness between nitride and silicon is characterized using the nanoindentation technique. Typical results are shown in Fig. 5. It can be observed that for a lower RTA temperature (e.g., specimen after 200°C RTA shown in Fig. 5a,b), the delamination zone is much larger than that experiencing RTA with higher peak temperature (e.g., specimen after 800°C RTA shown in Fig. 5c,d). The pop-in phenomenon observed from the load-penetration curves (i.e., Fig. 5 a,c) also reveals this argument. As a result, qualitatively speaking, the interfacial toughness is enhanced with increasing of RTA processing temperature. A more detail estimation of the relationship between interfacial toughness and RTA temperature converted by Eq.(3) is shown in Fig.6. It can be found that the toughness changes slowly from approximately 10J/m^2 until the temperature exceeding 400°C. For temperature reaching 800°C, no visual delaminations were detected and the interfacial toughness, therefore, reaches a very high level.

IV. Discussion

The fracture toughness of brittle ceramics such as silicon, oxide, and nitride is in the order of 1-2 MPa√m [3]. Therefore, the estimated fracture toughness 2.2 Mpa√m is reasonable. There is approximately 10-20% uncertainties in this calculated value (depends on the statistics model and definition of crack length uncertainty used) and it is most likely attributed to the unequal crack length in four corners of a Vickers indentation mark.

On the other hand, based on the information of the nanoindentation cracking measurement (i.e, by Fig. 5), it can be concluded that the interfacial strength between PECVD nitride and silicon can be significantly improved with an increase of RTA temperature (exceed 400°C). This is certainly important information for process and structural engineers. However, the increase of RTA temperature would also influence the material properties such as modulus, hardness, as well as the toughness, and the residual stress. These factors should also be considered simultaneously for an optimal RTA parameter setup.

Finally, it must be pointed out here that since only four RTA temperatures have been tested, some temperature-specified conclusion of this work (e.g., maximum tensile stress generated, transition temperature, etc) may vary once a more thorough characterization (i.e., using more RTA temperatures) is conducted. Nevertheless, the general tendencies and trends obtained from this work, together with the fracture toughness and the interfacial strength data, still provide useful information and reliability assessment for related MEMS and IC structure integrity design applications.

V. Summary and Conclusion

The toughness and interfacial properties of PECVD nitride subjected to thermal processing is traditionally an important concern for IC and MEMS applications. In this work, by combining both nano- and micro- indentation characterizations, the correlations between RTA processing parameters and thermal stress/fracture properties of PECVD silicon nitrides are investigated. In conjunction with the model proposed by Marshall and Lawn, and Zhang et al for data reduction, the fracture toughness of fresh nitride is obtained as 2.2 MPa√m based on a series of Vickers micro-indentation tests and this value is essentially unchanged if the RTA temperature is below 400°C. Further increase in RTA temperature would significantly enhance the fracture toughness. Meanwhile, by using nanoindentation in associated with the model proposed by Marshall and Evans, the interfacial strength between the nitride and silicon is determined as approximately $10 J/m^2$ for as-deposited nitrides and it could be significantly enhanced by RTA processes with temperature exceeding 400°C. The results provide a qualitative indication on the reliability assessment of MEMS or IC structures and should be very useful for the concerns of maintaining the structural integrity and improve fabrication performance in related applications.

Acknowledgement:

This work is supported by National Science Council (NSC) of Taiwan under contract no. NSC96-2628-E006-006 MY3. The comment from Prof. T-Y Chen of NCKU is also greatly appreciated.

References:

1. M. Ohring, *The Materials Science of Thin Films*, Academic Press, New York, (1992).
2. X. Zhang, X.,T. Y. Zhang, M. Wong, and Y. Zohar, *IEEE J. Micromechanical Systems* **7**, 356-364, (1998)
3. H. Y. Yan, SM Thesis, Department of Mechanical Engineering, National Cheng-Kung University (2007).
4. G. G. Stoney, *Proc. R. Soc. London, A*, **82**, 172, (1909).
5. J. W. Hutchinson and Z. Suo, *Advances in Applied Mechanics*, **29**, 63-191, (1992).
6. D. B. Marshall and A. G. Evans, *J. Appl. Phys.*, **56**, 2632-2638, (1984).
7. D. B. Marshall and B. R. Lawn, *ASTM STP* 889, 26-46, (1986).
8. T. Y. Zhang, L. Q. Chen, and R. Fu, *Acta Mater.*, **47**,3869-3878,(1999)

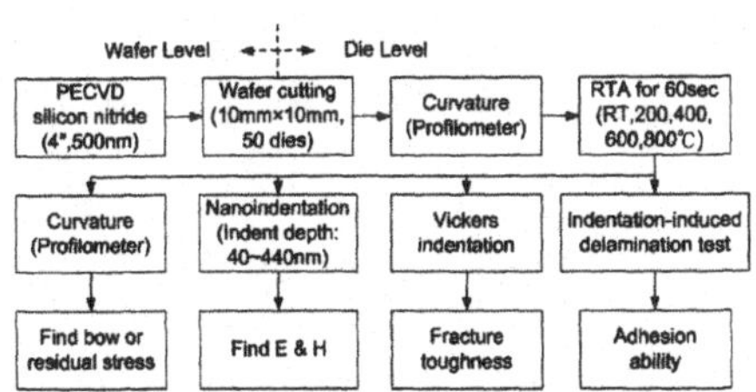

Fig.1 The overall characterization flow

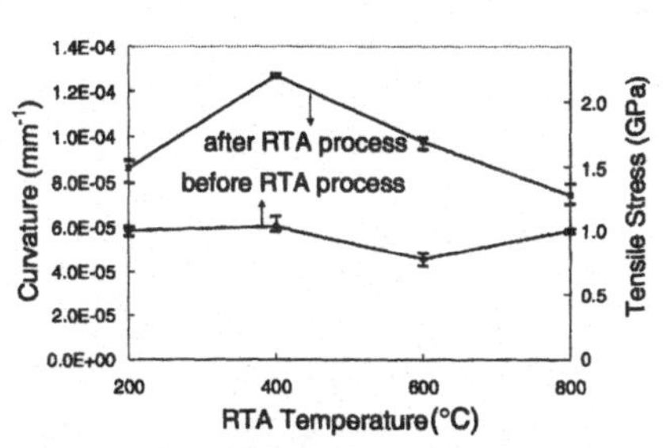

Fig. 2 The relationship between residual stress generation and RTA temperatures

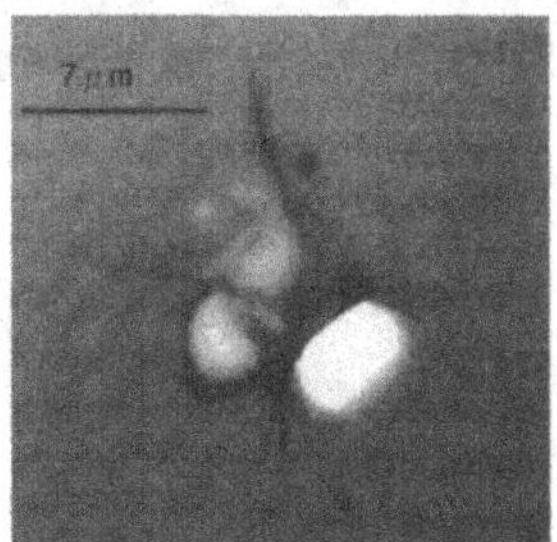

Fig. 3 A typical Vickers indentation on the silicon nitride film with a load of 245mN

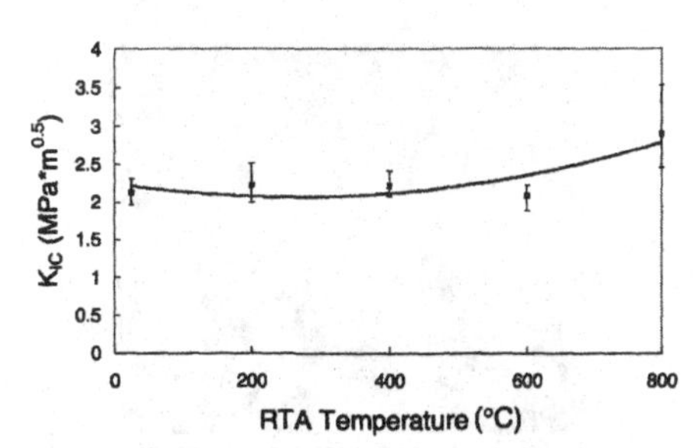

Fig. 4 The relationship between fracture toughness of silicon nitride and RTA temperatures

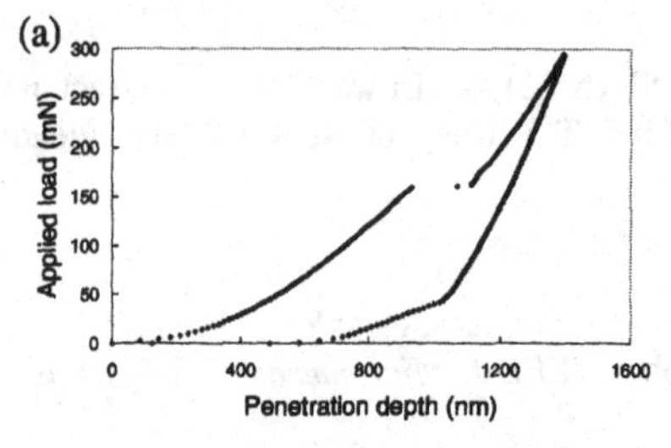

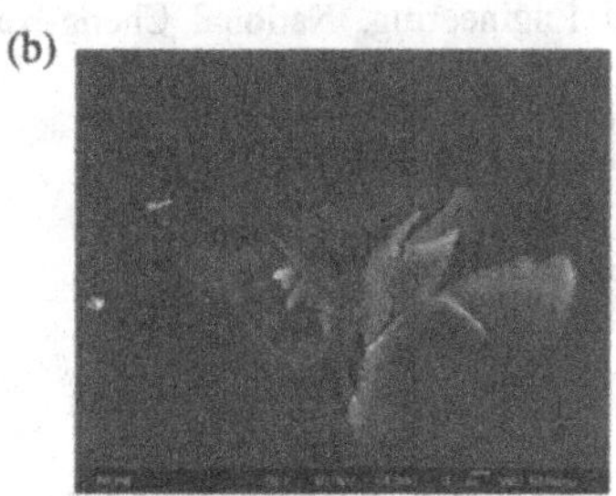

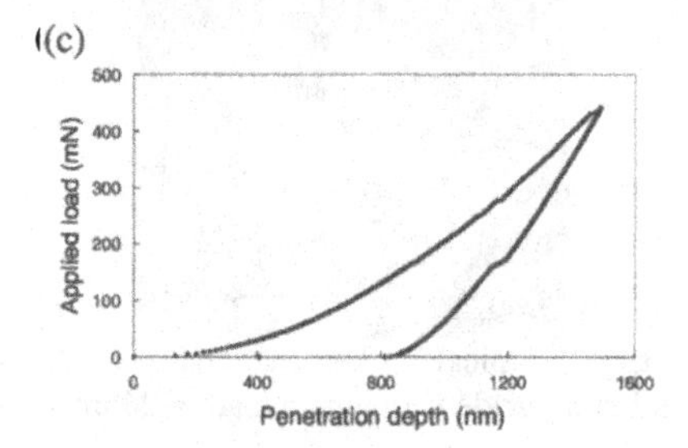

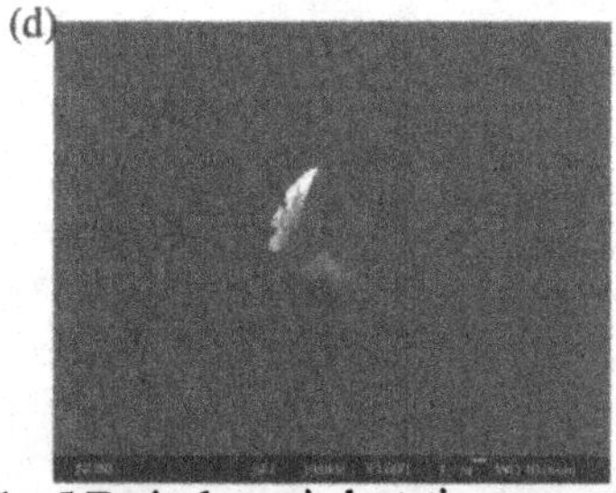

Fig. 5 Typical nanoindentation curves and images of nitride films after RTA (a) SEM micrograph and (b) indentation curve at 200°C. (c) and (d) SEM micrograph and indentation curve at 800°C

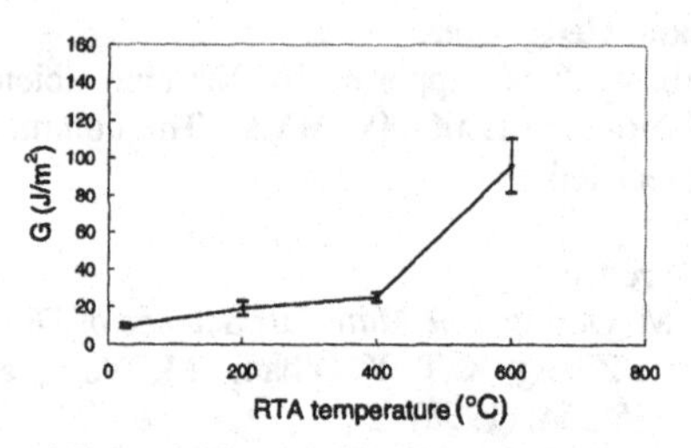

Fig. 6 The relationship between the interfacial toughness and RTA temperatures

Mater. Res. Soc. Symp. Proc. Vol. 1052 © 2008 Materials Research Society 1052-DD02-08

A Novel Micro Tensile Testing Instrument with Replaceable Testing Specimen by Parylene Passivation Technique

Yung-Dong Lau[1], Tso-Chi Chang[1], Hong Hocheng[1], Rongshun Chen[1,2], and Weileun Fang[1,2]
[1]Department of Power Mechanical Engineering, National Tsing Hua University, 101, Section 2, Kuang-Fu Road, Hsinchu, 30013, Taiwan
[2]Institute of NanoEngineering and MicroSystems, National Tsing Hua University, 101, Section 2, Kuang-Fu Road, Hsinchu, 30013, Taiwan

ABSTRACT

This study has successfully demonstrated a novel tensile testing approach to mount the thin film test specimen onto a MEMS instrument using microfabrication processes. The MEMS instrument consists of a thermal actuator, differential capacitance sensor, and supporting spring. The thermal actuator applies a tensile load on the test specimen to characterize the Young's modulus and the residual stress of the thin film. As compare with the existing approaches, the problems and difficulties resulting from the alignment and assembly of a thin film test specimen with the testing instrument can be prevented. Furthermore, the parylene passivation technique with the MEMS fabrication process allows the user to change the test materials easily. In application, the present approach has been employed to determine the Young's modulus and the residual stress of *Al* films.

INTRODUCTION

There are many thin-film mechanical testing approaches have been reported, for instance, the micro tensile test, the cantilever beam bending test, the nanoindentation test, and the bulge test [1-4]. However, it remains a challenge to precisely apply load as well as to determine the deformation during a thin-film material test. Moreover, it is not straightforward to prepare and place a thin film specimen during a test. In summary, the specimen preparation is one of the key sources of measurement error for thin-film mechanical testing.

In general, the conventional tensile testing machine contains a loading module, sensing module, and gripping mechanism. Because they match in size, load, and displacement, MEMS devices are appropriate for the mechanical testing of thin film materials. In [1], the micro fabrication process is only employed to prepare the test specimen, on-chip loading and displacement sensing is not available. MEMS actuators are used in [5] for micro tensile testing, but they may still have problems and difficulties from the alignment and assembly of a thin film specimen with the testing instrument. In [6], an on-chip tensile testing setup was fabricated and integrated using the MUMPs process. However, the test specimen is limited to the poly-silicon film only.

This study demonstrates a novel micro tensile testing instrument with replaceable testing specimen using a parylene passivation technique. Parylene is a very inert chemically to the various etches tested, it only had a significant etch rate in the oxygen plasma. Parylene is also biocompatible and has relatively low gas permeability [7]. Thus, parylene can be used to protect the test specimen during etching processes. Furthermore, the parylene passivation technique with the MEMS fabrication process allows the user to change the test sample easily using the same fabrication process.

EXPERIMENT

Concept and Design

The present micro tensile testing instrument illustrated in Fig. 1a consists of thermal actuator, differential capacitance sensor, supporting spring, and specimen holder. The two ends of thin film specimen are fixed to the specimen holder by means of micro fabrication processes. In addition, the shape of the test specimen is also defined by the fabrication processes. After applying input current to the instrument, the joule heat will be generated in the V-beam thermal actuator. Thus, the thermal expansion of silicon will lead to the in-plane buckling deformation of the V-beams. The displacement of the V-beam actuator will induce the in-plane and out-of plane deformations of the test specimen. The differential capacitance sensor is used to measure the in-plane displacement, and an optical interferometer is employed to measure the out-of-plane displacement of the test specimen. In addition to supporting the suspended components of the MEMS instrument, the supporting springs are also designed to direct the load applied on the test specimen only in the axial direction.

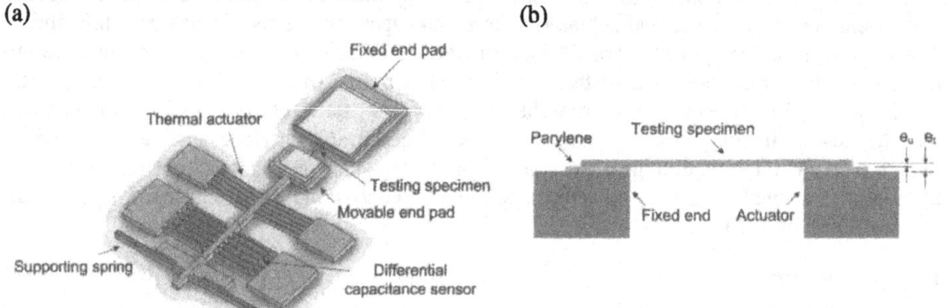

Figure 1. The schematic of the micro tensile testing instrument.

Figure 2 shows the finite element method (FEM) simulation models and results to predict the deformation of thin film specimen during tensile testing. The deformations of the test specimen resulting from external loads from the V-beam actuator were predicted. In the preliminary design, the mechanical properties from the bulk material were used in the FEM model. After that, the FEM analysis provided the design of the V-beam actuator arrays which can offer sufficient axial load on the specimen during the tensile test.

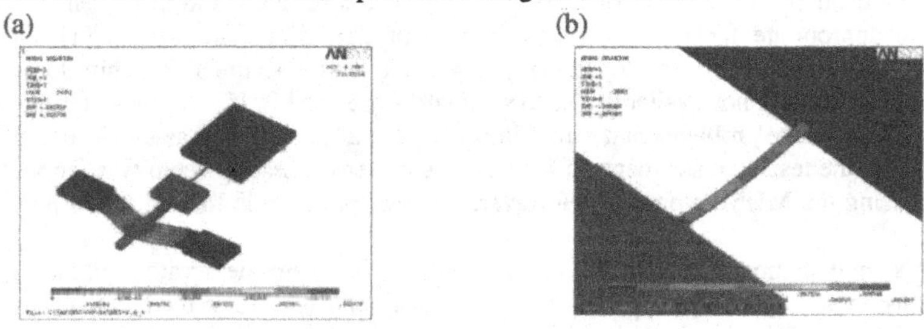

Figure 2. The typical FEM simulation models and results for load-deflection test, (a) the micro tensile testing instrument, and (b) zoom-in of the test specimen.

Theory of amended micro tensile testing equation

According to the characteristics of load-deflection regarding the present micro tensile testing, this study also established an analytical model to determine the Young's modulus and the residual stress of the test specimen. As shown in Fig. 1b, the boundary of the test specimen is only partially clamped to the holder [8], and the axial load F_t is applied on the specimen eccentrically. Thus, a bending moment is also produced during the tensile testing. As a result, the test specimen has both in-plane elongation and out-of-plane bending deformation $\delta_{atuator}$. Consider a prismatic test beam of length L and moment of inertia I, the out-of-plane bending deformation can be expressed as,

$$\delta_{actuator} = \frac{F_t \cdot e_t \cdot L^2}{8EI} \tag{1}$$

where e_t is eccentricity of axial load, F_t, and E is the Young's modulus of the test specimen. Moreover, the residual stresses also induce out-of-plane deformation after releasing the test specimen as a micro bridge. The residual stress includes both uniform stress and gradient stress. The tension state and compression state residual stress can be expressed as:

$$\delta_{tension} = \delta_{uniform} + \delta_{gradient} \tag{2}$$

$$\delta_{compression} = \delta_{uniform} + \delta_{gradient} + \delta_{buckling} \tag{3}$$

Where

$$\delta_{uniform} = \frac{F_u \cdot e_u \cdot L^2}{8EI} \tag{4}$$

$$\delta_{gradient} = \frac{E \cdot c \cdot A}{F_g}(1 - cos\theta) \tag{5}$$

$$\delta_{buckling} = \frac{F_u \cdot L^2}{4t \cdot A \cdot E} \tag{6}$$

In which F_u is the test specimen uniform residual stress, F_g is the test specimen gradient residual stress, e_u is the eccentricity of the test specimen, c is the distance from the neutral axis to the outer surface of the test specimen, A is the test specimen cross-sectional area, and t is the test specimen thickness.

Therefore, the test specimen total out-of-plane deformation during the tensile test, δ_{total}, is algebraic sum of $\delta_{actuator}$ and $\delta_{tension}$ or $\delta_{compression}$, respectively. We can rewrite the equation for F_t in terms of testing specimen total deflection, geometry, residual stress and material properties, yields the amended micro tensile testing equation:

$$F_t = -\frac{e_u}{e_t} F_u + \frac{8I \cdot \delta_{total}}{e_t \cdot L^2} E - \frac{8I \cdot e_u \cdot (1 - cos\theta)}{e_t \cdot L^2 \cdot \sigma_g} E^2 \tag{7}$$

$$F_t = -\frac{4e_u}{3e_t} F_u + \frac{8I \cdot \delta_{total}}{e_t \cdot L^2} E - \frac{8I \cdot e_u \cdot (1 - cos\theta)}{e_t \cdot L^2 \cdot \sigma_g} E^2 \tag{8}$$

Equation (7) is used when the residual stress is in a state of tension and equation (8) is used when the residual stress is in a state of compression. As a result, when apply the actuation force, F_t, the instrument can simultaneously determine the Young's modulus, E, and the uniform residual stress, F_u, of thin films.

Fabrication process and results

The fabrication process for the micro tensile testing instrument is shown in Figure 3. The substrate was a SOI wafer with a 50μm thick device layer. As shown in Fig. 3a, the parylene film was deposited and patterned to act as a bottom protection layer during the etching processes. In this study, the sputtered *Al* film was used as the test specimen, which was then patterned by a lift-off process. As illustrated in Fig. 3b, the second parylene film was deposited and patterned to act as the top protection layer. Thus, the test specimen (*Al* thin film) was fully covered by parylene films to protect the thin film specimen during the subsequent etching processes. As shown Fig. 3c, the PECVD silicon oxide film was deposited and patterned to define the planar shape of the present instrument. After applying the silicon nitride film to define the wet etching window, the test specimen was fully suspended using anisotropic bulk silicon etching (KOH), as sown in Fig. 3d. As illustrated in Fig. 3e, after removing the silicon nitride film, the silicon oxide film was used as the hard-mask for DRIE (deep reactive ion etching) to pattern the device silicon layer. As depicted in Fig. 3f, the silicon oxide film of SOI box-layer was isotropically etched with HF to release some of the device Si layer from the substrate. After release, the components of the test instrument such as the V-beam actuator and the differential capacitance sensor were free to move. Finally, the parylene passivation layer was removed isotropically by O_2 plasma to expose the test specimen.

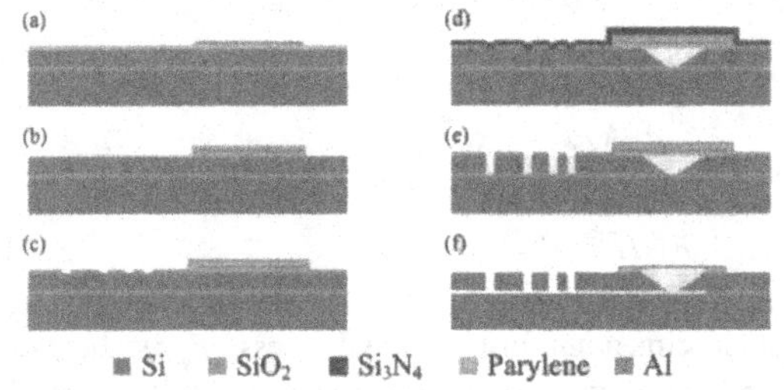

Figure 3. The fabrication process steps.

Figure 4 shows SEM micrographs of a typical micro tensile testing instrument after the process. The zoom-in SEM micrograph in Fig. 4b demonstrates the test specimen and the specimen holder. The etch release holes can also be clearly observed in this photo.

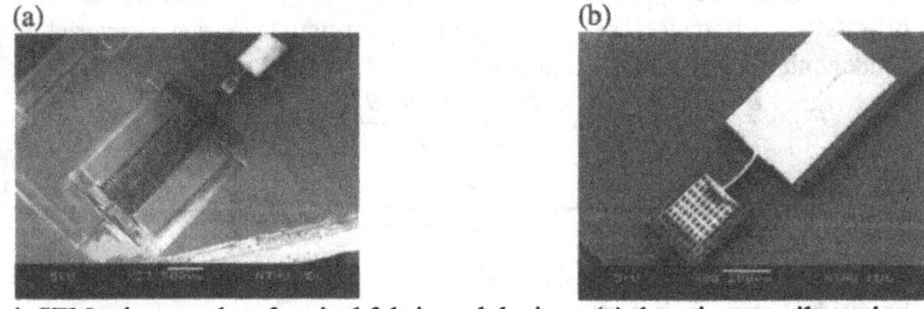

Figure 4. SEM micrographs of typical fabricated devices, (a) the micro tensile testing instrument, and (b) zoom-in of the test specimen.

In this case, the *Al* thin film specimen is 200 μm in length, 20 μm in width, and 0.4 μm in thickness. The thermal actuator is 2000 μm in length, 8 μm in width, and 1° in pre-bent angle. The supporting spring is 2000 μm in length and 10 μm in width. The thickness of the instrument is 50 μm which can be adjusted by varying the device layer thickness of the SOI wafer.

Experimental setup

Figure 5a shows a photo of a typical instrument chip after packaging, and the test setup in Fig. 5b-c was established to characterize the present instrument. The optical interferometer in Fig. 5b is used to measure the out-of-plane deformation of a test specimen during the tensile test. As shown in Fig. 5c, the in-plane deformation of the test specimen was measured by the differential capacitance change of the sensing electrode arrays. A voltage signal was provided after the capacitance-to-voltage conversion by a commercially available capacitance readout IC.

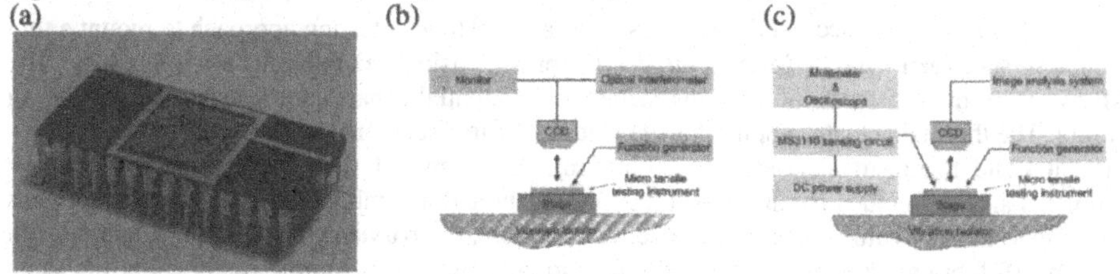

Figure 5. (a) The photos of completed micro tensile testing instrument chip and (b) out-of-plane deformation and (c) in-plane displacement experimental setup to characterize the micro tensile instrument.

DISCUSSION

Figure 7a shows a typical measured out-of-plane deflection profile of the released test specimen measured with a 3D interferometer before the tensile test. Figure 7b shows the typical test specimen out-of-plane deformation and in-plane displacement versus applied voltage.

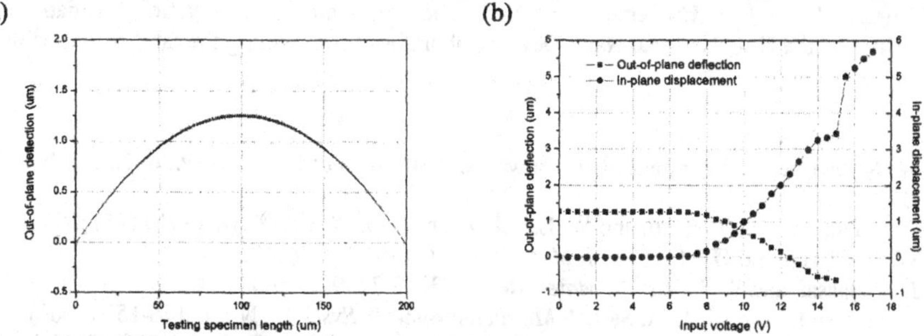

Figure 7. (a) The out-of-plane deflection profile by interferometer, (b) out-of-plane deformation and in-plane displacement of testing specimen versus applied voltage.

The Young's modulus and uniform residual stress were determined by calculating the slope of the actuation force versus total deflection curve using a least squares fit over the linear region. The measurement results show the Young's modulus and the residual stress of the thin *Al* film are 76 GPa and 58 MPa, respectively.

Table 1 shows the Young's modulus and the residual stress measured by different testing methods which the test specimen preparation is using the same fabrication process.

Table 1. Average of the Young's modulus and the residual stress measured by different testing methods.

	Young's modulus (GPa)	Residual stress (MPa)
This study	76	58
Nanoindentation test	82	-
Frequency resonant method	73	-
Diagnostic structure method	-	64

CONCLUSIONS

This study has successfully demonstrated a novel tensile testing approach to mount a thin film test specimen onto a MEMS testing instrument using microfabrication processes. The MEMS instrument consists of a thermal actuator, differential capacitance sensor, and supporting spring. The thermal actuator applies tensile load on the test specimen to characterize the Young's modulus and the residual stress of a thin film. As compared with existing approaches, the problems and difficulties resulting from the alignment and assembly of thin film test specimens with the testing instrument can be avoided. Furthermore, the parylene passivation technique with the MEMS fabrication process allows the user to change the test sample easily using the same fabrication process. According to the characteristics of load-deflection regarding the present micro tensile testing, this study also established an analytical model to determine the Young's modulus and the residual stress of test specimens. In application, the present approach has been employed to determine the Young's modulus and the residual stress of *Al* films.

ACKNOWLEDGMENTS

This research was sponsored in part by the NSC of Taiwan under grant of NSC-95-2221-E-007-068-MY3. The authors wish to appreciate the Center for Nano-Science and Technology of the National Tsing Hua University, the Nano Facility Center of the National Chiao Tung University and the NSC National Nano Device Laboratory for providing the fabrication facilities.

REFERENCES

1. W. N. Sharpe, B. Yuan, and R. L. Edwards, *J. Microelectromech. Syst.*, **6**, No. 3, 193-199 (1997).
2. G. L. Pearson, W. T. Read, and W. L. Feldmann, *Acta. Metall.*, **5**, 181-191 (1957).
3. W. C. Oliver, and G. M. Pharr, *J. Mater. Res.*, **7**, 1564-1583 (1992).
4. J. J. Vlassak and W. D. Nix, *J. Mater. Res.*, **7**, 3242-3249, (1992).
5. M. A. Haque, and M. T. A. Saif, *J. Microelectromech. Syst.*, **10**, No. 1, 146-152 (2001).
6. Y. Zhu, F. Barthelat, P. E. Labossiere, N. Moldovan, and H. D. Espinosa, *2003 SEM Annual Conference*, Session 77, Paper 155 (2003).
7. K. R. Williams, K. Gupta, and M. Wasilik, *J. Microelectromech. Syst.*, **12**, No. 6, 761-778 (2003).
8. W. Fang, and J. A. Wickert, *J. Micromech. Microeng.*, **6**, 301-309 (1996).

Poster Session

Mater. Res. Soc. Symp. Proc. Vol. 1052 © 2008 Materials Research Society 1052-DD03-01

Mechanical Stress Sensors for Copper Damascene Interconnects

Romain Delamare[1], Sylvain Blayac[1], Moustafa Kasbari[1], Karim Inal[1], and Christian Rivero[2]
[1]Centre de Microelectronique de Provence, GARDANNE, 13541, France
[2]STMicroelectronics, ROUSSET, 13106, France

ABSTRACT

We propose embedded microsensors to investigate the mechanical stress in copper damascene lines in a standard CMOS microelectronic technology. Those sensors are based on the silicon piezoresistive effect where strain in the active silicon is induced by orientated copper lines. The challenge is to correlate the electrical sensors signal directly to stress variation in lines.

We have performed electrical measurements of the structures as a function of temperature. A coupled analytical and finite element thermomechanical model of the structure was developed and a good agreement with measurements was obtained.

INTRODUCTION

It is well known that mechanical stress is developed during manufacturing process of CMOS microelectronic circuits [1]. This residual stress plays an important role in the devices reliability and often generates fabrication yield loss [2]. This phenomenon becomes of increasing importance as the complexity (integration density, number of metal levels, thermal cycling, packaging effects…) grows. It is therefore of major importance to correlate parameters with mechanical stress under systematic control along process to minimize yield losses. Thus, in response to industrial requests, we developed an embedded microsensor for monitoring the local stresses in Copper Damascene interconnects. In the present work, we report results obtained with a new test structure and demonstrates the link between electrical response of the sensor and stress variation in the first metal level.

The aim of this paper is to report the results obtained with this embedded structure and demonstrate its relevance as a copper stress sensor. The second part introduces the principles of this piezoresistive sensor. The third part details the electro-thermal finite-element model and shows the comparison with actual measurements.

SENSOR DESIGN

Sensor fabrication

The basic sensitive element is a silicon resistor underneath copper lines. Those lines induce an additional stress on the sensor through the piezoresistive effect (see Figure 1).

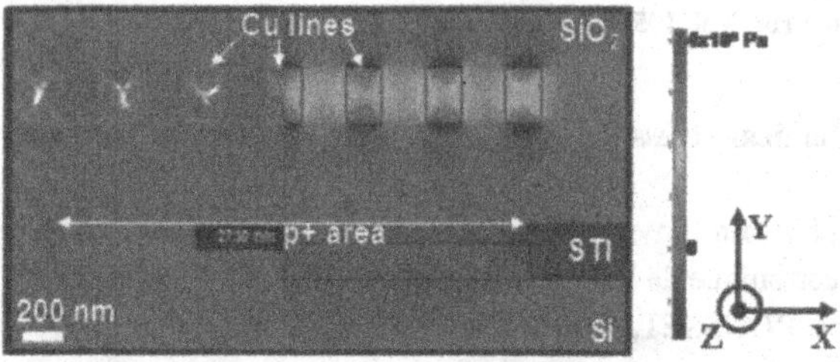

Figure 1: SEM cross sectional view of a stressed resistor with seven Copper lines (metal-1 level) and corresponding FEM simulation.

The cross section of the structure is made in the (110) plane perpendicular to the resistor main axis. Resistors (p+ doped silicon, [110] oriented, L = 27 μm and W = 2.7 μm) are surrounded by the SiO_2 shallow trench isolation (STI). In the silicon dioxide, above the resistor, seven copper lines are grown with a damascene process. The mean distance between silicon and copper lines is 550 nm. These lines are oriented along the long axis of the resistor (27 μm long) and their size is 0.2 μm large, 0.3 μm height (their aspect ratio L / W is 135). The first inter-metal dielectric oxide was etched, a thin barrier layer (25 nm Ta/TaN diffusion layer) was deposited, and the copper was electrochemically deposited and a furnace anneal at 400°C was performed for 30 minutes. After a chemical mechanical polishing, a silicon nitride capping layer was deposited on top. These lines are above the active resistor at the first metallization level only.

Electrical structure

A large literature exists on piezoresistive stress sensors which concerns mainly the measurement of externally imposed mechanical stress (for monitoring of packaging stress for example [3]). Usually, resistor rosettes are used, or more complex structures based on MOSFET devices [4]. In the present application, the source of stress is internal to the circuit stack. In this case, the main drawback is that the effective resistivity change may be due not only to stress but also to process variations. As was stated in [5], piezoresistive effect induces variations in the order of a few percent, as cumulative process variations, wafer to wafer or within wafer, which may induce overall resistance changes on the order of 2%. For this reason, a single resistor cannot be used here. The solution to circumvent this problem is to make a differential measurement [6], in other words to measure the resistivity change between two closely matched resistors: one standard (without copper lines), and the other with copper lines induced strain.

The electrical structure (Metal Wheatstone Bridge: MWB) is therefore based on a Wheatstone bridge including four resistors with two strained and two unstrained resistors. In Figure 2, the electrical representation of MWB sensor is presented. The stressed resistors are represented by top red lines. Special care was taken to obtain an optimum matching of the four resistor nominal values. The resistance change induced by the stress is noted as δR.

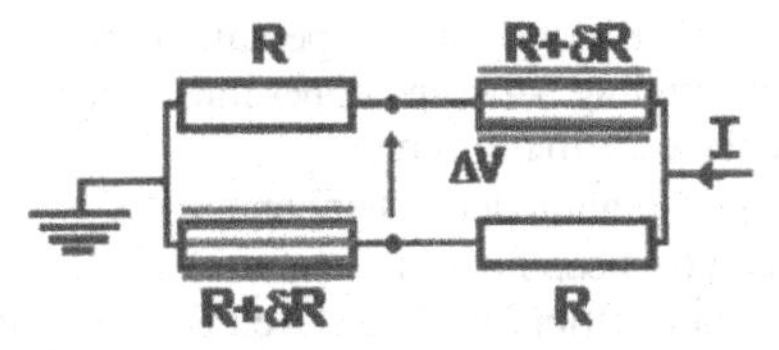

Figure 2: Electrical schematic representation of a metal Wheatstone bridge (MWB).

The offset voltage ΔV is measured as the current I is swept. The resistance variation δR can be expressed using the following equation:

$$\left(\frac{\Delta V}{I}\right)=\frac{R^2-(R+\delta R)^2}{4R+2\,\delta R} \qquad (1)$$

According to the piezoresistive effect, δR / R is on the order of a few percent and can be neglected. Equation (1) is reduced to:

$$\delta R=-2\left(\frac{\Delta V}{I}\right) \qquad (2)$$

Furthermore, the use of a Wheatstone bridge allows measurements free from contact error and then the measured offset voltage ΔV is proportional to the resistance variation induced by stress.

RESULTS AND DISCUSSION

Electrical characterization

All electrical measurements were performed on a dedicated high precision measurement system, consisting of a Keithley 4200 precision semiconductor parameter analyser and a Suss Microtec PA200 semi-automatic probe station.

In order to demonstrate the mechanical coupling between the copper lines and silicon resistor, a temperature sweep was performed to induce a strain variation due to the thermal expansion of the copper lines. I(V) measurements were carried out at a range of temperatures between 22 and 125°C.

Figure 4 compares I(V) curves obtained at different temperatures on a MWB structure. The linearity of the experimental data allows the measurement of the slope with high accuracy and calculation of the resistance change δR at each given temperature (equation 2).

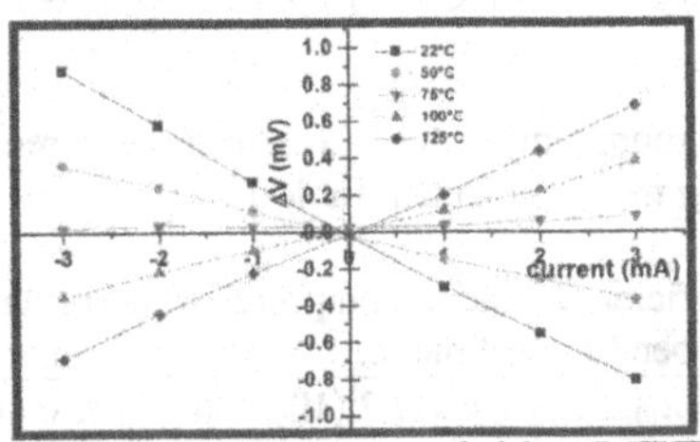

Figure 3: I-V curves evolution of a metal Wheatstone bridge (MWB) versus temperature.

A significant variation of MWB slope with temperature is observed. This result clearly demonstrates the mechanical coupling between copper lines and silicon. To the stress change induced bias, we should add the contributions of:

- Nominal resistance variation due to temperature;
- Ultimate mismatch between resistors.

To clearly quantify the different contributions, we developed analytical and finite elements models. The goal of the analytical model is to establish a correspondence between the stress field at the resistor level and the electrical response. The finite element model will allow making the relationship between the stress at the Copper lines level and the Copper-induced stress at the resistor

Analytical model

The orientation of the resistors on the wafer determines the piezoresistance term. The resistors are oriented along the [110] axis of silicon; the stress-induced resistance change can be written as follows:

$$\frac{\delta R}{R}=\left(\frac{\pi_{11}+\pi_{12}+\pi_{44}}{2}\right)\Delta<\sigma_L>+\left(\frac{\pi_{11}+\pi_{12}-\pi_{44}}{2}\right)\Delta<\sigma_T>+\pi_{12}\,\Delta<\sigma_y> \qquad (3)$$

Where $\sigma_L = \sigma_z$; $\sigma_T = \sigma_x$, $\Delta<\sigma_i>=<\sigma_i^{metal}>-<\sigma_i^{ref}>$, $<\sigma_i^{metal}>$ is the average stress in the resistor with copper lines in the i direction , $<\sigma_i^{ref}>$ is the mean stress in the resistor without lines in the i direction and π_{11},π_{12} and π_{44} are the three independent components of the piezoresistance tensor. The third term of equation (3) can be neglected as $\pi_{11}$and $\pi_{12} << \pi_{44}$. Using $\pi_L = (\pi_{11}+\pi_{12}+\pi_{44})/2$ and $\pi_T = (\pi_{11}+\pi_{12}-\pi_{44})/2$ we obtain the simplified piezoresistance term for resistors:

$$\frac{\delta R}{R}=\pi_L\,\Delta<\sigma_L>+\;\pi_T\,\Delta<\sigma_T> \qquad (4)$$

All the terms of (2) depend on temperature and the equation can be rewritten as:

$$\left(\frac{\delta R}{R}\right)_T=\pi_L(T)\Delta<\sigma_L(T)>+\;\pi_T(T)\Delta<\sigma_T(T)> \qquad (5)$$

From [7] we can approximate $\pi_i(T)$ to a linear function of the temperature in the range of 22 to 125°C. Then $\pi_i(T) = P(T).\pi_{i0}$ with $P(T)=1-2.5\times10^{-3}(T-25)$

The resistance change induced by copper lines at the temperature T can be written as:

$$\left(\frac{\delta R}{R}\right)_T=P(T)\left[\pi_{L0}\left(<\sigma_L^{metal}(T)>-<\sigma_L^{refl}(T)>\right)+\;\pi_{T0}\left(<\sigma_T^{metal}(T)>-<\sigma_T^{refl}(T)>\right)\right] \qquad (6)$$

The values of π_{L0} and π_{T0} at room temperature have been measured on our samples by using calibrated four point bending measurements (Table 1).

Table 1: Piezoresistive coefficient extraction on p+ resistors (bridge resistors) at room temperature with four point bending technique.

	$\pi_{L0} = (\pi_{11} + \pi_{12} + \pi_{44}) / 2$ (10^{-12} Pa^{-1})	$\pi_{T0} = (\pi_{11} + \pi_{12} - \pi_{44}) / 2$ (10^{-12} Pa^{-1})
Experiment	**- 273 ± 15**	**234 ± 10**

Once this electrical mechanical dependence at the resistor level is determined, it is necessary to investigate the mechanical coupling law between the copper lines and the active silicon.

Finite element model

COMSOL® software was used to develop a copper-silicon thermo-mechanical coupling model. We used a 3D description to avoid simplifications of the initial conditions and because of the relative small geometry size. The model geometry shown in Figure 1 has been reproduced from a SEM cross-section picture. Young's moduli and thermal expansion coefficients were assumed to be constant in this temperature range (E_{Cu} = 125 GPa, E_{Si} = 130.2 GPa, E_{SiO2} = 70 GPa, ν_{Cu} = 0.33, ν_{Si} = 0.28, ν_{SiO2} = 0.17, α_{Cu} = 16.85x10^{-6} K^{-1}, α_{Si} = 3.6x10^{-6} K^{-1}, α_{SiO2} = 0.65x10^{-6} K^{-1}). All materials (SiO2, Si and also Cu) were modeled as temperature-dependant elastic materials.

Given an initial stress state reported by Baldacci et al [8], a temperature variation was simulated. Baldacci et al. performed in-situ X-Ray Diffraction measurements during annealing of copper damascene lines with the same geometric ratio, making measurements from room temperature up to 500°C using synchrotron radiation. These results give at each temperature the complete stress tensor in copper lines. Figure 4 compares these experimental results with finite element analysis. A good agreement was obtained between these measurements and simulation results (Figure 4a).

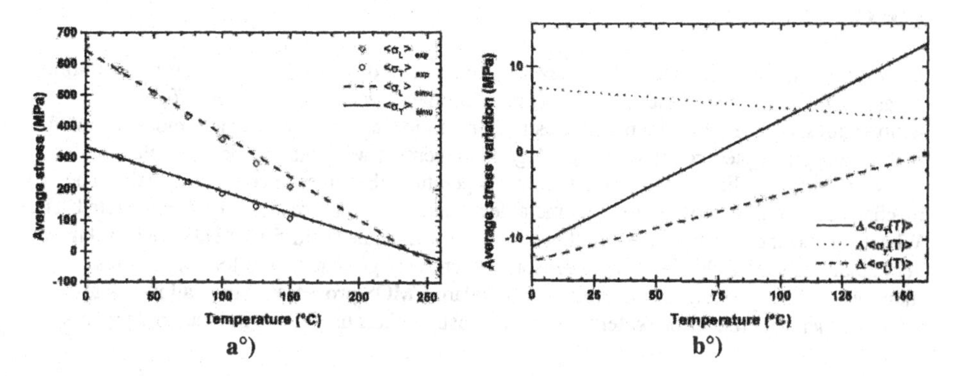

Figure 4: Simulated versus measured average stress components in copper lines (a), and transmitted components at the resistor (b).

As the copper line stress versus temperature is determined, the average copper-induced stress state in the resistor can be estimated ($\Delta <\sigma_i(T)>$). Figure 4b represents the stress variation $\Delta <\sigma_i(T)>$ versus the temperature.

Global model assessment

The electrical behavior of a MWB is determined by the knowledge of the stresses in the active silicon. Based on the silicon stress field in fig5b, we can translate the stress variation in a resistance variation by using equation (6): the temperature behavior of MWB is reported in Figure 5 both from simulation and electrical results.

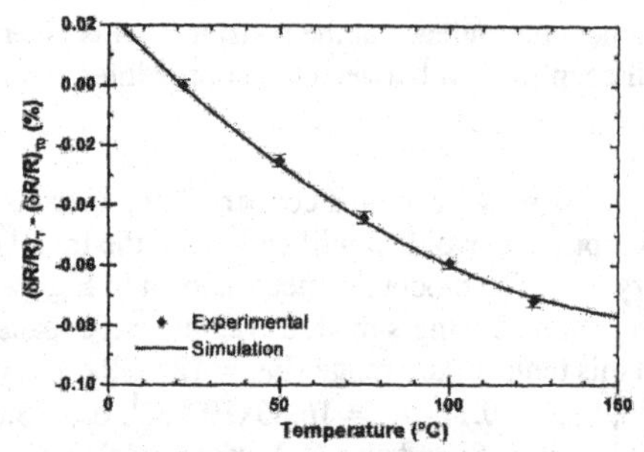

Figure 5: Comparison between the experimental evolution of a bridge (data Figure 4) and the calculation done by FEA using the model from equation (6).

The choice to represent the resistance change evolution in $(\delta R/R)_T - (\delta R/R)_{T0}$ has been done to avoid offset stress state resistance change $(\delta R/R)_{T0}$.

There is a good agreement between experimental and calculated MWB behavior which confirms that the proposed global model is realistic and that the main contributor in the MWB behavior is the mechanical stress induced by the copper lines.

CONCLUSION

We designed and characterized a piezoresistive sensor dedicated to copper stress monitoring. To assess the relevance of the sensor, I(V) measurements were carried out at different temperatures. We developed a purely elastic thermo-mechanical finite element model coupled with an analytical piezoresistive model. A good agreement was obtained between electrical measurements and the simulations. At room temperature, the stress in copper is triaxial. The mechanical coupling between the first metal level and the silicon was clearly demonstrated. The Wheatsone bridge structure used for this sensor showed high sensitivity and allowed tracking of very small resistance changes. These sensors present very good potential for copper stress monitoring. They are fully compatible with standard CMOS processes, which allows us to integrate them in production wafers and to demonstrate their usefulness in industrial process control.

ACKNOWLEDGMENTS

This research was supported by the « Communauté du Pays d'Aix », the « Conseil Général des Bouches-du-Rhônes » and the « Conseil Régional de PACA », through the focused research program called « ROUSSET 2003-2008 » in partnership with STmicroelectronics. Special thanks to Valérie Serradeil-Luton and Jean-Luc Liotard, STmicroelectronics Rousset team mangers for their help and interesting discussions. Thanks to Youssef Baltagi and Loïc Charrier for their help.

REFERENCES

[1] Hu, S.M., "Stress-related problems in silicon technology", J. Appl. Phys., vol. 70 (6), pp R53-R80, 1991.

[2] Flinn P.A., Gardner D.S. and Nix W.D, "Analysis technology for VLSI fabrication", IEEE Trans. On Elec. Dev., Vol 34(3), pp 689-699, 1987.
[3] Lwo B., Lin C., "Measurement of Moisture-Induced Packaging Stress With Piezoresistive Sensors", Advanced Packaging, IEEE transaction on, vol. 30,issue 3, p 393-401, 2007.
[4] Jaeger R. C. et al, « CMOS stress sensors on (100) silicon », IEEE journal of solid-state circuits, vol. 35, NO1, january 2000.
[5] Slattery O., « Sources of variation in piezoresistive stress sensor measurement », IEEE transactions on components and packaging technologies, 2004, pp 1-5.
[6] Kasbari M.et al, "Embedded Mechanical Stress Sensors for Advanced Process Control", ASMC proceedings, IEEE, 2007.
[7] Kanda Y., "A graphical representation of the piezoresistance coefficient in Silicon", IEEE Transactions on Electron Devices, vol 29, n°1, 1982.
[8] Baldacci A. et al, "Stresses in blanket films and damascene lines: Measurements and finite element analysis", P105 - 108 IEEE, 2004.

Mater. Res. Soc. Symp. Proc. Vol. 1052 © 2008 Materials Research Society 1052-DD03-02

Elastic and Viscoelastic Characterization of Polydimethylsiloxane (PDMS) for Cell-Mechanics Applications

I-Kuan Lin[1], Yen-Ming Liao[2], Yan Liu[1], Kuo-Shen Chen[2], and Xin Zhang[1]
[1]Manufacturing Engineering, Boston University, 15 Saint Marys St., Brookline, MA, 02446
[2]Mechanical Engineering, National Cheng Kung University, No.1, Ta-Hsueh Road, Tainan, 701, Taiwan

ABSTRACT

The mechanical properties of polydimethylsiloxane (PDMS) were characterized by using uniaxial compression, dynamic mechanical analysis (DMA), and nanoindentation tests as well as finite element simulation methods. A five-parameter linear solid model was used to emulate the behavior of PDMS. The study results indicated that the effect of viscoelasticity affected the PDMS pillar arrays significantly. The traditional approach for calculating the cell force basing on the linear elastic mechanics could result in considerable errors.

Keywords: PDMS, Viscoelasticity, Stress relaxation, finite element methods

INTRODUCTION

The mechanical interaction between cells and their neighboring extracellular matrix is believed to be of fundamental importance in various physiological processes such as division and it can be measured by soft material probes such as polydimethylsiloxane (PDMS) pillars [1]. Researches have constructed PDMS pillar arrays as extracellular matrices and the interaction forces can be characterized by converting the measured deflection of PDMS pillars, showing in Fig.1, using elastic mechanics of materials based on simple assumptions from mechanics of materials, where the elastic modulus of PDMS was assumed to be a constant [2]. However, PDMS is inherently a viscoelastic material and its elastic modulus changes with loading frequencies and elapsed time durations [3]. Neglecting the time- and frequency-dependent nature of PDMS could possibly result in significant errors in data interpretation and even on the mechanisms in cell responses. Unfortunately, to the best of our knowledge, no sophisticated constitutive models on PDMS have been reported. Therefore, it is important to perform a detail material characterization on PDMS materials for cell-mechanics applications.

This paper presents a detail mechanical behavior characterization for PDMS materials. As shown in Fig. 2, the entire study includes PDMS specimen fabrication, material and mechanical characterization in both macro and micro scales, and the corresponding finite element analyses (FEA) and simulations.

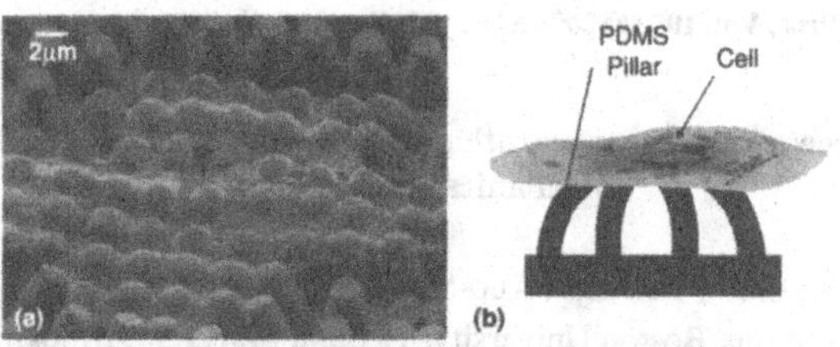

Fig. 1 A PDMS array for characterizing cell force

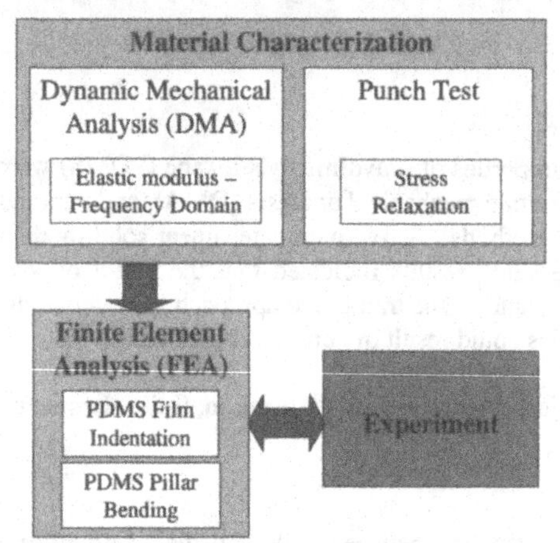

Fig. 2 The overall characterization flow

SPECIMEN FABRICATION AND EXPERIMENTS FOR PDMS VISCOELASTICITY BEHAVIOR

PDMS specimens in form of bulk and film were fabricated for dynamic mechanical analysis (DMA) and nanoindentation characterization, respectively. On the other hand, PDMS micro pillar arrays were also prepared by following the fabrication process demonstrated in Fig. 3. In order to form three types of PDMS sample: the 30mm × 30mm × 3mm metallic molds, 25mm×75mm glass slides, and patterned 40μm thick SU8 photoresist molds with 20×20 (μm^2) holes, were used. These PDMS samples were obtained by mixing PDMS prepolymer (Sylgard 184, Dow Corning) and a curing agent in a ratio of 10:1. The mixed PDMS prepolymer was poured on the three different molds and then put into a vacuum chamber for five minutes to remove the residual air bubbles in the film. Finally, the sample was baked at 65 °C for 90 minutes for the purpose of curing. After separately removing the PDMS and mold, these PDMS samples were completed.

The bulk PDMS specimens have been characterized to establish the material constitutive law by using a self-developed DMA system and a punch testing system shown in Fig. 4. By utilizing the punch equation developed in contact mechanics (with finite thickness corrections), the plane strain elastic modulus, E*, can be obtained by

$$E^{*} = \frac{E}{1-\nu^{2}} = 0.3\frac{P}{hD} \qquad (1)$$

where P, h, D, are the loading, thickness of the bulk PDMS, and the punch diameter, respectively. By Eq.(1), it is estimated that the initial modulus of the bulk PDMS at room temperature (25 °C) is approximately 1.45 MPa.

Since PDMS is inherently a viscoelastic material, a stress relaxation scheme based on the same punch test was also performed. The test data and the finite element simulation (using ABAQUS v.6.4) result by using a five-parameter linear solid model are all shown in Fig. 5(a). By using the Prony series approach, the viscoelastic behavior at RT of the PDMS can be modeled as

$$G(t) = 0.455(1-[0.08(1-e^{-t/0.165})+0.03(1-e^{-t/5})]) \quad \text{(MPa)} \qquad (2)$$

where G(t) is the shear modulus. Since the Poisson's ratio of polymer can be reasonably assumed to be approximately 0.5, the Young's modulus E(t) is therefore three times of G(t). The obtained viscoelasticity constitutive law would then be used for simulating the experimental data obtained by the nanoindentors. Next, a more comprehensive study on the PDMS viscoelasticity was also performed using a self-developed DMA system shown in Fig. 4(c). The test data and the finite element simulated results (using the frequency domain viscoelasticity model presented in Eq.(2) are both shown in Fig. 6. The results indicated that the simulation results essentially agree with the test data.

The obtained viscoelasticity model is then validated via nanoindentation process. A nanoindentation relaxation characterization by using a Hystron nanoindenter was performed with an indentation depth of 1000nm for 60s. In parallel, a finite element simulation model (i.e., Fig. 5(b)) is also constructed to emulate the indentation process by incorporating the material behavior (i.e., Eq.(2)) of the PDMS. Both the experimental and simulation data are plotted in Fig. 6(c). It can be observed that the tendencies of both set of data essentially agree to each other and this further validates the accuracy of the viscoelsticity model of the PDMS. Notice that quantitatively, there are still certain discrepancies between model and test data but this can be further improved by using a more sophisticated model. Nevertheless, in this primary work, we have chosen to concentrate on the qualitative characterization and have left detail modeling as our future work.

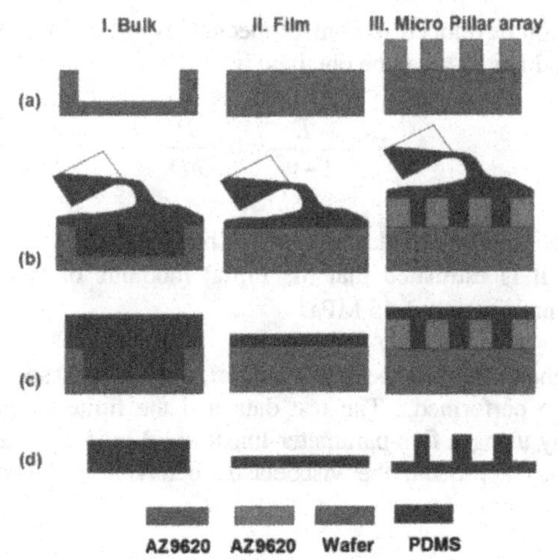

Fig. 3. PDMS sample fabrication (I.) bulk PDMS, (II.) PDMS film, (III.) PDMS pillar

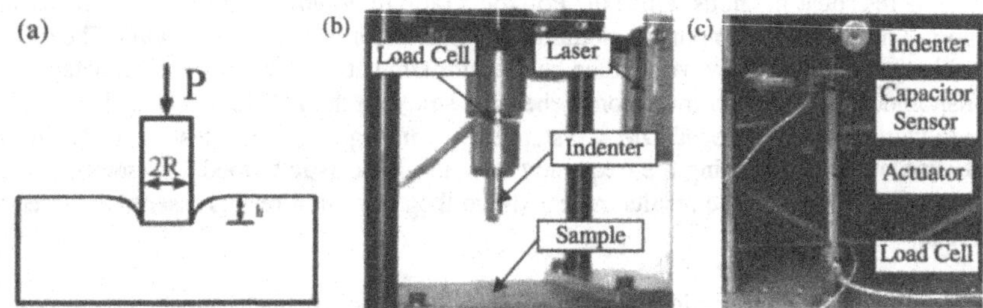

Fig 4. Punch problem and equipment setup:(a) Punch problem diagram,(b) Punch test system,(c) Dynamic mechanical analysis system

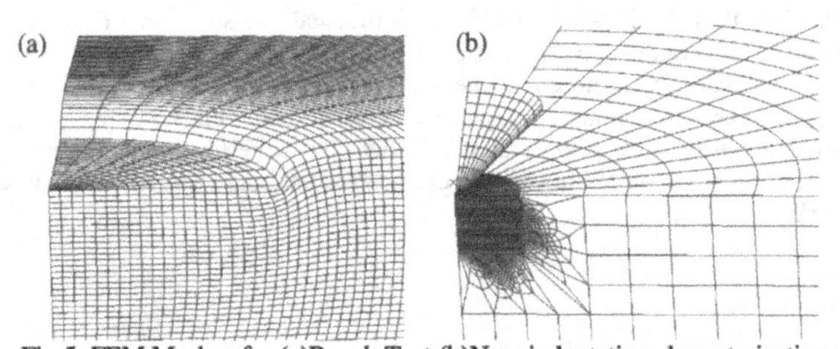

Fig 5. FEM Meshes for (a)Punch Test (b)Nanoindentation characterization

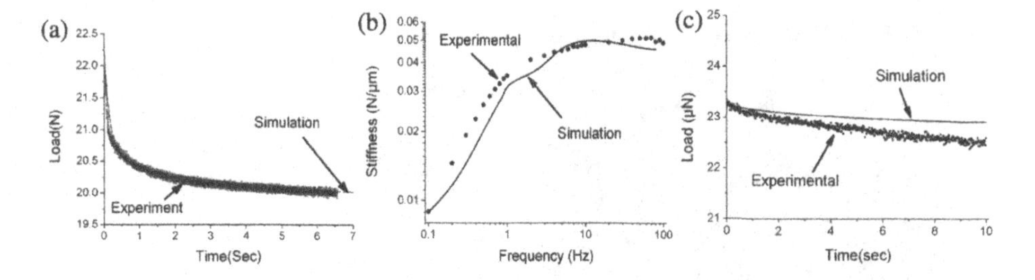

Fig. 6 Experimental and simulation results for (a) stress relaxation (b) DMA (c) Nanoindentation

BENDING BEHAVIOR OF PDMS PILLARS

The bending behavior of PDMS pillar was then performed by both finite element simulation and nanoindentation process. The picture of the test pillar array is shown in Fig. 7(a). Since the loading process involves the pre-indentation and the lateral scratch and the behavior is extremely complicated, a simple concentrated loading acted 3D finite element model, shown in Fig. 7(b), was used to qualitatively find the trend of the behavior. A typical force-deflection relation is shown in Fig. 8. From Fig. 8a, it can be found that the widely used simple cantilever beam model, the nonlinear finite element model based on linear elastic material, and the viscoelastic model actually have considerable differences.

Finally, the nanoindentation scratch experiment and the corresponding finite element simulation are correlated. The scratch was performed by pre-loading the nanoindenter vertically with a load of 10 μN for pre-loading. The lateral force was then applied to move 5μm within 12 seconds and followed by a relaxation period of 50 seconds. As shown in Fig. 8b, due to the problems faced in specimen preparation, experimenting skill such as centering, and the resolution (i.e., 3μN), the test data were highly scattered and the device results are poorly correlated with the FEA simulations. Nevertheless, it can be found that the trend of the simulation results agrees essentially with the test data. In addition, the essential solid mechanics tends to overestimate the pillar stiffness and it would result significant error. This primary result indicates that it is worth to perform an even detail characterization to clarify this concern in the future.

SUMMARY AND CONCLUSION

In this work, the mechanical properties of PDMS were characterized by using uniaxial compression, dynamic mechanical analysis (DMA), nanoindentation, as well as finite element simulation. A five-parameter solid model was used to emulate the behavior of PDMS. The study results indicated that the effect of viscoelasticity affected the bending behavior of PDMS pillar arrays significantly. The traditional approach for calculating the cell force basing on the linear elastic mechanics could result in errors.

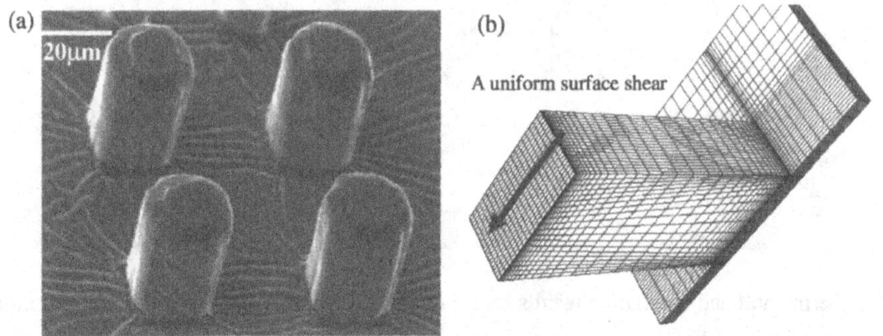

Fig. 7. The SEM micrograph and the corresponding finite element model of the PDMS pillar

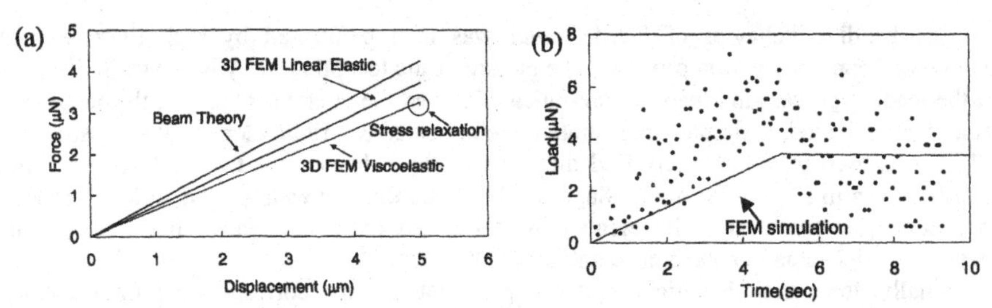

Fig. 8. A typical pillar bending characterization result (a) comparison between various model and (b) comparison between the test and the viscoelastic FEM simulation

ACKNOWLEDGMENTS

This work was supported by the National Science Foundation under grant number CMMI-0239163/0700688 and was also partially supported by National science council of Taiwan (NSC96-2628-E-006-006-MY3). The authors are grateful for the assistance of George Seamans and George Fraley in PDMS samples fabrication at Photonic Center, Boston University.

REFERENCES

[1] J.L. Tan, J. Tien, D.M. Pirone, et al., Proc. Natl. Acad. Sci., 100(2003) pp. 1484–1489.
[2] Y. Zhao and X. Zhang, Applied Physics Letters, 87(2005), pp. 144101.
[3] K. P. Menard, Dynamic Mechanical Analysis, (CRC press LLC, 1999) pp. 162-171.
[4] E. Riande, Polymer Viscoelasticity: Stress and Strain in Practice, (CRC press LLC, 1999), pp. 394

Mater. Res. Soc. Symp. Proc. Vol. 1052 © 2008 Materials Research Society 1052-DD03-06

A Direct Method of Determining Complex Depth Profiles of Residual Stresses in Thin Films on a Nanoscale - Mechanics of Residually Stressed Systems

Stefan Massl[1], Jozef Keckes[2], and Reinhard Pippan[1]
[1]Erich Schmid Institute, Austrian Academy of Sciences, Jahnstr. 12, Leoben, A-8700, Austria
[2]Department Materials Physics, University of Leoben, Jahnstr. 12, Leoben, A-8700, Austria

ABSTRACT

The basic ideas behind the calculation procedure of the developed ion beam layer removal method (ILR method) to determine complex depth profiles of residual stresses in thin films are explained. The mechanics of thin films on substrates in general and the effect of the substrate thickness on the stresses in particular are presented by means of a simple model system in order to improve the understanding and facilitate the application of the ILR method.

INTRODUCTION

The distribution of residual stresses in thin films and structures is essential for the mechanical performance and the lifetime of coated components. Therefore, the ion beam layer removal (ILR method) was developed to allow the determination of stress profiles on a nanoscale in amorphous and crystalline thin films [1]. The procedure starts with the fabrication of a cantilever from the initial system with a focused ion beam workstation. It consists of the substrate and the thin film and deflects due to residual stresses. The thickness of the cantilever is reduced step by step by removing the thin film gradually top-down with the ion beam. This affects the stress distribution and therefore the deflection of the cantilever. The mean stress that acted in the removed sublayer is determined for each step from the actual deflection measured from SEM images, the Young's moduli and the dimensions of the cantilever. The stress profile in the cantilever is determined by superimposing the previously removed sublayers and establishing the force and moment balances by means of the equations presented subsequently. The stress distribution in the initial system is calculated from the stress profile in the cantilever.

The ideas presented below are essential for the understanding of the mechanics of thin films as well as the ILR method. A simple model system is introduced to explain the two ways of describing coated systems and the corresponding equations. The same equations are necessary for the calculation procedure of the ILR method, where the sublayers removed previously from the cantilever are superimposed. The stress in the sublayer superimposed and the accumulated balanced stresses in the cantilever below are correlated with the actual curvature. Finally the procedure to determine the stress distribution in the initial system from the stress profile in the cantilever (and vice versa) is presented.

MECHANICS OF RESIDUALLY STRESSED THIN FILMS ON SUBSTRATES

In order to understand the mechanical behavior of residually stressed systems the basic ideas and equations [2] are explained by means of the cross section of a linear elastic model system sketched schematically in Figure (1). This system consists of three parts labeled $i = 0...2$: the substrate (t_0) with the Young's modulus E_0, the bottom of the coating (sublayer 1: t_1, E_{coat}) and the top of the coating (sublayer 2: t_2, E_{coat}). Sublayer 1 and sublayer 2 exhibit different stresses. In order to account for the 2D calculation procedure of a 3D system, the biaxial Young's moduli $E_b = E/(1-\nu)$ are used.

Initially the coating and the substrate are separated and balanced external forces are applied on both sides to induce homogeneous stress profiles as depicted in Figure 1 (a). The compressive stresses in the coating and tensile stress in the substrate adjust the lengths of both parts. In the presented example the standardized forces F_0, F_1 and F_2 induce the tensile stress σ_0 in the substrate as well as the compressive stresses σ_1 in sublayer 1 and σ_2 in sublayer 2. In the next step the coating and the substrate are joined and the external forces stay applied to prevent the system from bending (Figure 1(α)). The position of the neutral plane depends on the Young's moduli as well as the thicknesses of the three parts joined and is described subsequently. The pairs of forces or the corresponding moments that act on each side of the system serve two purposes: they straighten the system and induce the current stress distribution. This description –the joined system with the external forces or moments applied- is denominated "straightened" and is the preferred description in terms of the calculation procedure. Now the external forces are removed from the joined system. The system responds by curving and accommodating the internal stresses and is therefore denominated "relaxed" system.

To facilitate the calculation of the stress profile in this relaxed system, an initially stress free system of the same materials as well as dimensions is introduced (Figure 1(b)) and bending moments with the same magnitude but opposite sense as in the model system (α) are applied.

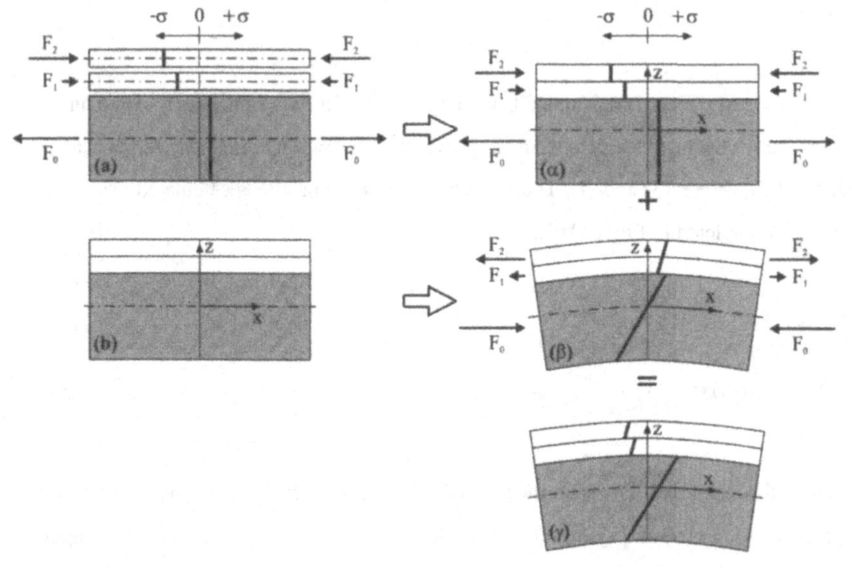

Figure 1. The straightened system (α) which is preferred for the calculation procedure and system (β) which is curved solely owing to external moments are added to obtain the relaxed system (γ) which describes the stresses in systems bent owing to internal residual stresses.

Then the position of the neutral plane is calculated. The neutral plane is the region in which the stresses induced by the bending moment are zero (Equation (1)). For both systems (α and β) it is at the same position, because it depends only on the thicknesses of the individual parts and the elastic moduli. The distance of the neutral plane from the bottom of the system h is calculated by solving Equation (2) for h, which has the advantage that the position can be calculated solely from the elastic moduli and the thicknesses of the corresponding parts. Equation (2) is a combination of Equation (1) and Equation (3), which is used for the determination of moment induced stresses from the corresponding curvature.

$$\int \sigma_M \, dz = 0 \tag{1}$$

$$E_{b,0} \int_{-h}^{-h+t0} z dz + E_{b,coat} \int_{-h+t0}^{-h+t0+t1+t2} z dz = 0 \tag{2}$$

$$\sigma_M = -E \kappa z \tag{3}$$

The bending moment is calculated from the stresses of the straightened system using Equation (4). Now the curvature of system (β) is determined from Equation (5), which is a combination of Equations (3) and (4). The corresponding stress profile is calculated from Equation (3) and depicted in Figure 1(β).

$$M = -\int \sigma z dz \tag{4}$$

$$M = \kappa E_{b,0} \int_{-h}^{-h+t0} z^2 \, dz + \kappa E_{b,coat} \int_{-h+t0}^{-h+t0+t1+t2} z^2 dz \tag{5}$$

Finally the stress distribution in the relaxed system (γ) is determined by superimposing the straightened system (α) and system (β) and adding the corresponding stresses according to Equation (6). The external forces and moments of the systems (α) and (β) cancel each other out and therefore, system (γ) curves solely owing to the internal stresses.

$$\sigma_\alpha + \sigma_\beta = \sigma_\gamma \tag{6}$$

As mentioned above, the relaxed system describes the stresses in the curved system, but the straightened system is preferred for the calculation procedure.

STRESSES IN THE CANTILEVER AND STRESSES IN THE INITIAL SYSTEM

The stresses in the straightened and the relaxed cantilever can be calculated from the stresses in the initial system and vice versa, because the material is assumed to be linear elastic. The procedure starts with the separation of the three parts of in the straightened initial system, i. e. with the applied external forces adjust the lengths and induce the stresses σ_i as depicted in Figure 2(a). For simplification the lengths in the straightened state are set to 1. Now the forces are removed so that the individual parts can contract or expand until they are stress free (Figure 2(b)). The length of the separated parts, $l_{i,stressfree}$, is calculated from Equation (7).

$$l_{i,stressfree} = 1/(\sigma_i/E_{b,i} + 1) \tag{7}$$

Now the thickness of the substrate is set to the desired value. In the presented example a cantilever substrate thickness of $t_{0,cant}$ is chosen and balanced external forces are applied to adjust the lengths of the three separated parts. Now the length of the cantilever and the balanced external forces that induce the stresses in the straightened cantilever are calculated. Therefore, four equations have to be solved in this example. Equation (8) establishes the force balance of the external forces that lead to compressive stresses in the coating and tensile stresses in the cantilever substrate. The three missing equations (Equation (9) with $i = 0...2$) calculate the forces necessary to adjust the lengths of the separated parts.

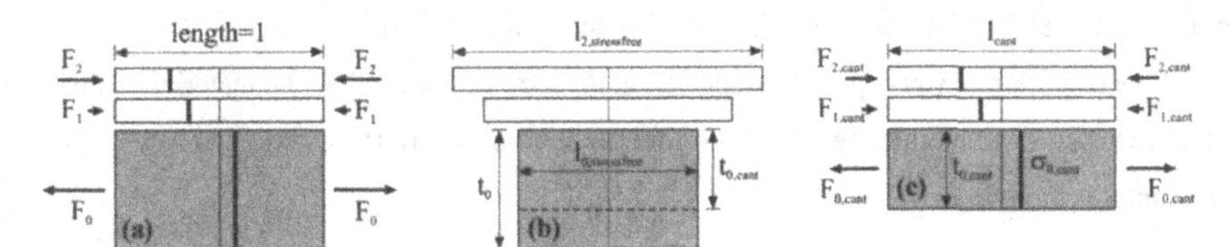

Figure 2. The stress profile in the straightened cantilever (c) is calculated by removing the external forces from the initial system (a) to obtain the unstressed parts of the system (b), reducing the substrate thickness and applying external forces to adjust the lengths of these parts.

$$F_{0,cant} + F_{1,cant} + F_{2,cant} = 0 \tag{8}$$

$$F_{i,cant} = t_{i,cant} E_{b,i} \frac{l_{cant} - l_{i,stressfree}}{l_{i,stressfree}} \tag{9}$$

The stresses in the straightened cantilever $\sigma_{0,cant}$, $\sigma_{1,cant}$ and $\sigma_{2,cant}$ are calculated by dividing the determined forces by the corresponding thicknesses. The stress profile in the relaxed cantilever is calculated in principle like the stresses in the relaxed initial system using Equations (1) to (6) to determine the neutral plane, the moments and the stresses.

The presented calculation procedure can be performed similarly for an arbitrary number of sublayers of different thicknesses and material combinations. It has to be taken into account that the thickness of the cantilever substrate influences the stress profile in the cantilever essentially. Depending on the Young's moduli and the stress profile in the initial system, a change from compressive stresses to partly tensile stresses in the curved cantilever is possible and then has to be considered (Figure 3).

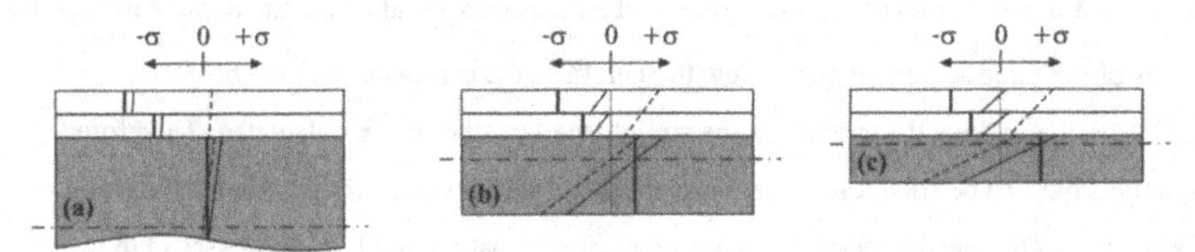

Figure 3. The effect of the substrate thickness on the stresses in the initial system (a) and two corresponding cantilevers of different substrate thicknesses (b) and (c) is demonstrated. The stresses in the straightened systems (thick line), the initially unstressed systems which curve only owing to the external bending moment (dashed line) and in the relaxed systems (thin line) are drawn in. The slope of the stress profiles depends on the Young's moduli of the materials. The materials are assumed to be linear elastic, therefore the stresses of (a), (b) and (c) can be converted into eachother.

ACKNOWLEDGMENTS

The authors thank the Austrian Science Foundation FWF for supporting this work and G. Dehm, FD. Fischer, T. Antretter, and KJ. Martinschitz for helpful discussions.

REFERENCES

1. S. Massl, J. Keckes and R. Pippan, *Acta Mater.* **55**, 4835 (2007).
2. J.M. Gere, *Mechanics of Materials*, 6th ed. (Brooks/Cole, Belmont, 2004) pp.300.

Mater. Res. Soc. Symp. Proc. Vol. 1052 © 2008 Materials Research Society 1052-DD03-10

Modification of Conductivity and of Mechanical Properties of Electroactive Polymer (EAP) Thin Films by Titanium Ion Implantation

Muhamed Niklaus[1], Samuel Rosset[1], Massoud Dadras[2], Philippe Dubois[1], and Herbert R. Shea[1]
[1]LMTS, EPFL, Neuchâtel, 2002, Switzerland
[2]IMT, University of Neuchâtel, Neuchâtel, 2002, Switzerland

ABSTRACT

We present a study of the influence on Young's modulus, stress and electrical conductivity of poly-dimethylsiloxane membranes implanted with titanium ions at energies from 5 to 35 keV, with doses up to $8x10^{16}$ ions/cm^2. The motivation for this study was to find the optimum implantation conditions to create electrodes for microfabricated dielectric electroactive polymer actuators, which must combine low resistivity with low stiffness. Two implantation techniques are used, Filtered Cathodic Vacuum Arc (FCVA) and the more conventional Low Energy broad beam Implanter (LEI). Of the two, it is found that the FCVA implanter is much better suited to create compliant electrodes.

INTRODUCTION

Dielectric electroactive polymer (DEAP) actuators have very large percentage displacements, often exceeding 100%, and therefore require compliant electrodes. Macro-scale (cm and larger) DEA devices, often referred to as artificial muscles, have electrodes made of carbon or metal powder or of conductive grease [1]. This electrode fabrication method is not applicable to micro-scale devices for which electrodes patterned with micron resolution are required. Simple evaporation or sputtering of metal electrodes is not suitable due to the 5 orders of magnitude larger Young modulus of the metals compared to the elastomers, and due the fracture of metal films at very small strains.

We have shown [2, 3, 4] that it is possible to create compliant electrodes by Filtered Cathodic Vacuum Arc (FCVA) implantation of metal ions into poly-dimethylsiloxane (PDMS) elastomers. Low energy (2 t 35 keV) implantation creates a thin conductive layer a few nm below the surface of the polymer without forming a continuous metallic layer, and thus without the stiffening expected from a metal layer. Electrical conductivity of a few kΩ/ square can be reached with less than 1 MPa increase in Young's modulus for doses of order 10^{16} ions/cm^2 with a FCVA. However, large FCVA equipments are not readily available, and ours can only implant small surfaces (1 cm^2). Therefore, we conducted a series of implantations with a Low Energy broad beam Implanter (LEI) at 10 and 35 keV in order to obtain implanted layers whose parameters (dose, uniformity and energy) are better defined.

The field of ion implantation in polymers is largely unexplored. The higher the energy and the higher the atomic number of the ions, the longer their stopping distance in the polymer and the larger their distribution or straggle. A numerical simulation [5] of Ti ion distribution in PDMS for three different implantation energies (5, 10 and 35 keV), ignoring the effect of metal ions accumulated in the film, shows increasing average penetration depth D and increasing straggle s with increasing energy. At 5 keV, D=13 nm, s= 10 nm FWHM, while at 35 keV,

D=60 nm and s=60 nm FWHM. Lower implant energies are therefore preferred to obtain a sufficient concentration, and hence conductivity, with the lowest possible dose.

Conduction mechanisms in thin films have been studied for many years [6, 7], and several models, such as percolation, are probably applicable to thin films implanted in polymers. In this work we present the influence of the implantation technique, ion energy and dose on the mechanical and electrical properties of PDMS membranes and films implanted with Titanium.

EXPERIMENT

Samples for resistivity measurements consist of 25 μm thick PDMS films on Si chips. Resistivity is measured on a 5x1 mm^2 zone between gold electrodes. For the stress (σ) and Young modulus (E) measurements a "bulge test" setup was used that measures the deflection of a PDMS membrane due to an applied gas pressure. Samples consisted of 25 μm thick square and circular PDMS membranes of lateral dimensions 2.5 mm. Roughness measurements were performed with AFM and an optical profiler.

Two implantation techniques were used to fabricate electrodes on PDMS: FCVA and LEI. Ti was implanted at 5 keV with the FCVA, at 10 keV and 35 keV with LEI. Doses at 10 keV and 35 keV were chosen to be comparable with doses that produced conductive and compliant electrodes with the FCVA technique.

RESULTS

In the three figures that follow, Young's modulus (E) and one additional parameter are plotted vs. Ti ion dose for implantation at energies of 10 keV and 35 keV. As will be clear below, there are two main regimes: at lower doses (below $2x10^{16}$ at/cm^2) no cracking of the membranes is observed, while at higher doses cracks are observed, which must be taken into account when interpreting the values of E and σ which are an effective value for the complete membrane.

Figure 1 presents the evolution of E and stress σ vs. dose. Starting from an initial unimplanted value of 0.7 ± 0.1 MPa, E increases monotonically to reach a peak value of 64 MPa (at an ion dose of $1.8x10^{16}$ at/cm^2) and 173 MPa (dose of $1x10^{16}$ at/cm^2) for 10 keV and 35 keV respectively. At higher doses, E decreases rapidly, then for 10 keV increases again.

For both energies, except for the high doses of 35 keV where cracks dominate, the magnitude of the stress follows the same trend as the Young's modulus. The evolution of E and σ is clearly different for the two energies. For 10 keV the initial tensile stress, with a value of 53 ± 5 kPa, becomes strongly compressive, resulting even in a buckling of the membrane up to 120 μm, while for 35 kV the stress has a much smaller absolute magnitude and remains tensile.

The roughness of the implanted membranes is presented in figure 2, and follows the same trend as E. At low doses the roughness reaches relatively high values, approaching 130 nm and 170 nm RMS for 10 and 35 keV respectively, while at the high doses it remains low (around 20 nm RMS). For the samples presenting cracking, roughness has been measured in uncracked regions.

Figure 3 plots surface resistivity vs. dose. The membranes are initially insulating. As the implanted dose increase the resistivity drops. At lower doses there is initially the formation of electrically separated Ti clusters. When the percolation limit is reached, the resistivity drops by orders of magnitude as a continuous conductive path is established along the electrode surface. For 10 keV and for 5 keV (Figure 4), this resistivity drops exactly in the region where the Young's modulus rises, indicating a link between the mechanical properties of the film and the percolation threshold.

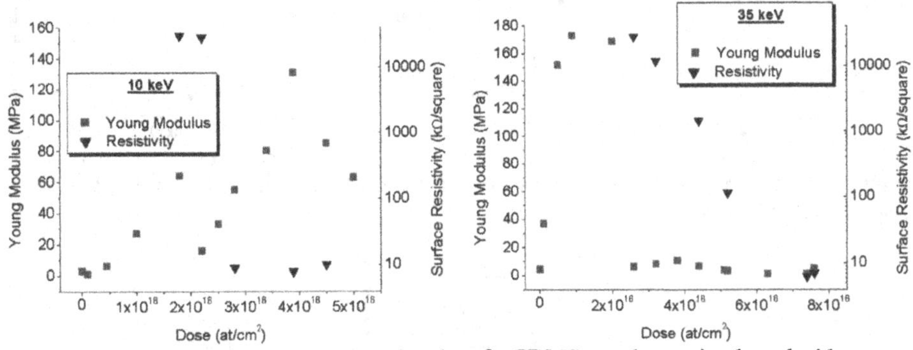

Figure 1. Young's modulus and stress vs. ion dose for PDMS membranes implanted with Titanium at 10 keV (left) and 35 keV (right) by LEI. For both energies, samples present cracking above $2x10^{16}$ at/cm^2, see figure 5 below.

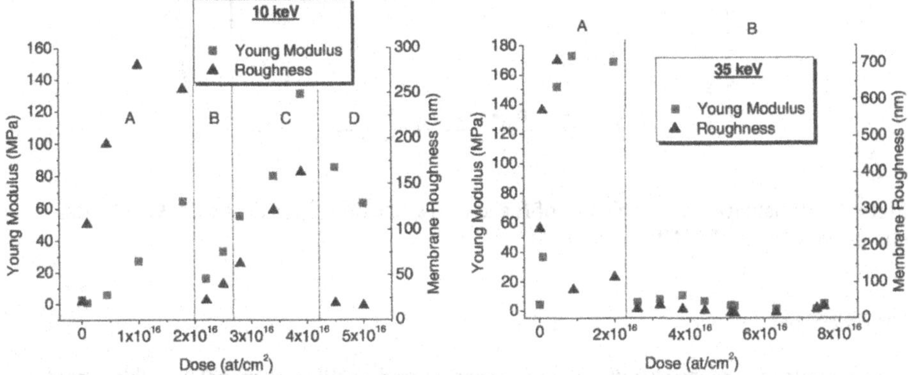

Figure 2. Young's modulus and membrane roughness vs. Ti ion dose, for 10 keV (left) and 35 keV (right). Dash dotted vertical lines delimit different zones of cracking (A and C is uncracked, while B and D are explained in the Discussion section).

The mechanical and electrical properties of PDMS membranes implanted with 5 keV Ti ions from an FCVA implanter are shown in figure 4. A continuous increase in E with dose is seen. The stress behaves just as for 10 keV: overall it inversely follows E. Unlike for 10 keV and 35 keV, no cracking of the films are seen at 5 keV. The maximum value of E for the FCVA samples was 2.3 MPa at $4x10^{16}$ at/cm^2, 57 times less than the maximum E at 10 keV.

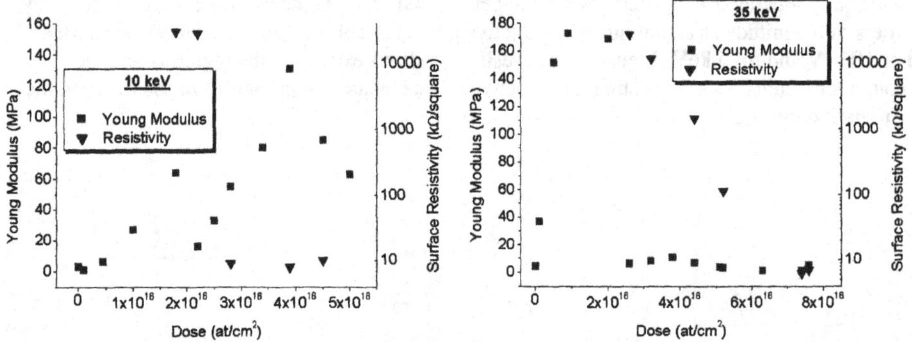

Figure 3. Surface resistivity and Young's modulus vs. ion dose for the 10 keV and 35 keV Ti implanted membranes.

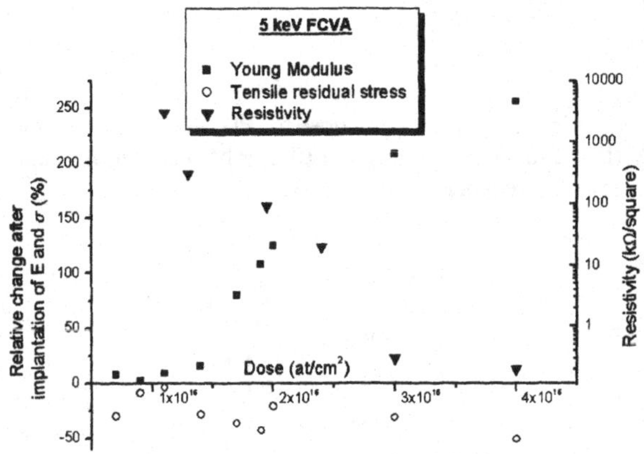

Figure 4. Resistivity, relative change of E and σ as a function of ion dose for 5 keV Ti ions implanted with the FCVA technique.

DISCUSSION

While the change in E with dose may appear surprisingly non-monotonic, this is readily explained by the cracks that appear in 10 keV and 35 keV samples at doses greater than $2x10^{16}$ atoms/cm^2. In regime A of figure 6, the films are uncracked. In regions B and D the films all show evidence of cracking which occurs during the deformation of the membrane.

Once the films cracks, the E and σ values obtained for the whole films by bulge test are dominated not by the stiff implanted layer, but by the virgin PDMS in the cracked regions, which explains why in figure 1 at 35 keV the values of E and σ are within 50% of those for virgin films.

The surface resistivity of the 10 keV and 35 keV samples was measured for over 6 months. These measurements show a very good stability for all doses: for 10 keV and 35 keV we observed a relative increase of resistivity of 25% and 35% after 187 days. In contrast, the 5 keV samples have unstable conductivity at low doses. Storing those samples in vacuum greatly extends their lifetime. 5 keV FCVA samples with Au or Pd ions exhibit stable resistivity in air.

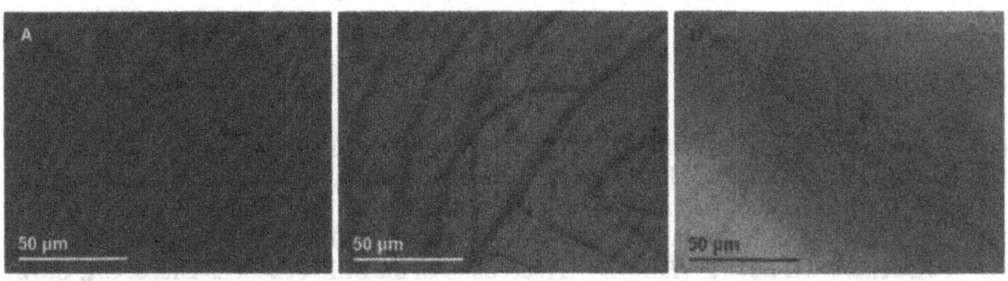

Figure 5. Membrane surface corresponding to the different zones presented in the figure 2 for 10 keV (left) and 35 keV (right).

The differences in cracking and in the increase of Young's modulus between 5 keV FCVA and 10 keV LEI stems probably from the very different implantation conditions: in addition to the obvious difference in ion energy, the FVCA produced a pulsed beam of ions, with a large energy spread in each pulse. Average beam current is 3 mA/cm^2 for FCVA compared to 0.4 µA/cm^2 for LEI at 10 keV. This leads to very different charging and heating effects. In addition of its lower impact on the membranes' stiffening, FCVA at 5keV produced layers that were much more conductive at comparable dose. For example, at $4x10^{16}$ at/cm^2, the surface resistivity for 5keV, 10keV and 35 keV are respectively 150 Ω/square, 10 kΩ/square, and 1 MΩ/square. At low energy, the penetration depth of incoming ions is much more limited, which increases the ion's spatial density. Unfortunately, it is not possible to work with LEI at such a low energy.

CONCLUSION

This study aimed at determining the impact on Young's modulus and resistivity of PDMS samples implanted in conventional implanters with Ti ions at 10 keV and 35 keV, and at comparing those results with data for Ti ion implanted with an FCVA implanter at 5 keV. It is found that the FCVA implanter is much better suited than conventional implanting equipment to create compliant electrodes for artificial-muscle type electroactive actuators, require lower doses for stable conduction, and showing much less stiffening and no cracking. While the films deposited at higher energy had excellent stability in time, they resulted in dramatic (factor of 100) increase in the Young's modulus of the elastomer membranes, followed by cracking at higher doses. We are currently comparing different ion species (Au, Pd) implanted with the FCVA tool and the LEI. Ongoing research focuses on TEM observations in order to image the resulting microstructure, and modeling of this microstructure to relate microstructure to electrical and mechanical properties.

ACKNOWLEDGMENTS

This research was supported by the Swiss National Science Foundation grant # 200021-111841 and the EPFL. The LEI implantations were performed by the Center for Application of

Ion Beams in Materials Research at the Forschungszentrum Dresden-Rossendorf (FZR), Germany. We also would specially like to thank Mr. I. Winkler for his work with LEI, as well as Dr. M. Doebeli (PSI, Ion Beam Physics) for the RBS measurements.

REFERENCES

[1] S. Ashley, *Scientific American* **289**(4), 52–59 (2003).

[2] P. Dubois, S. Rosset, S. Koster, J. Stauffer, S. Mikhaïlov, M. Dadras, N.-F. de Rooij and H. R. Shea, *Sensors and Actuators A: Physical* **130-131**, 147-154 (2006).

[3] S. Rosset, M. Niklaus, P. Dubois, M. Dadras, and H. R. Shea in *Electroactive Polymer Actuators and Devices (EAPAD)*, (Proceedings of the SPIE **6524**, San Diego, CA, USA, May 2007), pp. 652410.

[4] S. Rosset , M. Niklaus, P. Dubois, H. R. Shea, Submitted to *Sensors and Actuators*, (September 2007).

[5] J. F. Ziegler and J. P. Biersack, *SRIM - The Stopping and Range of Ions in Matter*, software at: http://www.srim.org/.

[6] E. Simanek, *Solid State Communications* **40**, (Pergamon Press Ltd. 1981) pp. 1021-1023.

[7] Scott Kirkpatrick, *Solid State Communications* **12**, 1279 - 1283 (1973); *PHYSICAL REVIEW B* **74**, 174201 (2006).

Mater. Res. Soc. Symp. Proc. Vol. 1052 © 2008 Materials Research Society 1052-DD03-11

C-Axis Oriented ZnO Film by RF Sputtering and Its Integration With MEMS Processing

Sudhir Chandra, and Ravindra Singh
Centre for Applied Research in Electronics, Indian Institute of Technology Delhi, Hauz Khas, New Delhi, 110016, India

ABSTRACT

In the present work, we report a new fabrication process to integrate the "c-axis oriented" ZnO films with bulk-micromachined silicon diaphragms. ZnO films are very sensitive to the chemicals used in the micro-electro-mechanical systems (MEMS) fabrication process which include acids, bases and etchants of different material layers (e.g. SiO_2, chromium, gold etc.). A Si_3N_4 layer is incorporated to protect the ZnO film from the etchants of chromium and gold used for patterning the electrodes. A mechanical jig is used for protecting the front side (ZnO film side) of the wafer from ethylenediamine pyrocatechol water (EPW) during the anisotropic etching of silicon. The resistivity measurement performed on the ZnO film integrated with micro-diaphragm shows the reliability of the fabrication process proposed in this work.

INTRODUCTION

During last several years, an extensive research has been carried out on zinc oxide (ZnO) thin films because of their applications in surface acoustic wave (SAW) devices, bulk acoustic wave (BAW) resonators, optical waveguides, transparent conducting coatings, light emitting diodes (LED), photodetectors and electroluminescence devices [1-5]. These films can also be used in MEMS as a sensing material because of their piezoelectric properties [6]. More recently, piezoelectric based MEMS have shown great potential for bio-sensor applications [7]. For applications of ZnO films based on its piezoelectric properties, it is a requirement that the films should be "c-axis oriented" and have high resistivity. Furthermore, ZnO films deposition at comparatively low temperature is of great interest for realization of MEMS in post-CMOS processing steps. Among the various deposition techniques, sputtering is the preferred method as oriented and uniform ZnO films can be obtained even at relatively low substrate temperatures. The integration of ZnO films with MEMS processing is a very challenging task because these films are very sensitive to most of the chemicals used in MEMS fabrication process such as H_2SO_4, hydrofluoric acid (HF), buffered HF (BHF), silicon anisotropic etchants (e.g. KOH, EPW), metal (e.g. chromium, gold, aluminum) etchants etc. [6, 8-9]. The subsequent fabrication processes after ZnO deposition could degrade or damage ZnO film if it is not protected properly. The present work focuses on the techniques to protect the ZnO film against degradation in subsequent micromachining processes. A new fabrication process has been demonstrated to integrate the ZnO films with bulk-micromachined diaphragms in which ZnO film is safely protected from the different chemicals during the etching processes.

EXPERIMENTAL DETAILS

Preparation of c-axis oriented ZnO films

The c-axis oriented ZnO films have been prepared on Pt (150 nm) / Ti (20 nm) / SiO_2 (1 µm) / Si substrate by RF magnetron sputtering. For preparing c-axis oriented ZnO film, the optimization of various sputtering parameters such as RF power, sputtering gas, sputtering pressure, target-to-substrate spacing and substrate temperature has been reported elsewhere [10]. The films prepared without external substrate heating at 100 watt RF power, 50 mm target-to-substrate spacing, 10 mtorr sputtering pressure in the mixture of argon-oxygen (1:1) were found to be highly c-axis oriented. The crystallographic properties of the ZnO film were investigated using PW 3040 Philips X'Pert X-ray diffractometer (CuKα radiation, λ=1.5405Å). For the measurement of resistivity of the films, aluminum electrodes of area 0.442 mm^2 were formed by thermal evaporation using a sheet metal mask. The dc resistivity was measured using a computer interfaced Keithley 2410, source meter.

Integration of ZnO film with bulk-micromachined diaphragms

The schematic view of the diaphragm structure for demonstrating the integration of ZnO film is shown in Fig. 1. The diaphragm consists of a heavily boron doped silicon (P^+-silicon) as structural layer, and ZnO thin film sandwiched between two electrodes. A composite layer of platinum and titanium is used as bottom electrode of ZnO while chromium-gold layer is used for the top electrode. A Si_3N_4 layer is incorporated to protect the ZnO film from the etchants of chromium and gold used for patterning the top electrodes. The square shaped diaphragms of different dimensions (e.g. 300 µm × 300 µm, 500 µm × 500 µm, 1 mm × 1 mm etc.) were fabricated using standard IC fabrication techniques and bulk micromachining process. A mechanical jig is used for protecting the front side (ZnO film side) of the wafer from EPW during the anisotropic etching of silicon.

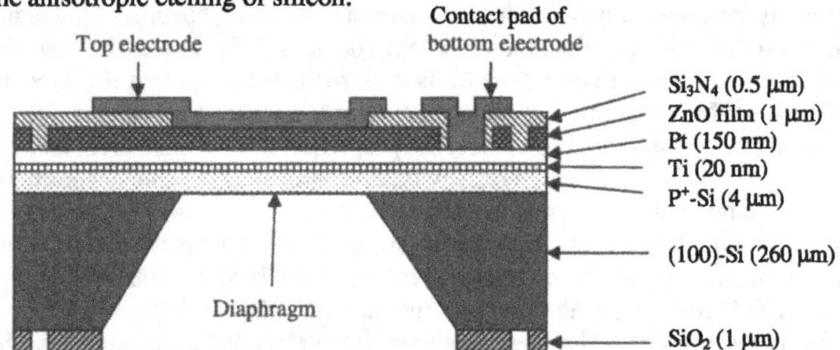

Fig. 1. Cross-sectional schematic view of the proposed structure (diaphragm) for demonstrating the integration of ZnO film.

The cross-sectional schematic view of the mechanical jig used for shielding the one side of the wafer during anisotropic etching of silicon is shown in Fig. 2. The jig is designed to process a 2-inch silicon wafer. It consist of two Teflon parts, two stainless steel (SS) plates, one neoprene gasket, two silicone O-rings, one SS pipe and eight pairs of nuts-bolts. The silicon wafer to be etched is sandwiched between the two Teflon parts and sealed with the help of inner silicone O-ring and neoprene gasket as shown in the Fig. 2. The front-side of the wafer, which is required to be protected during anisotropic etching process, is kept towards Teflon part # 1 so that it does not come in contact with the etching solution. The outer O-ring is used for sealing the two Teflon parts so that the etching solution does not seep-in. The outer O-ring is pressed with the help of SS plate # 1 and four SS bolts-nuts while the inner O-ring is pressed with SS plate # 2 using the remaining four bolts-nuts. A portion of Teflon part # 2 between the two O-rings is machined to 0.5 mm thickness which allows the flexibility for individual pressing of the two O-rings. To release the air from the closed cavity between wafer and Teflon part # 1 during etching process, a SS pipe is fixed in a hole made in Teflon part # 1. The SS pipe opens outside the etching solution in the air during etching process. The photograph of the assembled jig is shown in Fig. 3. Using this jig, only one side of the wafer on which the etching is required, comes in contact with the etching solution and the other side is completely protected during the etching process.

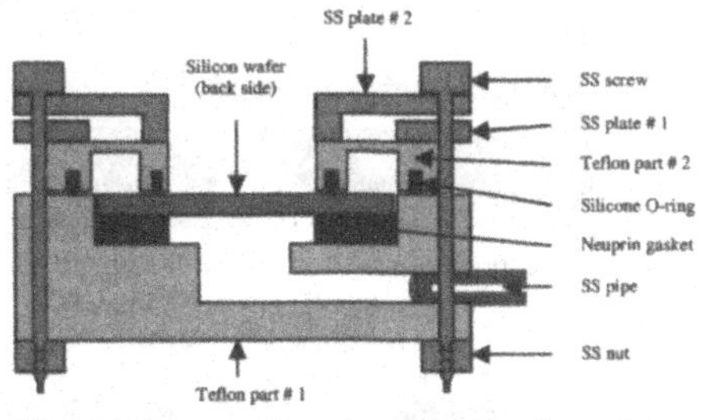

Fig. 2. Cross-sectional schematic view of the mechanical jig used to protect the front side of the processing wafer during silicon anisotropic etching.

Fig. 3. Photograph of the mechanical jig used to protect the front side of processing wafer during silicon anisotropic etching.

The complete fabrication process steps for the realization of proposed structures, as shown in Fig. 1, are schematically presented in Fig. 4. Initially, both-side polished and cleaned wafers were thermally oxidized at 1150 °C to grow about 1 μm silicon dioxide (SiO_2) layer. The positive photoresist (Shipley, AZ-1400-27) was then spin coated on one side (referred to as back side) of the oxidized wafer. Thereafter, the wafer is hard baked at 120 °C for 30 min and SiO_2 on the other side (referred to as front side) of the wafer was etched out in buffered hydrofluoric (BHF) acid. Now, for structural layer (P^+-silicon) formation, boron pre-deposition was carried out at 1050 °C using $Boron^+$ solid source followed by drive-in step out at 1100 °C. To obtain the thickness of P^+-silicon of about 4 μm, pre-deposition and drive-in were performed for 8 and 4 hrs respectively. Thereafter, the front-to-back alignment marks are defined on the back-side using photolithography with sheet metal mask and etching of SiO_2 [11]. Now, for defining the squared shaped windows of various dimensions in SiO_2 on the back-side, photolithography (mask # 1) followed by etching of SiO_2 was carried out. In the next step, titanium (Ti), platinum (Pt) and ZnO layers were sequentially deposited on the front-side of the wafer by RF sputtering method. Thereafter, the front-to-back alignment marks are defined on the front side of the wafer in ZnO film. The etching of ZnO was carried out in 1% HCl solution. Photolithography (mask # 2) was employed to open the windows in ZnO for taking the contact from Pt/Ti bottom electrode. In the next step, a 0.5 μm thick Si_3N_4 layer was deposited by RF sputtering. To open the contact windows, Si_3N_4 layer was patterned using lithography (mask # 3) and reactive ion etching (RIE) processes. The RIE of Si_3N_4 was performed in SF_6 plasma at 60 watt RF power and 2×10^{-2} torr

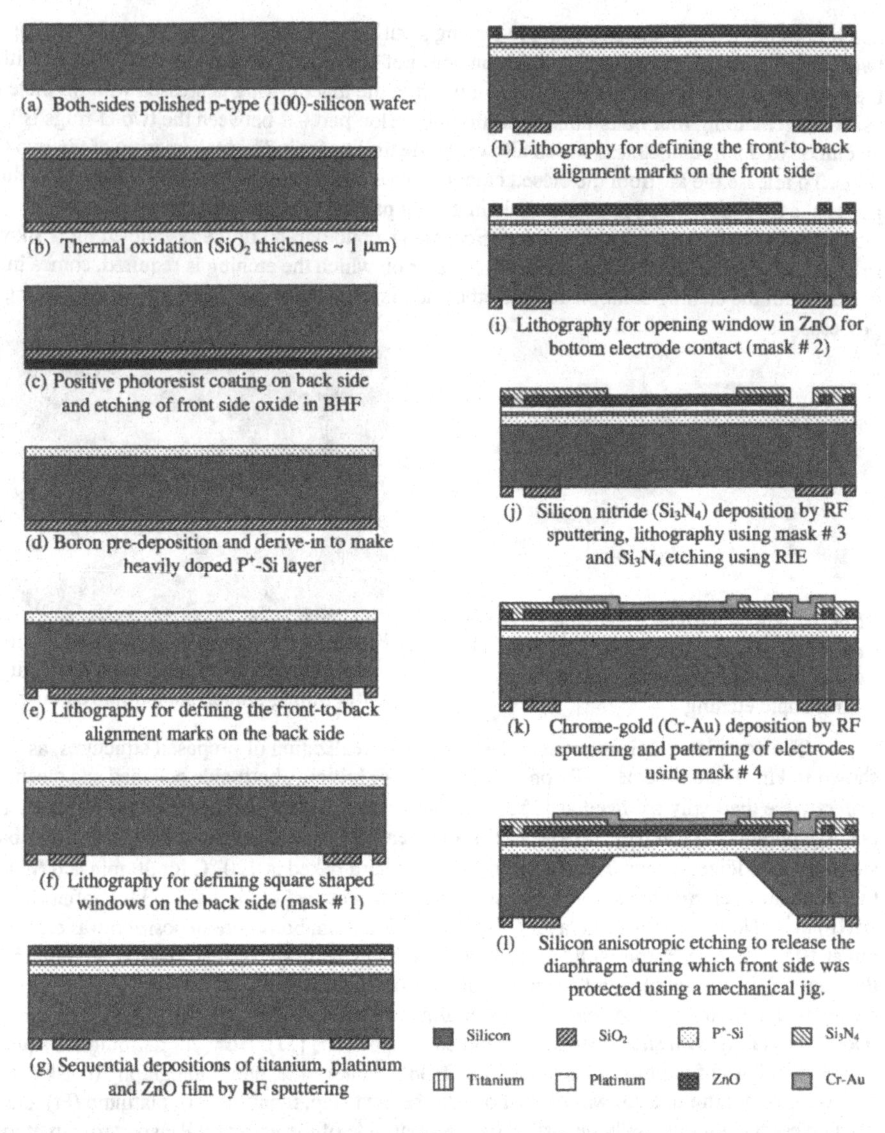
(a) Both-sides polished p-type (100)-silicon wafer
(b) Thermal oxidation (SiO_2 thickness ~ 1 μm)
(c) Positive photoresist coating on back side and etching of front side oxide in BHF
(d) Boron pre-deposition and derive-in to make heavily doped P^+-Si layer
(e) Lithography for defining the front-to-back alignment marks on the back side
(f) Lithography for defining square shaped windows on the back side (mask # 1)
(g) Sequential depositions of titanium, platinum and ZnO film by RF sputtering
(h) Lithography for defining the front-to-back alignment marks on the front side
(i) Lithography for opening window in ZnO for bottom electrode contact (mask # 2)
(j) Silicon nitride (Si_3N_4) deposition by RF sputtering, lithography using mask # 3 and Si_3N_4 etching using RIE
(k) Chrome-gold (Cr-Au) deposition by RF sputtering and patterning of electrodes using mask # 4
(l) Silicon anisotropic etching to release the diaphragm during which front side was protected using a mechanical jig.
Silicon
SiO_2
P^+-Si
Si_3N_4
Titanium
Platinum
ZnO
Cr-Au

Fig. 4. Schematic cross-sectional view of the process steps to fabricate the proposed structure shown in Fig. 1.

pressure. For top electrode, Cr-Au layers were deposited by RF sputtering. The top electrodes and contact pads of bottom electrode were then patterned using lithography (mask # 4) and etching process. Finally, the silicon anisotropic etching was performed in EDW. During this process, the front side of the wafer (ZnO film side) was protected using the mechanical jig presented in the previous paragraph.

DISCUSSION

Fig. 5 shows the XRD pattern of c-axis oriented ZnO film prepared on optimized sputtering parameters. It can be seen from this figure that the film has only a single peak at 34.3° corresponding to the c-axis orientation. I-V characteristic of the film is shown in Fig. 6. The DC resistivity of the film, calculated from I-V measurements, was found to be of the order of 10^{11} Ω-cm at low electric fields (~ 10 kV/cm), which is very good for piezoelectric based applications.

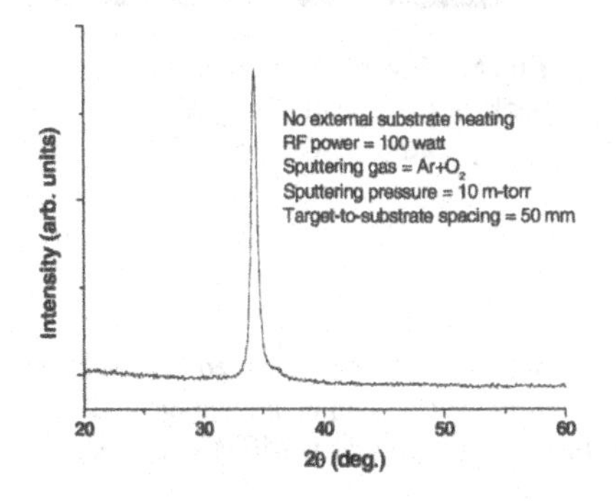

Fig. 5. XRD pattern of c-axis oriented ZnO film.

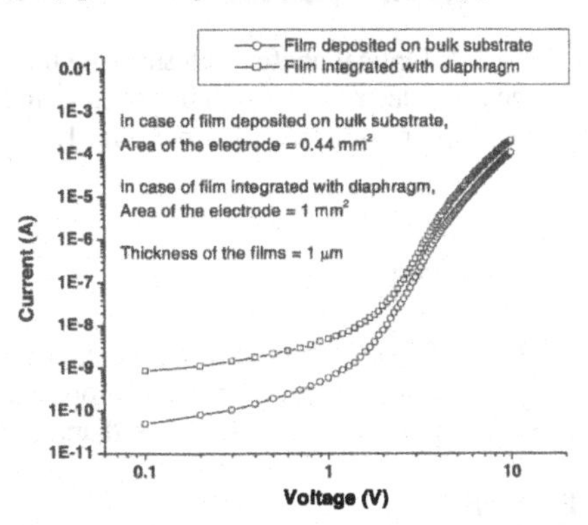

Fig. 6. I-V characteristics of the ZnO films.

Fig. 7 shows the optical photograph of Au/Cr top electrode and contact pad of the bottom electrode after fabrication of the diaphragm structure as shown in Fig. 1. The SEM photograph of the diaphragm (from the back-side) incorporating ZnO film is shown in Fig. 8. This diaphragm is composed of a P^+-silicon structural layer and the ZnO film sandwiched between Pt/Ti and Au/Cr electrode layers. It is evident from the photograph that a well-defined cavity has been formed with smooth and sloped side-walls of (111) planes. No buckling is visible in the diaphragm, which indicates very low stress in the diaphragm. The aim of the present work was limited to developing the fabrication process to integrate ZnO film with MEMS based diaphragms. However, for showing the reliability of the fabrication process proposed in this work, it is important to perform a basic electrical measurement on the ZnO film integrated with diaphragm. Therefore, we have also performed the I-V measurement on ZnO film integrated with diaphragm. This I-V characteristic is also shown in Fig. 6 and found to be almost similar to that observed for the film deposited on bulk substrate. On the basis of above results, it can therefore be concluded that the fabrication process demonstrated in the present work can be advantageously used to integrate the ZnO films with diaphragms for various MEMS applications.

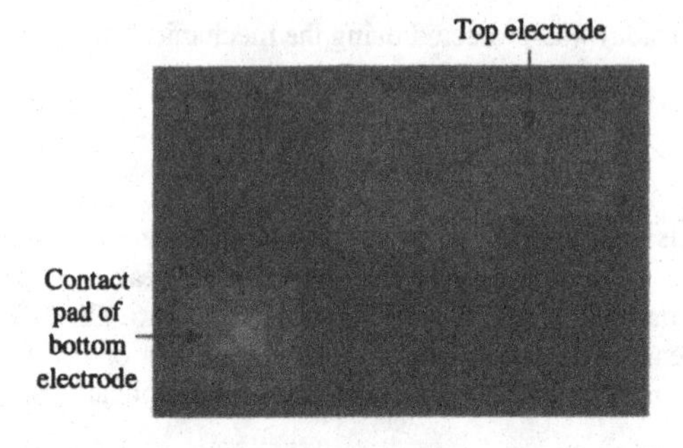

Fig. 7. Optical photograph showing the contact electrodes after fabrication of the diaphragm structure shown in Fig. 1.

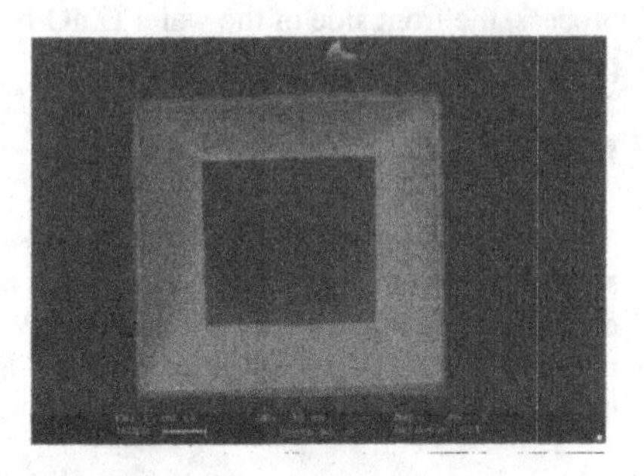

Fig. 8. SEM photograph showing the back side of the diaphragm incorporating ZnO film.

CONCLUSIONS

A new fabrication process has been presented to integrate the RF sputtered ZnO film with bulk-micromachined diaphragms. The proposed fabrication process can be advantageously applied to integrate the ZnO films with diaphragms for developing various MEMS devices.

REFERENCES

1. L. Jian-quo, W. Lei, Y. Zhi-zhen and Z. Bing-hui, *J. Func. Mat. Dev.* 8 (3) (2002) 303.
2. D.L. DeVoe, *Sens. Actuat. A* 88 (2001) 263.
3. T. Yamamoto, T. Shiosaki and A. Kawabata, *J. Appl. Phys.* 51 (1980) 3113.
4. T. Soki, Y. Hatanaka and D.C. Look, *Appl. Phys. Lett.* 76 (2000) 3257.
5. S.S. Lee and R.M. White, *Sens. Actuat. A* 71 (1998) 153.
6. T. Xu, G. Wu, G. Zhang and Y. Hao, *Sens. Actuat. A* 104 (2003) 61.
7. G.Y. Kang, G.Y. Han, J.Y. Kang, I.-H. Cho, H.-H. Park, S.-H. Paek and T.S. Kim, *Sens. Actuat. B* 117 (2006) 332.
8. M.J. Vellekoop, C.C.G. Visser, P.M. Sarro and A. Venema, *Sens. Actuat. A* 23 (1990) 1027.
9. H. Maki, N. Ichinose, T. Ikoma, I. Sakaguchi, N. Ohashi, H. Haneda and J. Tanaka, *J. Ceram. Soc. Jpn.* 110 (2002) 395.
10. R. Singh, M. Kumar and S. Chandra, *J. Mater. Sci.* 42 (12) (2007) 4675.
11. P. Pal, Y.-J. Kim and S. Chandra, *Sens. Lett.* 4 (2006) 1.

Mater. Res. Soc. Symp. Proc. Vol. 1052 © 2008 Materials Research Society 1052-DD03-12

Optimization of the Geometry of the MEMS Electrothermal Actuator to Maximize In-Plane Tip Deflection

Edward S. Kolesar, Thiri Htun, Brandon Least, Jeffrey Tippey, and John Michalik
Department of Engineering, Texas Christian University, TCU Mail Stop 298640, Tucker Technology Center, 2840 Bowie Street West, Fort Worth, TX, 76129

ABSTRACT

Several microactuator technologies have been investigated for positioning individual elements in large-scale microelectromechanical systems (MEMS). Electrostatic, magnetostatic, piezoelectric and thermal expansion represent the most common modes of microactuator operation. This investigation optimized the geometry of the asymmetrical electrothermal actuator to maximize its in-plane deflection characteristics. The MEMS polysilicon surface micromachined electrothermal actuator uses resistive (Joule) heating to generate differential thermal expansion and movement. In this investigation, a 3-D model of the electrothermal actuator was designed, and its geometry was optimized using the thermo-mechanical finite-element analysis (FEA) capabilities of the ANSYS computer program. The electrothermal actuator's geometry was systematically varied to establish optimum values of several critical geometrical ratios that maximize tip deflection. The value of the ratio of the length of the flexure component relative to the length of the hot arm was discovered to be the most sensitive geometrical parameter ratio that maximizes tip deflection.

INTRODUCTION

The single-hot arm electrothermal actuator depicted in Figure 1 is a widely used positioning component in MEMS devices. This actuator is popular because it is capable generating the large tip forces and deflections required to position various MEMS components compared to electrostatic and piezoelectric devices. Polysilicon electrothermal actuators can operate in the conventional IC current/voltage regime, and their fabrication process is compatible with that of semiconductor devices. The magnitude of the actuator's tip force and deflection is the primary performance metric associated with its design and application. Many studies have focused on the analysis of the dimensions of the different components of the electrothermal actuator to maximize tip deflection, and numerous researchers have proposed different models.

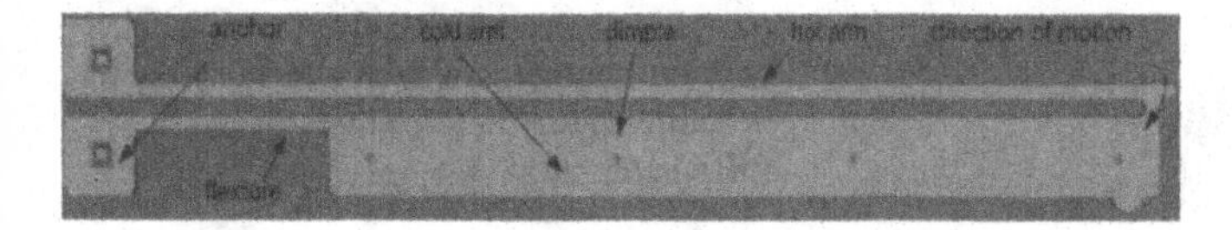

Figure 1. Single-hot arm polysilicon electrothermal actuator.

Guckel *et al* [1] proposed the original flexure-based electrothermal actuator in 1992. Huang *et al* [2] were first to develop an analytical model to predict the response of the Guckel electrothermal actuator. Hickey *et al* [3] then proposed a model that included the effect of thermal conduction through the device's narrow air gap to the substrate. Mankame *et al* [4] presented a model that accounted for all the modes of heat transfer, and it included the temperature dependence of the thermophysical properties. Pan *et al* [5] presented a modified version of the Guckel actuator that permitted different adjacent arm lengths, but they had to have the same cross sections. Lee *et al* [6] developed another modified model of the Guckel actuator that incorporated different arm lengths, but now they were allowed to have different cross sections. Chen *et al* [7] proposed a modified actuator design that included a gold layer deposited on the cold arm. Kolesar *et al* [8,9] studied the various designs of the electrothermal actuator, and using finite-element analysis (FEA) along with experimental performance, reported actuator geometries that yielded large tip deflections and forces.

This investigation was focused on optimizing the geometry of the single-hot arm electrothermal actuator to maximize tip deflection. Using the thermo-mechanical module of ANSYS®, a 3-D model of this actuator design was developed. This investigation determined the optimum ratios of the different geometrical elements shown in Figure 2, such as (L_3/L_1), (L_2/L_1), (t_1/L_1), (t_2/L_1), (t_3/L_1), and (t_4/L_1), that would maximize tip deflection. It is noted that this investigation was initiated by using the set of dimensions suggested by Kolesar *et al* [8,9].

THEORY

As depicted in Figure 1, the single-hot arm electrothermal actuator is composed of two parallel in-plane cantilever arms, and their free ends are connected to each other. One arm has a narrower cross section than the other. The actuator is anchored to its substrate at the fixed ends of the arms. The electrothermal actuator uses resistive (Joule) heating to generate thermal expansion and movement. When electrical current is passed through the actuator from one anchor to the other, the higher current density in the narrower arm ("hot" arm) causes it to heat and expand along its length more than the wider arm ("cold" arm). Since both arms are joined at their free ends, the difference in length of the two arms causes the electrothermal actuator tip to move in an arc-like pattern about the flexure element incorporated at the anchor end of the cold arm, as illustrated in Figure 3. Removing the electrical current from the device allows it to cool and return to its equilibrium state. The polysilicon MEMS version of this electrothermal actuator has been experimentally shown to produce incremental in-plane tip deflections that span 0 – 10 μm, while also generating force magnitudes greater than 10 μN [8,9].

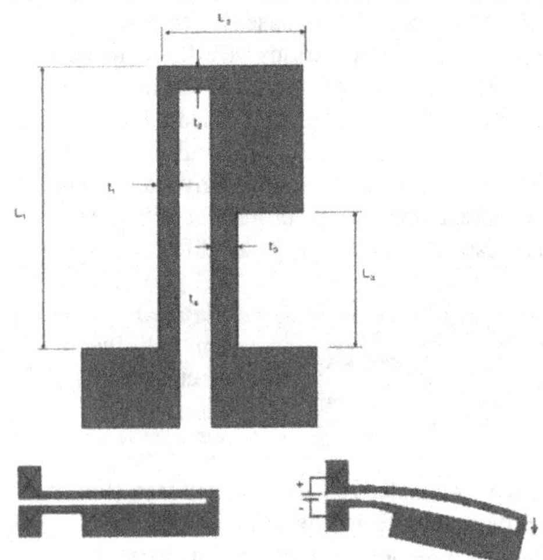

Figure 2. Schematic of the single-hot arm polysilicon electrothermal actuator labeled with the geometrical parameters used to accomplish the finite-element analysis (FEA) computations.

Figure 3. Schematic of the single-hot arm polysilicon electrothermal actuator's dynamic operation.

NUMERICAL MODELING

The finite-element analysis (FEA) of the MEMS electrothermal actuator was accomplished with the thermo-mechanical module of ANSYS® 10 program. Since the most popular material used to fabricate this actuator is polyslicon [2-9], it was adopted in this investigation, and its basic properties are summarized in Table 1.

To create a 3-D model of the electrothermal actuator for the ANSYS® 10 thermo-mechanical FEA optimization, the SOLID98 element geometry was used. As shown in Figure 4, this tetrahedral element has a total of 10-nodes. Each node has 3 degrees-of-freedom in the *x*-, *y*-, and *z*-directions. The SOLID98 element is useful for performing 3-D magnetic, thermal, electrical, piezoelectric, and structural FEA computations.

Table 1. Material Properties for Polysilicon (room temperature) [10].

Material Property	Value
Young's modulus	169 GPa
Poisson's ratio	0.22
Resistivity	2.3×10^{-5} ohm-m
Coefficient of thermal expansion (CTE)	2.9×10^{-6}/°K
Thermal conductivity	150 W/m°K
Melting Point	1410 °C

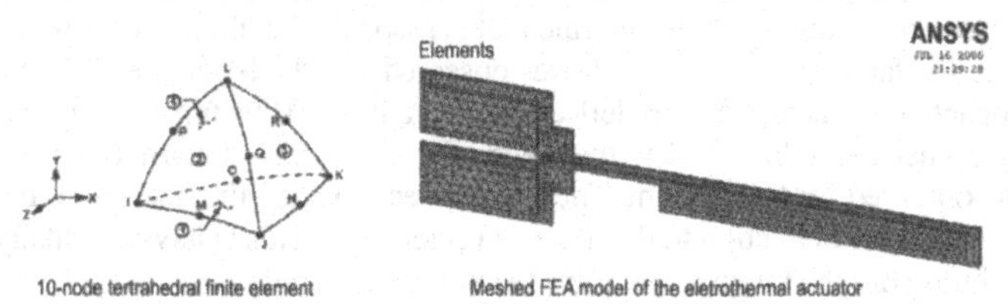

Figure 4. Tetrahedral finite-element node and FEA meshed model of the single-hot arm polysilicon electrothermal actuator.

A 5 VDC boundary condition was applied to the electrical pad connected to hot arm. The voltage set on electrical pad connected to cold arm was 0 VDC (ground). The temperature of the undersides of the contact pads was set to 23°C (room temperature) for the thermal boundary condition. To fix the actuator's anchors, all displacements in the *x*-, *y*-, and *z*-directions were set to zero. Figure 4 shows the meshed finite element model of the electrothermal actuator created in ANSYS® 10. Since Mankame *et al* [4] reported that the dominant mode of heat transfer for this actuator design is thermal conduction, this investigation restricted its focus to this mode.

OPTIMIZATION PARAMETERS

In order to perform the FEA optimization computations, the design variables, state variables, and objective function needed to be specified. Figure 2 describes the geometrical parameters that were used as the design variables for the electrothermal actuator. The numerical values associated with the design variables are summarized below:

$L_1 = 200\ \mu m$ (constant)

$1\ \mu m < L_2 < 50\ \mu m$

$1\ \mu m < L_3 < 180\ \mu m$

$$0.1\ \mu m < t_1 < 15\ \mu m$$
$$0.1\ \mu m < t_2 < 10\ \mu m$$
$$0.1\ \mu m < t_3 < 15\ \mu m$$
$$0.01\ \mu m < t_4 < 15\ \mu m$$

The state variable in this study was the melting point of polysilicon, which is 1410 °C (Table 1). Finally, the objective function of this investigation was to determine the actuator's maximum tip deflection relative to the practical range of geometrical parameters and the applied voltage.

SENSITIVITY ANALYSIS

The values for the design variables of the electrothermal actuator that were used to initiate the FEA computations were taken as reported in [8], and they include: $L_1 = 200\ \mu m$, $L_2 = 19\ \mu m$, $L_3 = 40\ \mu m$, $t_1 = 2\ \mu m$, $t_2 = 5\ \mu m$, $t_3 = 2\ \mu m$, and $t_4 = 2\ \mu m$. It was observed that as the value of L_2 increases, the magnitude of the actuator's tip deflection also increases. But after L_2 is increased to 25 µm, the gradient of the actuator's tip deflection decreases significantly. Changing the value of L_3 was observed to have a significant effect on the actuator's tip deflection. In this case, the maximum tip deflection was found when the value of L_3 was 50 µm. As depicted in Figure 5, decreasing the value of parameter t_1 from its initial value had the most significant effect on the value of the actuator's tip deflection. As the width of t_1 decreases to 0.5 µm, the actuator's tip deflection dramatically increases. This behavior is likely due to the reduction in the stiffness of the actuator's hot-arm as the width of t_1 decreases. However, the minimum width of t_1 will be limited by the minimum linewidth design rule that is associated with the polysilicon material and the foundry used to fabricate this device. It was observed that the t_2 parameter did not have a significant impact on the actuator's tip deflection. Since it would be functionally impractical for the width of parameter t_3 to be less than the width of the actuator's hot arm (t_1), the maximum tip deflection was observed for $t_3 = 2.9\ \mu m$. The gap between the hot arm and the cold arm, t_4, causes the actuator's tip deflection to decrease as t_4 increases. This behavior is likely due to the enhanced flexibility of the structure that joins the hot arm and cold arm.

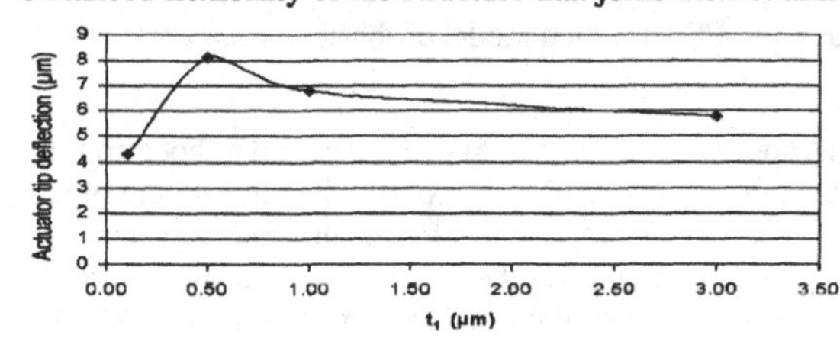

Figure 5. Plot of the FEA computed tip deflection of the single-hot arm polysilicon electrothermal actuator versus the "t_1" geometry variable (Figure 2).

RESULTS AND DISCUSSION

It was observed in the sensitivity analysis that the actuator's tip deflection significantly increased as the initial value of t_1 decreased. However, due to foundry fabrication linewidth limitations, and the consequence of having a hot arm whose width is less than 2 µm, it would be expected that significant amounts of heat would be generated, and thus, destroy the actuator. Therefore, the minimum width of t_1 was set at 2 µm, as reported in [8].

The flexure element should be as narrow as possible. A narrower flexure allows more of the force generated by the thermal expansion of the hot-arm to cause movement at the tip of the actuator. If the flexure were to be narrower than the hot arm, the temperature of the flexure element would be greater than the hot arm, and it could be destroyed by excessive heat [9].

Furthermore, there is a trade-off between the values for L_3 and t_3. As L_3 is held at its optimum value, ($L_3/L_1 = 1/20$), as observed in the sensitivity analysis, t_3 needs to decrease to maximize actuator tip deflection. Consequently, the optimum minimum flexure width, t_3, was subsequently restricted to be the same as the width of t_1.

Another important parameter is the flexure length, L_3. The flexure element needs to be sufficiently long so that it can be elastically deflected by the thermally-induced length expansion of the hot-arm. However, if the flexure is too long, movement of the actuator's tip will be significantly reduced [9]. Large L_3, for example, $L_3 = \frac{3}{4} L_1$, results in small tip deflections. When L_3 is made even smaller relative to L_1, the actuator's tip deflection dramatically increases, because the cold arm area increases, and this situation results in a larger temperature difference between the hot arm and the cold arm. However, after the (L_3/L_1)-ratio exceeds 1/20, the actuator's tip deflection decreases, because the short length of the flexure and its increased stiffness acts to limit the deflection of the actuator's tip. The sensitivity analysis of the computations confirmed that the optimum value of the (L_3/L_1)-ratio was 1/20, and it was also observed that the actuator's tip deflection could be maximized by using the design parameter ratios summarized in Table 2.

Table 2. Optimized ratio values of the design variable parameters for the electrothermal actuator.

Parameter Ratio	Optimized Parameter Ratio Value
(L_3/L_1)	1/20
(L_2/L_1)	1/8
(t_1/L_1)	1/100
(t_2/L_1)	1/40
(t_3/L_1)	1/100
(t_4/L_1)	3/400

The electrothermal actuator similar to that depicted in Figure 1 was designed subject to the optimized design parameter ratios summarized in Table 2 with $t_1 = 2$ μm. The electrothermal actuator was designed using the MEMSPro® CAD software and the material system of the PolyMUMPS foundry [10]. The ANSYS 10® FEA program was then used to compute the tip deflection characteristics of this optimized design, and Figure 6 depicts these results compared to a prior non-optimized design where $L_1 = 200$ μm.

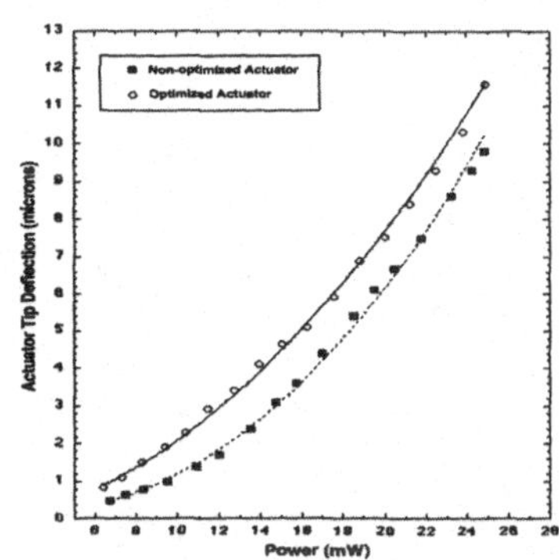

Figure 6. Plots of the ANSYS 10® FEA computed tip deflections (d) versus activation electrical power (mW) for an optimized [design parameter ratios summarized in Table 2 (with $t_1 = 2$ μm)] and a non-optimized geometry device. The smooth curve corresponds to the Levenburg-Marquardt non-linear least-squares curve fit for an equation of the form: d (μm) = k (μm) x [W(mW)/(W_o)] ^{m} where $W_o = 1$ mW (unit normalization); (k = 0.005, m = 2.344 and Corr. Coef. = 0.987 for the non-optimized actuator and k = 0.028, m = 1.875 and Corr. Coef. = 0.978 for the optimized actuator).

CONCLUSIONS

The most important part of this investigation was the sensitivity analysis. After performing the sensitivity analysis with ANSYS® 10, several important design parameters were determined that can be scaled to increase the actuator's tip deflection. The important parameters are the hot arm width (t_1), cold arm width (t_3), and the flexure length (L_3). The actuator's other design parameters also influence the magnitude of tip deflection, but their influence is not as significant as the t_1, t_3, and L_3 parameters. As result, a set of optimum ratios for these parameters was also determined. These findings will be valuable to other investigators who will design specific electrothermal actuator for their MEMS applications. An electrothermal actuator was designed subject to the optimized design parameter ratios using the MEMSPro® CAD software and the material system of the PolyMUMPS foundry. This design will be fabricated in the near future, and it experimental performance will be reported in a subsequent publication.

ACKNOWLEDGMENTS

This work was supported by the National Science Foundation, grant CTS-9601283; the Lockheed Martin Corporation, Tactical Aircraft Systems Division, Fort Worth, TX, grant N-549-SOW-7009357.

REFERENCES

1. H. Guckel, et al, "Thermo-magnetic metal flexure actuators," *Technical Digest of IEEE Solid-State Sensor and Actuator Workshop* (1992) p. 73.
2. Q. Huang and N. Lee, *J. Micromech. Microeng.,* **9**, 64 (1999).
3. R. Hickey, M. Kujath, and T. Hubbard, *J. Vac. Sci. Technol. A,* **20**, 971 (2002).
4. N.D. Mankame and G.K. Ananthasuresh, *J. Micromech Microeng.,* **11**, 452 (2001).
5. C.S. Pan and W. Hsu, *J. Micromech. Microeng.,* **7**, 7 (1997).
6. C.C. Lee and W. Hsu W, *Microsystem Technologies,* **9**, 331 (2003).
7. R.S. Chen, C. Kung, and G.B. Lee, *J. Micromech Microeng.,* **12**, 291 (2002).
8. E.S. Kolesar, S.Y. Ko, J.T. Howard, P.B. Allen, J.M. Wilken, N.C. Boydston, M.D. Ruff, and R.J. Wilks, *Thin Solid Films*, **377**, 719 (2000).
9. E.S. Kolesar, et al, *Thin Solid Films*, **420**, 530 (2002).
10. J.C. Tucker, Guide For Designing Micro-electromechanical Systems in MUMPS, http://www.ece.ncsu.edu/erl/tutorials/mumps/

Mater. Res. Soc. Symp. Proc. Vol. 1052 © 2008 Materials Research Society 1052-DD03-15

Aligned Low Temperature Wafer Bonding for MEMS Manufacturing: Challenges and Promises

Viorel Dragoi[1], Thorsten Matthias[2], Gerald Mittendorfer[1], and Paul Lindner[1]
[1]Technology, EV Group, DI Erich Thallner Str. 1, St. Florian am Inn, Austria
[2]Technology, EV Group Inc., 7700 South River Parkway, Tempe, AZ, 85284

ABSTRACT

The increased complexity of current generations of MEMS devices imposes new requirements for wafer bonding. Among these can be mentioned low process temperature (<400°C), precise optical alignment of substrates, ability to bond a large variety of substrates and the possibility to bond with defined intermediate layers. An important aspect in aligned wafer bonding is that alignment accuracy needs to be correlated to the type of bond process. Especially in case of processes using bonding layers the post-bond alignment accuracy will be given by the behavior of the bonding layers. This paper aims to review the main criteria to be considered in defining aligned wafer bonding processes. Particularly bonding of substrates containing electronics (e.g. CMOS wafers) is currently of high technological interest.

INTRODUCTION

During the last decade MEMS (Micro Electro Mechanical Systems) or MOEMS (Micro Opto Electro Mechanical Systems) became more and more important in our daily life. Safety features on cars, portable GPS navigation systems, multimedia equipment, mobile phones, game pads... Who can imagine today life without the presence of all these items? And on the other hand, how many people know that many of these are based on various types of small dimensions devices named microsystems?

Having their roots in pressure sensors development in early 80's, microsystems gained an increasingly important role in automotive applications and more recently in consumer products.

MEMS bring the major benefit of combining on a single device the electronic signal processing functions and the signal detection functions. The assembly of different modules accomplishing the above mentioned functions leads to a particular feature of MEMS, different from standard IC devices: tri-dimensional architectures vs. bi-dimensional structures.

MEMS manufacturing process requires combination of different technologies and integration of new manufacturing techniques in the process flow. New techniques were developed in order to accommodate the new requirements. Among them the most representative are Deep Reactive Ion Etching - DRIE, Surface Micromachining, Chemical Mechanical Polishing and Wafer Bonding.

Reported in mid 80's as an approach for thick SOI manufacturing [1, 2], wafer bonding was constantly raising interest from industry. Although low and medium volume manufacturing activities included wafer bonding in the process flow (e.g. anodic bonding for pressure sensors, glass frit bonding) the real potential of wafer bonding was discovered only during last decade.

One of the main limiting factor for wafer bonding applications was the high process temperature (e.g. standard silicon fusion bonding requires a high temperature annealing step above 1000°C). Due to their specific requirements as dissimilar materials combination or bonding of "electronic" wafers (e.g. CMOS) MEMS manufacturing was a major driving force for low temperature wafer bonding process development (for wafer bonding, "low temperature" refers to temperature of maximum 400°C). In case of dissimilar materials, difference in thermal expansion coefficients will result in a mechanical stress proportional with temperature. Bonding of substrates with electronic structures is more and more used in MEMS manufacturing, CMOS became in last years an interesting "MEMS substrate". The use of metal interconnects requires temperature limitations to 400°C.

In the low temperature area developments were carried into two main directions: wafer bonding using intermediate bonding layers (adhesive bonding with polymers layers [3], eutectic bonding [4], metal diffusion bonding or thermocompression bonding [5]) and direct bonding methods using special surface preparation techniques [6].

Direct bonding process is based on molecular bonds established between molecules from the two substrates' surfaces. Typically the process consists of two main steps: one room temperature step in which weak adhesion is established between the substrates (e.g. due to van der Waals forces) followed by a thermal annealing meant to provide the energy needed to transform the weak reversible bonds in strong permanent bonds (covalent).

The surface preparation methods developed for low temperature processes are based on surface activation resulting in different types of bonds at room temperature (compared to the non-activated situation) and requiring less energy during thermal annealing step (lower annealing temperature and annealing time).

WAFER-TO-WAFER ALIGNMENT

Wafer bonding process is not limited to bonding of blank wafers, most of applications require patterned wafers alignment and bonding. Wafer to wafer alignment technology is a very dynamic field as the demands in terms of alignment accuracy are continuously increasing. For each application the right combination wafer bonding process/alignment technology has to be carefully selected for the best result.

Two main types of wafer-to-wafer alignment methods are currently used:

1. Mechanical alignment:

The main flats/notches of the two wafers are aligned one to each other mechanically. Typically a setup of special pins is used to accomplish this task. For SEMI standard wafers the alignment accuracy is ±50μm (wafer diameter up to 300mm).

In MEMS manufacturing such alignment method is used for applications were a cap wafer is bonded to a device wafer (e.g. for some pressure sensors applications)

2. Optical alignment:

The two wafers are precisely aligned using optical alignment equipment. In this case, high alignment accuracy is expected typically in the range of 1 ÷ 10μm. For this type of alignment special alignment keys have to be placed on the wafers during patterning. Alignment keys size and positioning across the wafers depend mainly on the type of alignment required.

Due to the different nature of the two processes (wafer-to-wafer alignment and wafer bonding), precise *in situ* alignment in the bond chamber is not possible.

For this reason, the two processes are separated in the process flow in two steps (Figure 1.)

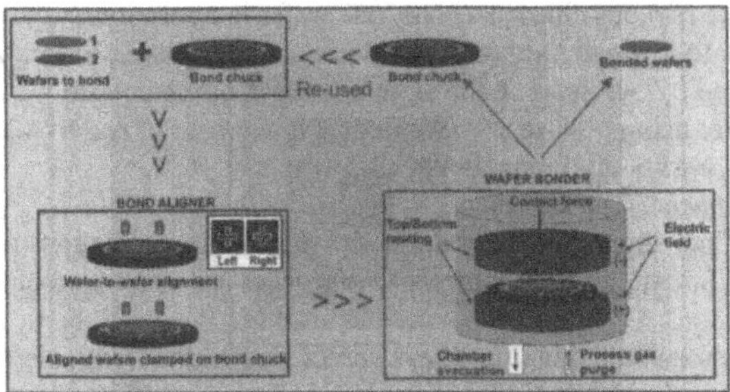

Figure 1. Process separation flow for aligned wafer bonding.

In the process separation, the flow is the following:
- Wafers are loaded to the wafer-to-wafer aligner (known also as bond aligner).
- Wafers are aligned (manually or automatically) and mechanically fixed on the bond chuck. If controlled atmosphere is required at the bonding interface (inside devices) the two wafers are mechanically secured separated by three spacers.
- Bond chuck is loaded to the bond chamber. The bond chamber can be evacuated down to high vacuum range and bond is performed after spacers removal by applying contact force and heating from two sides.

There are two types of optical alignment processes: direct alignment (if at least one wafer is transparent to light - visible or IR - the alignment keys from both wafers are observed in live mode) and indirect alignment methods (wafers are not transparent and one wafer is aligned using as reference the digitized image of second wafer).
The most used wafer-to-wafer optical alignment methods are shown in Figure 2.

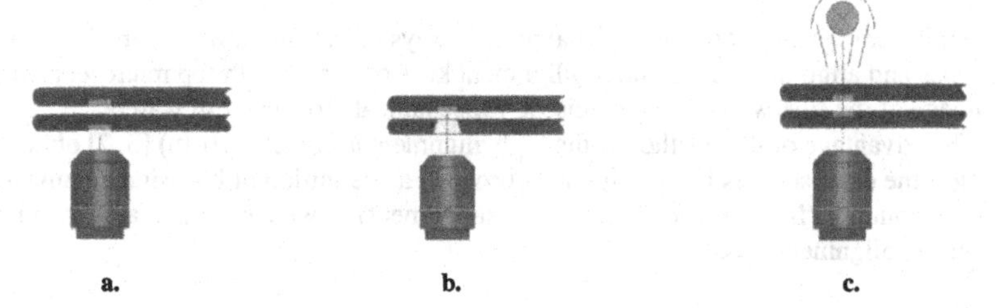

Figure 2. Principle of back side wafer-to-wafer alignment: a.- backside alignment, b.- backside alignment with glass wafer, and c.- IR alignment.

a. *Backside alignment (indirect alignment method):*
In this approach, one wafer has alignment keys defined on the front-side, while the second has alignment keys defined on the backside (Figure 2.a.).
Alignment is achieved by storing the image of first loaded wafer and aligning second wafer to the digital image of first wafer. The alignment keys of bottom wafer are overlapped to the alignment keys on the digitized image of first wafer using precise x-y mechanical movement.
b. Backside alignment with transparent wafer:
This alignment approach is using a similar flow with the previous method, but in this case the transparency of bottom wafer offers a technological advantage (no need to digitize the image of

first wafer). There are two possible situations: one wafer is transparent to visible light (e.g. glass wafer, Figure 2.b) or to infrared light (e.g. low doping silicon wafer, Figure 2.c).
Alignment is achieved by overlapping the alignment keys of the bottom wafer to the alignment keys on the first wafer using precise x-y mechanical movement of wafers while observing the live images of the two pairs of alignment keys.

c. Face-to-face alignment (figure 3):

An original face-to-face alignment method was developed for applications requiring very high alignment accuracy: SmartView® alignment. This process uses two pairs of microscopes, one for each alignment key.
In this alignment approach the alignment keys are located on the front sides of the two wafers and will be enclosed in the bond interface after bonding.

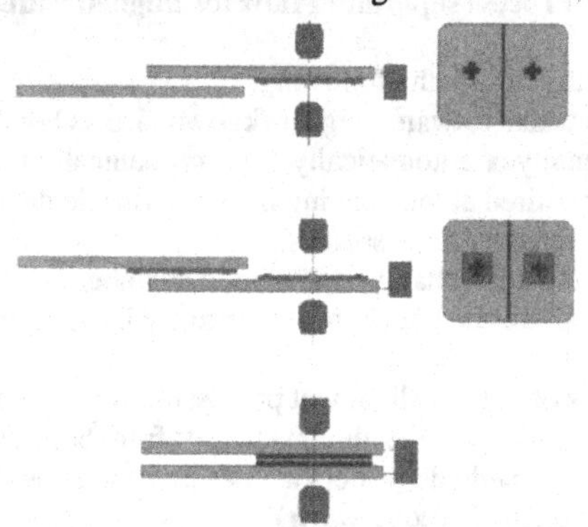

Figure 3. Principle of SmartView® face-to-face wafer-to-wafer alignment.

Alignment is completed by observing the alignment keys of the first wafers using bottom microscopes and align the second wafer (alignment keys observed with top microscopes) to the digital image of the first wafer using precise x-y mechanical movement of wafers.

The advantage of this method is the high alignment accuracy (<1μm) [3, 7] obtained by removing some error sources in the alignment process, as definition of backside alignment keys (with corresponding "front-to-back" induced misalignment) as well as large z-axis travel of wafers during alignment process.

DISCUSSION

In order to be sure the desired process benefits of all advantages of the wafer-to-wafer alignment method, few items have to be considered.

First of all, wafers design has to consider the alignment method which will be used. It is important to know from the design stage the alignment method which will be used, in order to include the dedicated bond alignment targets in wafer layout design.

The choice of the wafer alignment method is made based on substrates type and specific features, as well as on requested alignment accuracy.

Alignment keys selection is a very important step in process definition. For a specified alignment method the alignment accuracy and the alignment process reliability depend on the alignment keys shape, dimensions and keys positions on the wafer.

Wafer bonding process is strongly influencing the alignment accuracy. Table I summarizes the main wafer bonding processes used for aligned wafer bonding and shows main process feature with impact on wafer-to-wafer alignment.

Table I. Main wafer bonding features with impact on wafer-to-wafer alignment.

	Bonding temperature	Interface (bonding)	Interconnection type	Alignment accuracy
Direct (fusion) bonding*	20°C–400°C	Solid	Mechanical	0.4μm
Anodic bonding	350°C–450°C	Solid	Mechanical	1μm
Metal diffusion bonding	300°C–450°C	Solid	Mechanical+electrical	0.6μm
Eutectic bonding	200°C–400°C	Liquid	Mechanical+electrical	1μm
Adhesive bonding	50°C–300°C	Liquid	Mechanical	1 – 5μm
Glass frit	400°C-450°C	Viscous	Mechanical	>5μm

* Low temperature fusion bonding (e.g. surface activated by plasma [6])

In table I column "Interface (bonding)" refers to physical status of interface during wafer bonding process and column "Interconnection type" refers to the type of connection between substrates which can be achieved for the particular process.
Some of the processes listed only as providing "mechanical connection" are used also for applications in 3D Interconnects, where electrical interconnects are typically realized after wafer bonding step by creating vias and filling them with metal.

Wafer-to-wafer alignment accuracy is influenced by various factors during the bond process. Among them the most significant factors can be mentioned compression of intermediate bond layers which induces shifts (for bonding with intermediate layers as for eutectic bonding, adhesive bonding or glass frit bonding), different thermal expansion of the two bonding partners which induces run-out-type errors and the z-travel range of the wafers when brought in contact (e.g. given by thickness of the spacers used in the bond setup). For such bond processes typically two alignment specifications are typically defined: post-alignment accuracy (accuracy provided exclusively by the optical alignment equipment) and post-bonding accuracy (the final accuracy measured after bonding, when the bonded interface is already rigid). With respect to the above factors, direct (fusion) bonding methods are compatible with high alignment accuracy.

From the processes listed in table I three are extensively used for the application which brought wafer bonding into the mainstream IC industry: 3D integration by wafers stacking. The 3D chips could be successfully demonstrated using adhesive wafer bonding [8], metal diffusion bonding [5] or fusion bonding [7].

A process gaining more and more importance due to the high alignment accuracy and low process temperature is aligned wafer bonding using plasma activation and SmartView® alignment. Such a process is used for various application based on wafer bonding of CMOS wafers. In such a process flow the CMOS wafer needs first to be planarized by deposition of a CVD silicon oxide followed by Chemical Mechanical Polishing (CMP). After the high quality structure is achieved by CMP (microroughness <1nm, measured by AFM on 2x2μm^2) the CMOS wafer is activated in a plasma chamber. After alignment the two wafers are brought into contact inside the aligner and then annealed at a temperature of 250°C - 350°C.

The above described low temperature process was proven not to affect the CMOS performance. Plasma activated aligned wafer bonding is an important process for 3D integration as well as for MEMS devices manufacturing.

CONCLUSIONS

Being a wafer scale process, wafer bonding brings to MEMS manufacturing some major advantages: improved throughput, increased yield, allows devices testing at wafer scale (prior to dicing). Wafer bonding is of high interest in MEMS devices packaging by enabling costs decrease through Wafer Level Packaging (current MEMS packaging cost may be as high as 70% of the total device costs). The low temperature bonding processes are considerably expanding wafer bonding field of applications.

Combined with high accuracy wafer-to-wafer alignment methods as face-to-Face SmartView® alignment wafer bonding technology allows fabrication of complex devices by allowing precisely aligned combination of substrates processed using different technologies with yield and reliability which recommends it for high volume manufacturing activities.

The novel applications in the field of 3D integration benefit from the technology developed in the past mainly for MEMS. CMOS wafers planarization, wafers thinning and wafer bonding are nowadays processes belonging to the mainstream IC industry.

Due to the lack of standardization in MEMS generally and particularly in wafer bonding sometimes major confusions are generated. This has to be considered when selecting wafer bonding technology for specific applications.

REFERENCES

1. J. B. Lasky, S. R. Stiffler, F. R. White and J. R. Abernathey, *IEDM Tech. Dig.* 684, 1985.
2. M. Shimbo, K. Furukawa, K. Fukuda and K. Tanzawa, *J. Appl. Phys.* 60 (8), pp. 2987, 1986.
3. V. Dragoi, T. Glinsner, G. Mittendorfer, B. Wieder, P. Lindner, SPIE Proceedings, vol. 5116, pp. 160, 2003.
4. K. T. Turner, R. Mlcak, D. C. Roberts and S. M. Spearing, MRS Proc. vol. 687, pp. B. 3.2.1., 2002.
5. C. H. Tsau, S. M. Spearing, and M. A. Schmidt, *J. of Microelectromech. Systems* 11 (6), p. 641, 2002.
6. V. Dragoi and P. Lindner, ECS Transactions 3 (6), pp. 147, 2006.
7. Lea Di Cioccio, ECS Transactions 3 (6), 2006, p. 19.
8. Y. Kwon, A. Jindal, J. McMahon, J.-Q. Lu, R.J. Gutmann and T.S. Cale, MRS Proc. vol. 766, pp. E5.8.1., 2003.

Mater. Res. Soc. Symp. Proc. Vol. 1052 © 2008 Materials Research Society 1052-DD03-20

Compressive Stress Accumulation in Composite Nanoporous Gold and Silicone Bilayer Membranes: Underlying Mechanisms and Remedies

Erkin Seker[1,2], Ling Huang[1], Matthew R. Begley[3,4], Hilary Bart-Smith[3], Robert G. Kelly[4], Giovanni Zangari[4], Michael L. Reed[2], and Marcel Utz[3]
[1]Department of Chemistry, University of Virginia, Charlottesville, VA, 22904
[2]Department of Electrical and Computer Engineering, University of Virginia, Charlottesville, VA, 22904
[3]Department of Mechanical and Aerospace Engineering, University of Virginia, Charlottesville, VA, 22904
[4]Department of Materials Science and Engineering, University of Virginia, Charlottesville, VA, 22904

ABSTRACT

This paper outlines a simple method to fabricate a bilayer membrane consisting of a thin nanoporous gold layer infused with uncured polydimethylsiloxane. The fabrication technique offers excellent adhesion due to mechanical interlocking between porous layer and elastomer, and excellent electrical conductivity of $1x10^{-6}$ Ω *-m* at small strains and $3x10^{-5}$ Ω *-m* at strains of 25% despite a very low effective elastic modulus (~1.35 MPa) due to cracks in the embedded gold layer. Initially freestanding circular membranes displayed significant out of plane buckling, and created difficulties in extraction of membrane mechanical properties. The underlying mechanisms of compressive stress accumulation that lead to membrane buckling and remedies to prevent it are discussed.

INTRODUCTION

Integration of metal films with elastomers has recently received significant attention as the demand grows for the development of macroelectronics [1] and chemo-mechanical sensors [2] that rely on surface functionalization of the metal surface. Such applications require a relatively low elastic modulus for optimal device performance. Gold has been traditionally used for its high conductivity, chemical inertness, and ease of performing thiol-chemistry for surface functionalization [3]. However, gold does not adhere well to most polymers without adhesive layers such as titanium and chrome that increase the overall stiffness of multilayers [4]. This problem has been circumvented by using patterned electrodes with concentric circle patterns [5] or mechanical cracking of metal layers [6]. Several groups have mixed conductive particles (e.g., gold, carbon, etc.) with polymers to increase conductivity [7]; while others attempted ion implantation of conducting species [8,9]. These efforts concentrated generally on either decreasing elastic modulus or increasing conductivity, but not both. Recently we demonstrated a new approach [10] by infusing a nanoporous gold (np-Au) leaf with uncured polydimethylsiloxane (PDMS) to obtain a composite bilayer membrane with desirable properties such as: (i) superior adhesion between gold and elastomer due to mechanical interlocking; (ii) very low effective elastic modulus of ~1.35 MPa; (iii) low specific resistivity of $1x10^{-6}$ Ω *-m* at small strains and $3x10^{-5}$ Ω -m at strains of 25%. At first, this technique was utilized to fabricate freestanding composite bilayer membranes on PDMS substrates; however, accumulation of compressive stress during fabrication resulted in significant membrane buckling. This paper will discuss the underlying mechanism of compressive stress accumulation and recommend solutions to circumvent this problem.

EXPERIMENT

Fabrication

A commercially available 12-karat $Au_{35}Ag_{65}$ "Monarch" white gold leaf (10 cm x 10 cm) (Sepp Leaf Products, New York) was dealloyed [11] by floating it on concentrated (65%) nitric acid in a petri dish, as seen in figure 1(a). The leaf was dealloyed for approximately 45 minutes until it became transparent-brown and did not change color with additional etching. The dealloying step dissolved silver in the leaf and resulted in the formation of a nanoporous network of gold. The acid was removed with a transfer pipette, and the leaf was rinsed with deionized (DI) water several times. The petri dish was filled with DI water for a last time, and a 2-inch silicon wafer coated with 2 μm photoresist was placed underneath the floating foil. The liquid was aspirated, and the leaf coated the surface of the wafer uniformly, as shown in figure 1(b). The leaf was dehydrated on a hot plate at 50°C to remove residual liquid.

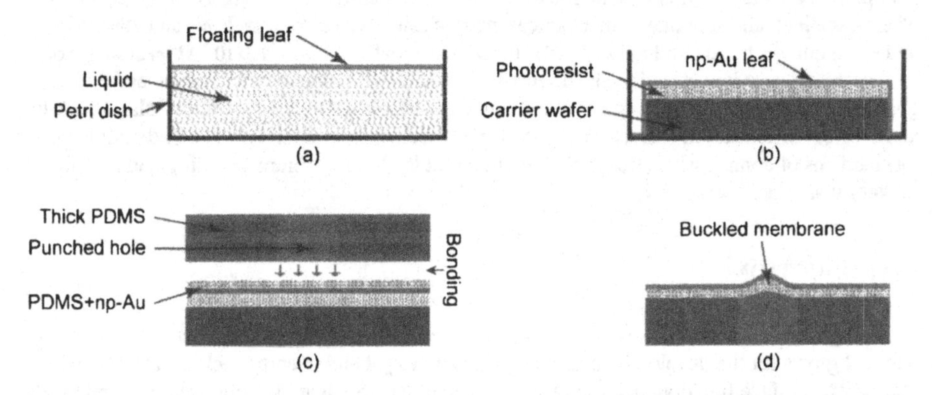

Figure 1. Fabrication process of a freestanding np-Au and PDMS bilayer membrane. (a) A gold-silver leaf is dealloyed by floating over nitric acid and rinsed with deionized water. (b) Np-Au leaf is collected with a photoresist coated silicon wafer and dehydrated. (c) Np-Au is spin coated with PDMS, placed in a vacuum chamber to promote infusion of PDMS into the np-Au matrix, bonded to a thick PDMS sheet with punched holes, and cured. (d) The membranes were released in methanol by dissolving the sacrificial photoresist layer.

A mixture of Sylgard 184 PDMS (Dow Corning, Michigan) was prepared by mixing and degassing 1 part curing agent and 10 parts elastomer base. The np-Au leaf (~140 nm-thick) on the wafer was spin coated with the PDMS mixture by spinning at 4000 rpm for 1 minute. The thickness of the PDMS layer was approximately 22 μm as measured with a white light interferometer. The spin-coated sample was subsequently transferred to a vacuum chamber to remove the trapped air in the pores, and simultaneously fill them with PDMS by capillary action. A study of mass transport of liquids through np-Au films can be found elsewhere [12]. While the sample was kept under vacuum, holes of various diameters were punched in a 250 μm-thick thick PDMS film using sharpened copper tubes of 2 and 4 mm diameters, and mechanical pencils with 0.3 to 0.9 mm diameters. Following the formation of holes, the thick PDMS was cleaned with isopropanol, dried with nitrogen, and treated with oxygen plasma at 100W for 1 minute. The thick PDMS was then placed onto the PDMS-infused np-Au leaf as shown in figure 1(c). The thick PDMS substrate with holes served as the pedestals for the freestanding membranes. Finally, the bonded sample was cured in a convection oven at 60°C for 10 hours, and the composite structure was released by dissolving photoresist in a methanol bath, as illustrated in figure 1(d). The released membranes displayed out of plane buckling. The underlying mechanism for buckling will be discussed later.

Characterization

For membrane characterization only the 2 mm-diameter membranes were used, since smaller membranes increased the difficulty of alignment during deflection measurements and 4 mm-diameter membranes displayed irreproducible buckling profiles. A stereoscopic optical microscope was used to examine membranes. Figure 2(a) shows typical top views of 2 and 4 mm membranes. Note that the rim of the membrane is slightly wrinkled due to the ragged edges of the punched hole.

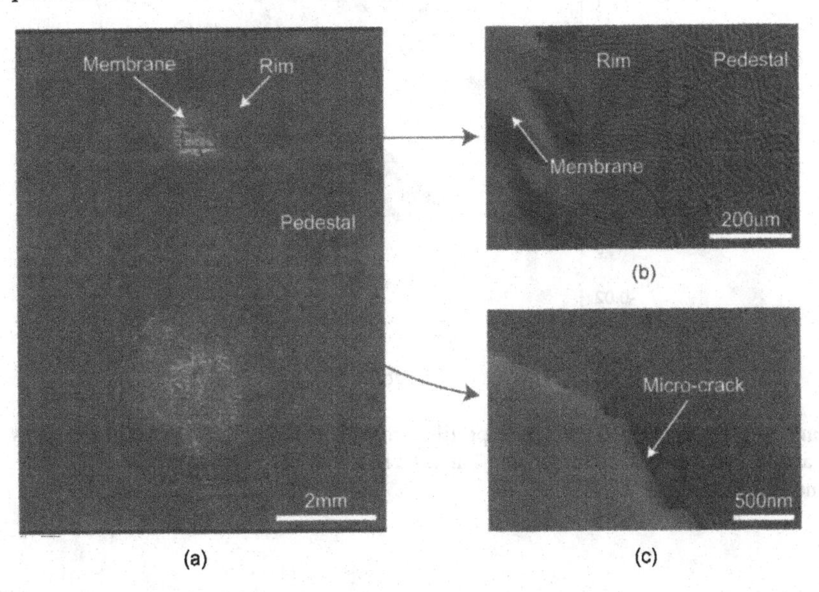

Figure 2. Micrographs of the membranes at various magnifications. (a) Optical micrograph of 2 and 4 mm-diameter membranes. (b) Close-up of a region around the membrane. (c) A high magnification SEM image of the top surface of a freestanding membrane.

The membranes were observed with a scanning electron microscope (SEM) and the porosity was determined by image analysis [13]. The percent porosity was ~30%, while the average pore radius was ~11 nm. The compressive stress in the PDMS film resulted in microbuckling patterns on the pedestal region, as seen in figure 2(b). The buckling disappeared closer to the membrane. The chip was placed on a silicon wafer during the SEM study which sealed the membrane, capturing air in the membrane cavity. When the SEM chamber was pumped down, the air in the cavity inflated the membrane and strained it. Therefore, the buckling patterns were no longer visible on the membrane. Micro-cracking due to bulging of the membrane was observed in the high magnification image of the membrane surface in figure 2(c); however, the np-Au film did not delaminate.

The membranes buckled out of plane after release, suggesting compressive stress build up in the composite film, as seen in the inset of figure 3. Buckling profiles of ten membranes were examined using a white light interferometer. For analysis, 2D profiles of the membranes were captured along the center-line where the maximum buckling was observed. The line profiles of ten randomly picked membranes were averaged as seen in figure 3. The average maximum buckling amplitude (i.e., the distance between the lowest point on the rim and the highest point on the membrane) was 112±12 μm. It was observed that the membrane shapes were very similar from chip to chip, and the process was repeatable.

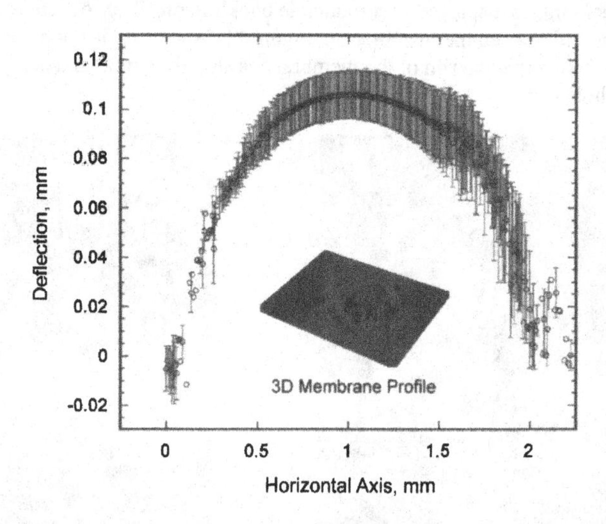

Figure 3. Average buckled membrane profile from ten randomly picked membranes show repeatability in membrane fabrication. The inset illustrate a typical 3D profile of a 2 mm-diameter membrane.

DISCUSSION

The compressive stress accumulation in the bilayer membrane is due to drastically different coefficients of thermal expansion (CTE) of the thick PDMS substrate and the carrier silicon wafer. Note that the thick PDMS substrate was bonded to a layer of *uncured* PDMS/np-Au layer deposited on a photoresist-coated silicon wafer. Following the bonding step, the sample was cured in a convection oven at 60°C. We have observed that it takes more than 3 hours for PDMS to fully cure at this temperature and form a permanent bond with the thick PDMS substrate. However, the thick PDMS experiences rapid thermal expansion as it is placed in the oven since it is not yet bonded to the uncured PDMS. In other words, the thick PDMS floats on the uncured PDMS due to the slippery interface. Since the spun PDMS is attached to the rigid silicon wafer, the limit of its thermal expansion is dictated by the thermal expansion of the silicon wafer. As the PDMS cures, a firm bond is established between the spun PDMS layer and the thick PDMS substrate. Finally, as the PDMS is removed from the oven, it cools down, and the thick PDMS contracts two orders of magnitude more than the silicon wafer (and the spun PDMS temporarily bonded to the wafer), putting the composite layer in compression. The compressive thermal strain in the membrane is calculated to be approximately 1% using equation 1:

$$\varepsilon_{thermal} = \Delta\alpha\Delta T \,, \tag{1}$$

where $\Delta\alpha$ is the difference in CTE between silicon (~3 ppm/°C) and PDMS (~300 ppm/°C), and ΔT is the difference between room (~25°C) and curing (~60°C) temperatures. Using this strain value, a buckling amplitude of ~65 μm was calculated with equation 2 based on the approximation of a simple sinusoidal deformation of the centerline of a membrane due to elongation:

$$\delta_{max} = \frac{D\sqrt{\varepsilon_{thermal}}}{\pi} \,, \tag{2}$$

where δ_{max} is the buckling amplitude and D is the diameter of the membrane (~2 mm). The calculated buckling amplitude is slightly lower than the experimental observation. This is thought to be due to additional strains introduced to the bilayer membrane during handling and deflection measurements. However, we believe that this diagnosis captures the physical basis of the membrane buckling and micro-buckling patterns on the substrate, as shown in figure 2(b).

The compressive stress accumulation was remedied by two different methods. Instead of a punched PDMS substrate, a drilled glass slide was used. Similar CTEs of silicon and glass avoided the thermal strain associated with thermal mismatch of the two materials. Another method was to first cure the spun PDMS, and then bond it to a glass or PDMS substrate after oxygen plasma activation of the bonding surfaces. Both of these methods prevented membrane buckling after release, and produced devices compatible with pressure-bulge experiments to extract mechanical properties of the bilayer composite membranes [10]. The outlined fabrication method can be customized to produce a variety of freestanding structures, by molding precise channels in PDMS substrates and/or by micromachining glass or silicon substrates.

CONCLUSIONS

A simple fabrication has been developed to produce adherent, conductive, and flexible bilayer membranes by infusing a np-Au leaf with PDMS. The freestanding membranes produced using the outlined technique resulted in compressive stress accumulation and consequent buckling. The underlying mechanisms for the stress accumulation are due to a combination of slippery boundary between the substrate and the uncured membrane, and the difference in the thermal expansion coefficients of silicon and PDMS. This problem was prevented by using glass slides as the substrate to prevent thermal mismatch, or postponing the bonding step until curing the PDMS/np-Au layer. This composite material is expected to find use in the development of conductive electrodes and sensitive chemo-mechanical sensors.

ACKNOWLEDGMENTS

This work was supported by the National Science Foundation, award DMI 0507023.

REFERENCES

1. S. P. Lacour, J. Jones, Z. Suo, S. Wagner, IEEE Electr. Device. L. **25**, 179 (2004).
2. M. R. Begley, M. Utz, U. Komaragiri, J. Mech. Phys. Solids **53**, 2119 (2005).
3. S-H Lim, D. Raorane, S. Satyanarayana, A. Majumdar, Sensor. Actuat. B-Chem **119**, 4 (2006).
4. M. R. Begley, J. Micromech. Microeng. 15, 2379 (2005).
5. R. Pelrine, R. Kornbluh, J. Joseph, R. Heydt, Q. Pei, S. Chiba, Mater. Sci. Eng. **C11**, 89 (2000)
6. S. P. Lacour, D. Chan, S. Wagner, T. Li, Z. Suo, Appl. Phys. Lett. **88**, 204103 (2006).
7. J. Engel, J. Chen, N. Chen, S. Pandya, C. Liu, IEEE 19th MEMS Conference, Istanbul, Turk January 2006, pp. 246-249
8. Y. Wu, T. Zhang, H. Zhang, X. Zhang, Z. Deng, G. Zhou, Nucl. Instrum. Methods Phys. R Sect. B **169**, 89 (2000).
9. P. Dubois, S. Rosset, S. Koster, J. Stauffer, S. Mikhailov, M. Dadras, N-F. de Rooij, H. Sh Sensor. Actuat. A-Phys. **130**, 147 (2006).
10. E. Seker, M. L. Reed, M. Utz, M. R. Begley, *Appl. Phys. Lett.* (in review).
11. Y. Ding, Y. Kim, J. Erlebacher, Adv. Mater. **16**, 1897 (2004).
12. E. Seker, M. R. Begley, M. L. Reed, M. Utz, *Appl. Phys. Lett.* (in review).
13. E. Seker, J. T. Gaskins, H. Bart-Smith, J. Zhu, M. L. Reed, G. Zangari, R. G. Kelly, M. Begley, *Acta Mater.* **55**, 4593 (2007).

Mater. Res. Soc. Symp. Proc. Vol. 1052 © 2008 Materials Research Society 1052-DD03-21

Variation in Dislocation Pattern Observed in SCS Films Fractured by Tensile Test: Effects of Film Thickness and Testing Temperature

Shigeki Nakao[1], Taeko Ando[1], Shigeo Arai[2], Noriyuki Saito[2], and Kazuo Sato[1]
[1]Department of Micro-Nano Systems Engineering, Nagoya University, Furo-cho, Chikusa-ku, Nagoya, 464-8603, Japan
[2]High Voltage Electron Microscope Laboratory, Nagoya University, Furo-cho, Chikusa-ku, Nagoya, 464-8603, Japan

ABSTRACT

This paper reports a transition in the fracture behavior of micron-sized single-crystal-silicon (SCS) film in a MEMS (microelectromechanical systems) structure for various film thicknesses and ambient temperatures. The mean fracture toughness of 4-μm-thick SCS films was 1.28 MPa$\sqrt{m}$ at room temperature (RT), and the value increased as the film thickness decreased, reaching 2.91 MPa$\sqrt{m}$ for submicron-thick films. The fracture toughness of 4-μm-thick film did not change for ambient temperatures ranging from RT to 60°C. However, it increased at 70°C and reached 2.60 MPa$\sqrt{m}$ at 150°C. Enhanced dislocation activity in the SCS crystal near the fracture surface was observed on 1-μm-thick film at RT and 4-μm-thick film at 80°C by high-voltage electron microscopy (HVEM). This change in dislocation activity seemed to correlated with the transition in fracture behavior.

INTRODUCTION

Mechanical characterization of micron-sized structures plays an important role for microelectromechanical systems (MEMS) devices. Single-crystal silicon (SCS) has high mechanical strength at room temperature (RT) and is widely used in MEMS physical sensors and moving mechanical elements. In recent years, however, demands for MEMS that can operate at elevated temperatures have been increasing. We need a better understanding of the mechanical behaviors of micron-sized SCS films in MEMS structures at elevated temperatures.

The brittle-to-ductile transition (BDT) of bulk SCS has been studied for a long time [1-4]. It is well known that SCS shows a sharp transition from brittle to ductile in a narrow temperature range above 600°C in fracture toughness measurements. However, micro- and nanometer-sized SCS films were recently reported to show some plasticity in their crystals in a low temperature range around 100°C [5-6]. There should be a size effect on the BDT temperature and dislocation activity in silicon crystals, but this has not yet been clarified.

In this paper, we present experimental results for fracture toughness measurements of SCS films in the temperature range from RT to 500°C for a film thickness of 4μm and in the thickness range from submicron to 4.5 μm at RT. After the toughness measurements, the dislocation patterns existing near the fracture surface were observed by high-voltage electron microscopy (HVEM). HVEM is a suitable method for analyzing the dislocations in SCS films with a thickness of a few microns because of its high acceleration voltage of 1 MV. These experimental results determined the temperature and size effects on the fracture toughness and the dislocation activity of SCS within the micron range.

EXPERIMENT

"On-chip" tensile test

We carried out the fracture toughness measurement using an "on-chip" tensile test [7]. This method uses a special test device that consists of a thin film tensile specimen and a loading system, as shown in Figure 1. The loading system is composed of torsion bars, a load lever, and a support frame. When a normal load is applied to the load lever, the lever rotates around the axis of the torsion bars, and the specimen is uniaxially stretched (Figure 1(b)). The applied load balances the tensile force on the specimen and the repulsive force of the torsion bars before the specimen fractures. The tensile force is calculated from the applied load by subtracting the restoring force of the torsion bars, which can be measured after the specimen has fractured (Figure 1(c)). The device was fabricated on a SOI wafer by photolithography and etching processes.

The test device was fixed on the heating stage by vacuum chucking. The heating stage contained cartridge heaters and a thermocouple, and the stage temperature was kept constant to within ±2°C of the desired test temperature by manually controlling the voltage applied to the cartridge heaters. Details of the setup and experimental procedure were reported previously [8].

Specimens

SCS tensile specimens with a single notch in one edge were used for the fracture toughness measurements. The fracture toughness K_C can be calculated from the fracture stress σ_f and the notch length a according to the theoretical formula

$$K_C = Y\sigma_f\sqrt{\pi a} \quad (1)$$

where Y is a factor related to the crack configuration given by

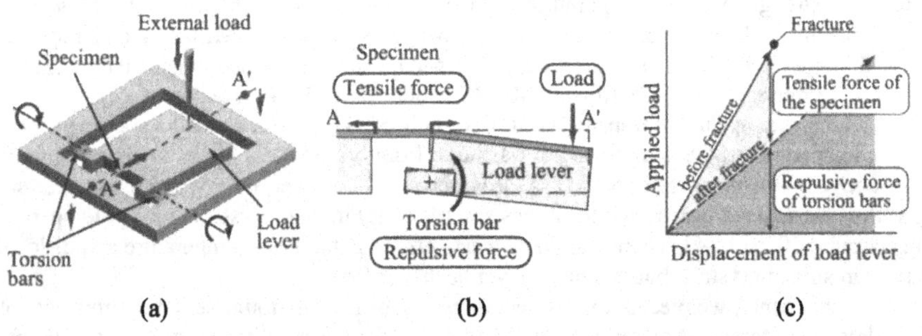

Figure 1. Schematics of "on-chip" tensile test device. (a) Overall view of the device consisting of a thin film tensile specimen and a loading system. (b) A–A' cross section of the device showing the principle of the tensile test. (c) Relationship between applied load and displacement of the load lever. The tensile force is calculated from the applied load by subtracting the repulsive force.

$$Y = 1.12 - 0.231\lambda + 10.55\lambda^2 - 21.72\lambda^3 + 30.39\lambda^4 \quad (2)$$

$$\lambda = a / W \quad (3)$$

where W is the width of the specimen.

The surface orientation was (001), and the tensile axis was aligned in the <110> direction. The straight portion of the specimen was 50 or 100 μm long and 45 μm wide. The film thickness ranged from 0.17 to 4.7 μm. The notch was introduced on one side of the straight portion by using a focused ion beam (FIB). The notch length ranged from 1 to 2 μm, and the tip radius was less than 30 nm. The notch fabricated by FIB was not an atomically sharp crack, and the large tip radius will effectively lead the large fracture toughness values [11]. However, we think that the amount of increase value is negligible with nanometer-scale tip radius. The FIB process was carried out using an SMI-2050 (SII NanoTechnology, Inc.) under an acceleration voltage of 30 kV.

RESULTS

Relationship between fracture toughness and film thickness

The fracture toughness of SCS film was evaluated for film thicknesses ranging from 0.17 μm to 4.7 μm at RT. The relationship between fracture toughness and film thickness is plotted in Figure 2. For thickness over 4 μm, the fracture toughness was about 1.32 MPa$\sqrt{\text{m}}$, which is a little higher than that of bulk silicon reported elsewhere [9-10]. For thicknesses below 4 μm, the fracture toughness increased as the film thickness decreased, reaching about 2.91 MPa$\sqrt{\text{m}}$ for submicron-thick films. One reason for this change in fracture toughness can be regarded as the increase in the opportunity for dislocations to be induced in silicon crystal due to the thickness reduction. We discuss the behavior of dislocations in silicon film below.

Relationship between fracture toughness and test temperature

We carried out the tensile tests on notched SCS films with thicknesses of 4 and 1 μm at

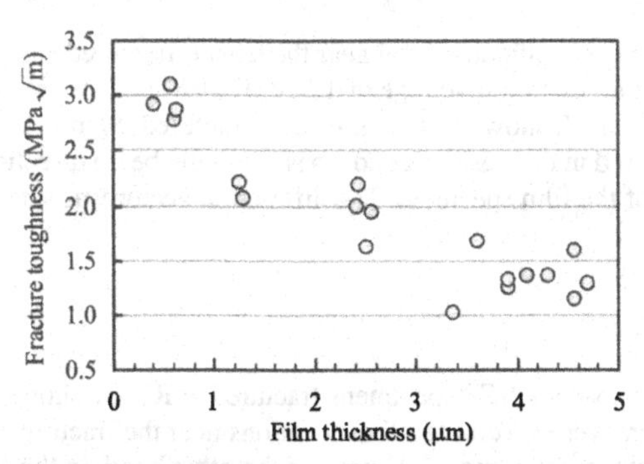

Figure 2. Relationship between fracture toughness and film thickness at RT. Fracture toughness increased as the film thickness decreased.

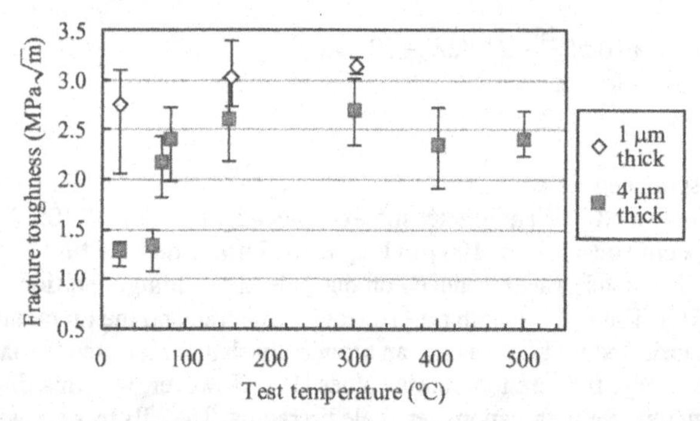

Figure 3. Relationship between mean fracture toughness of 4-μm-thick SCS films and test temperatures. Fracture toughness showed a steep increase at 70°C.

temperatures ranging from RT to 500°C. The relationship between fracture toughness and test temperature is plotted in Figure 3. For 4-μm-thick SCS films, the fracture toughness was almost constant from RT to 60°C. However, it increased drastically at 70°C, reaching 2.60 MPa$\sqrt{m}$ at 150°C, which is twice that at RT. At higher temperatures, it saturated. This increase in fracture toughness is similar to the phenomenon observed at the BDT temperature in bulk silicon above 600°C [1-2]. The present results suggest that the fracture behavior of 4-μm-thick silicon film clearly changed, even at 70°C. Dislocations in micron-thick silicon film were presumably induced more easily than in bulk materials, thus resulted in BDT transition at such a low temperature.

On the other hand, the fracture toughness of 1-μm-thick film was high even at room temperature, as mentioned above. And it increased moderately with increasing test temperature up to 300°C. It did not show a transition in this temperature range. Considering the 4-μm results, we speculate that there may be a transition point at a temperature below 0°C for 1-μm-thick SCS film.

DISCUSSION

We observed the dislocation image in silicon crystal near the tensile-fractured surface by HVEM (H-1250ST, Hitachi, Ltd) at an acceleration voltage of 1 MV. HVEM images are high-quality ones for a relatively thick film, which show the specimen as it fractured. Figure 4 shows bright field images of SCS films fractured under various conditions when the beam direction was [001], which is normal to the surface of the film specimen. The diffraction vector was set to [220].

Thickness dependence

HVEM images of the 4- and 1-μm-thick SCS specimens fractured at RT are shown in Figures 4(a) and (b), respectively. There were several small dislocations near the fracture surface of the 4-μm-thick specimen; their lengths were at most 200 nm. On the other hand, in the 1 μm-thick specimen, large dislocations with lengths from 1 to 2 μm penetrated from the fracture surface in the tensile direction. The fracture toughnesses of these specimens were 1.58 and 2.06 MPa$\sqrt{m}$, respectively. This dislocation growth with thickness reduction corresponded to the increase in fracture toughness. It seems that the short distance between the top and bottom surfaces of the film specimen made dislocation motion easier by allowing dislocations to escape from the surfaces.

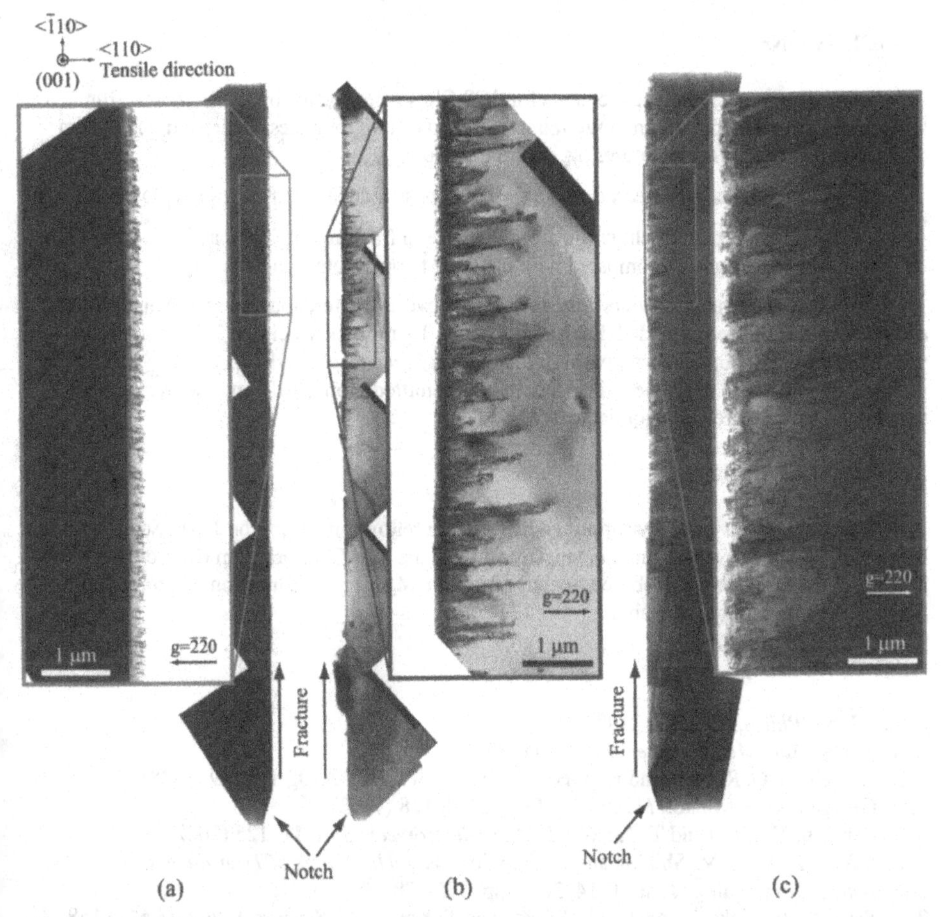

Figure 4. HVEM images of SCS film specimens under various conditions. (a) 4-μm-thick specimen fractured at RT (fracture toughness: 1.58 MPa$\sqrt{m}$). (b) 1-μm-thick specimen fractured at RT (fracture toughness: 2.06 MPa$\sqrt{m}$). (c) 4-μm-thick specimen fractured at 80°C (fracture toughness: 2.73 MPa$\sqrt{m}$).

Temperature dependence

An HVEM image of a 4-μm-thick specimen fractured at 80°C, after the steep increase in fracture toughness, is shown in Figure 4(c). The grown dislocations could be observed near the fracture surface. The length in the tensile direction was over 2 μm, which is more than 10 times as large as that at RT. Extended dislocations like these should result from the sharp transition in fracture behavior of SCS film even at 70°C. The mechanism of dislocation formation, which should be related to fracture morphology, needs to be analyzed in future work.

CONCLUSIONS

Fracture toughness measurements for SCS films were carried out with various film thicknesses and test temperatures. We found that the fracture toughness depended on the film thickness and ambient temperatures, as described below.

- Fracture toughness increased as the film thickness reduced from 4 μm to below 1 μm.
- 4-μm-thick specimens showed a steep increase in fracture toughness at 70°C, which is a very low temperature compared with the BDT temperature of bulk silicon.

HVEM observations evinced the relations between fracture toughness and dislocation activities. We could see extended dislocations on the 1-μm-thick specimen at RT and on the 4-μm-thick ones at 80°C, which shows high toughness. These results indicate that in SCS film, as the film thickness was decreased, dislocations were emitted more easily and the transition temperature of fracture behavior decreased.

ACKNOWLEDGMENTS

This work was partially supported by research fellowships from the Japan Society for the Promotion of Science for Young Scientists, and from the 21st COE program (Micro- and Nano-Mechatronics for Information-based Society) from the Ministry of Education, Culture, Sports, Science, and Technology, Japan.

REFERENCES

1. C. StJohn, *Philos. Mag.* **32**, 1193 (1975).
2. M. Brede, *Acta Metall. Mater.* **41**, 211 (1993).
3. J. Samuels, S. G. Roberts and P. B. Hirsch, *Mater. Sci. Eng.* **A105-106**, 39 (1988).
4. A. George and G. Michot, *Mater. Sci. Eng.* **A164**, 118 (1993).
5. T. Namazu, Y. Isono and T. Tanaka, *J. Microelectromech. Syst.* **11**, 125 (2002).
6. S. Nakao, T. Ando, M. Shikida and K. Sato, in *Dig. Tech. Papers of Transducers '07 conference*, Lyon, France, June 10-14, 2007, pp. 375-378.
7. K. Sato, T. Yoshioka, T. Ando, M. Shikida and T. Kawabata, *Sensors Actuators* **A70**, 148 (1998).
8. S. Nakao, T. Ando, M. Shikida and K. Sato, *J. Micromech. Microeng.* **16** 715 (2006).
9. F. Ebrahimi and L. Kalwani, *Mater. Sci. Eng.* **A268**, 116 (1999).
10. Y. Tsai and J. Mecholsky, *J. Mater. Res.* **6**, 1248 (1991).
11. P. Lowhaphandu and J.J. Lewandowski, *Scripta Mater.* **38**, 1811 (1998).

Mater. Res. Soc. Symp. Proc. Vol. 1052 © 2008 Materials Research Society 1052-DD03-27

Micro-Topography Enhances Directional Myogenic Differentiation of Skeletal Precursor Cells

Yi Zhao
Department of Biomedical Engineering, The Ohio State University, 294 Bevis Hall, 1080 Carmack Road, Columbus, OH, 43210

ABSTRACT

Skeletal muscle tissues were constructed by differentiating *in vitro* cultured skeletal myoblasts using an array of closely spaced linear microstructures. The adaptation of skeletal myoblasts has been characterized with immunoflurescence microscopy during cell proliferation and differentiation. In particular, the dependence of the alignment efficiency on the dimensions of the microstructures was studied. The morphology difference of the myotubes in the three-dimensional tissues was reported. This paper holds the promise of efficient on-chip fabrication of skeletal muscle tissues and has an important implication in direct muscle repair and muscular mechanics.

INTRODUCTION

Skeletal muscle accounts for 48% of the human body weight and is the major consumer of body fuels [1], largely responsible for the voluntary control and active movement. The prototypic skeletal muscle is composed of striated myotubes that are arrayed in parallel with one another along a common axis. Each multinucleated myotube functions as a single cell and spans the entire length of the muscle. Such striated organization is essential for generation of contractile forces sufficient for body movement at the expense of ATP consumption. In cases of inherited muscular dystrophies, accidents, or removal of tumor tissues, the striated structures of the skeletal muscles are damaged to certain extents, which directly affect the physiological and contractile performance of the muscles. There has been a long-standing research interest to develop striate muscle implants that mimic the architectural organization and physiological function of intact muscles for the repair and replacement against muscular abnormities [2-4].

Particularly, the skeletal precursor cell is a relatively accessible cell source with myogenic potential, and can be directed into musculoskeletal tissues. Therefore, skeletal myoblasts transplantation has been developed as an attempt to substitute the segments of the damaged muscles against muscular abnormities [2, 3]. In this practice, muscles with highly organized structures and aligned myotubes are preferred as they mimic the natural muscular morphology. This is expected to help the fusion between the implanted tissues with the reception regions, and to enhance the physiological functions.

To this end, a variety of engineering approaches have been developed to regulate the alignment of skeletal myotubes, including the use of aligned collagen gel system [5], microcontact printing of cell-adhesive domains [6], and cyclic mechanical loadings [7]. In particular, topographic modulation is an effective method for regulating cell growth and proliferation. Various studies have indicated that it is possible to align the skeletal myoblasts and myotubes using micro-scale topographies, such as nano fibers [8], micro- grooves [9, 10], micro-cantilevers [11], and continuous wavy micropatterns [12]. Nonetheless, most of these studies were focused on the alignment effect, while little information is provided on the how individual cells adapted to the micro-scale structures during proliferation and differentiation, and its impact in subsequent construction of three-dimensional muscles. The objective of this study was to develop a microfabricated platform for investigation of cell adaptation to the microstructures during growth, proliferation and differentiation, which could be used to facilitate on-chip fabrication of engineering skeletal muscles.

EXPERIMENTAL DETAILS

Fabrication

We used a revised replica molding process to fabricate microscale PDMS lines array (Figure 1). The master template was first fabricated using a glass substrate. The glass substrate was spin coated with a layer of thick positive photoresist AZ 9620. The thickness of the photoresist was in the range of a few micrometers and controlled by the spin rate. The substrate was then patterned with closely spaced lines array using a mask aligner equipped with an optical filter. After developing and post exposure bake, the complimentary features were formed on the glass substrate.

PDMS prepolymer (Sylgard@ 184) was prepared by mixing the base solution and the curing agent at 10:1. After degassing the liquid mixture to remove the microbubbles, the prepolymer was poured on the aforementioned master template, and was cured under an elevated temperature (65 °C) for about 1 hour. After the prepolymer was fully solidified, we dissolved the photoresist template by rinsing the substrate using acetone solution. The closely spaced microscale lines array was transferred to the PDMS substrate. Since this method uses a sacrificial layer as the master template which is dissolvable after the replication process, the interfacial stress between the polymer microstructures and the master template was avoided. As a result, the structural failure due to the mechanical interactions with the rigid master template was avoided. Closely spaced PDMS microstructures with a aspect ratio above 3 can be easily fabricated with a high yield. The PDMS template was sterilized by immersing into 70% ethanol for 30 minutes, which followed by three-time rinse using 1X PBS solution. Finally, the substrate was rinsed and kept in the culture medium until the cell seeding.

Cell Preparation

C2C12 myoblasts (ATCC, VA, USA) were cultured according to a previously described protocol and seeded on the PDMS thin film. Growth medium consisted of Dulbecco's Modified Eagle Medium (DMEM) (Invitrogen, CA) supplemented with 10% fetal bovine serum and 1% penicillin/streptomycin. Differentiation media consisted of DMEM supplemented with 2% horse serum and 1% penicillin /streptomycin. The cell adaptation to the linear PDMS micro lines array was observed every 24 hrs from the start of culture. The differentiation media was applied when the confluence was reached.

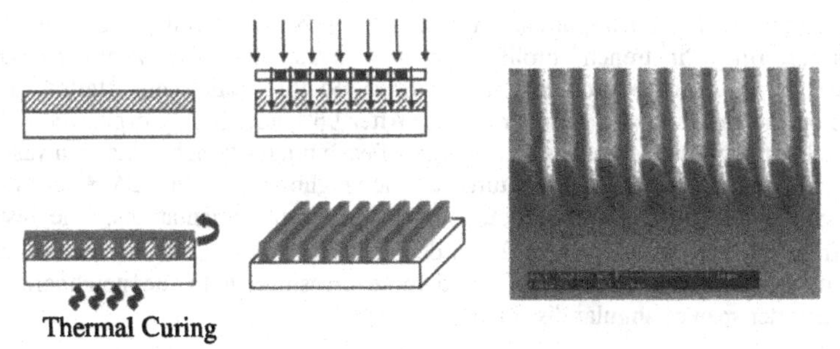

Figure 1: Fabrication of microscale lines array. The positive photoresist are first spin coated on a wafer and patterned, followed by pattern transfer to a PDMS polymeric substrate using replica molding. The PDMS substrate is then peeled off the wafer. SEM micrograph shows the line width and the spacing between neighboring lines are both 2 μm.

RESULTS AND DISCUSSION

Cell Adaptation during Proliferation

Careful examination of the microscopic images showed that at the beginning of the proliferation, the skeletal myoblasts seemed to prefer the bottom of the grooves rather than the ridges, different from the previous reports. This may due to the relatively smaller widths and higher aspect ratios of the microstructures, which form parallel and narrow trenches with subcellular widths. When 7 μm PDMS microlines were used, most cells were forced to accommodate themselves within the trenches as they spread on the surface. The cells were thus elongated along the longitudinal axles of the micro- trenches. Although the width of the micro-trench is smaller than the characteristic dimensions of a C2C12 myblast, the majority of each cell was "stuffed" into a single trench spontaneously without substantial contact to the neighboring trenches. Excellent angular control of individual myoblasts was thus achieved.

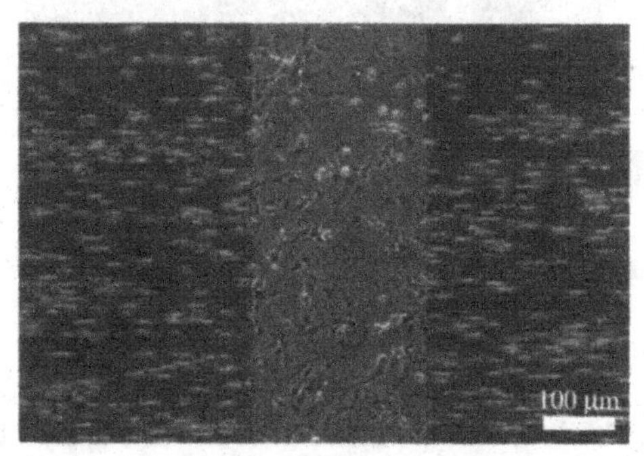

Figure 2: PDMS microlines array lead to directional myoblasts proliferation.

This alignment has an important implication in subsequent cell proliferation, where the myoblasts performed "in trench" proliferation to generate new cells within the same micro-trench. As the proliferation proceeds, it was observed that many cells started to grow and proliferate on the ridges of the microstructures. After 96 hours, highly organized cell cascades were formed in the groove as well as on the ridge of each micro- trench. This is in vast difference from the neighboring control group cultured on the neighboring plain PDMS substrates, where the cells grown and proliferated without obvious preferable orientations. The result is also different from the experiments where 2 μm deep trenches were used. In cases of the shallow trenches, it was observed that the cells had more crosslinking to the neighboring trenches, leading to a wider span of angular distribution.

<u>Directional Cell Differentiation</u>

Once the confluence is reached, the differentiation media was supplied to initiate cell fusion. When 7 μm high microstructures were used, individual C2C12 myoblasts in the cascades fused into long, multinucleated myotubes, aligning along the longitudinal axles of the microstructures (Figure 3). It is interesting to note that although the highly oriented cell growth and proliferation initiated from the bottom of the grooves, the skeletal myotubes seemed to mainly reside on the ridges. In particular, the nuclei of the myotubes were well confined by the ridges, having the same width as the ridges.

To fabricate manipulable muscle tissues, additional layers of myotubes were allowed to grow on top of the first layer. Observation showed the secondary and additional layers also exhibited highly oriented organization. Nevertheless, the alignment in these layers is morphologically different from that in the primary muscle layer, due to the fact that the cells in the additional layers did not have direct contact with the linear microstructures. In particular, it was observed that the widths of these myotubes were much larger than those contacting with the micro- ridges. The nuclei in these multinucleated myotubes were not elongated. This can be clearly seen by nuclei staining, where the round nuclei cascades of the cells in the secondary layers and the elongated nuclei cascades of the cells in the primary layer are in different focal planes. To examine the thickness of the muscle and the morphology difference between layers, z-

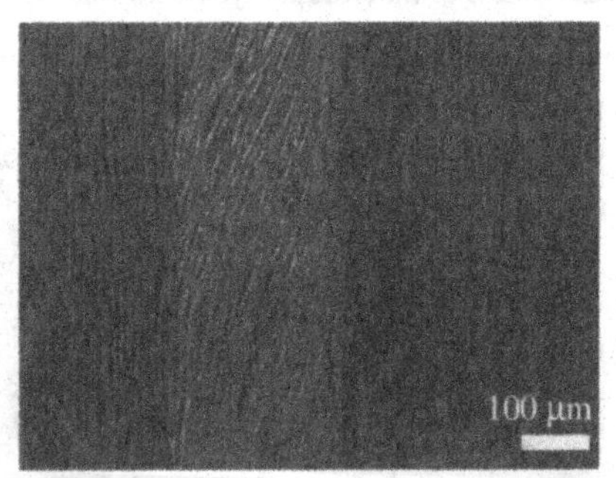

Figure 3: The directional proliferation of the skeletal myoblasts leads to organized myotubes upon differentiation, along the longitudinal axis of the microlines array.

axis scanning was performed (Figure 4). The results showed that skeletal muscle after two days culture was about 40 μm thick. The intensity traces indicated that the cell nuclei in the bottom surface have an obvious period along the transverse axles of the microstructures. Such period, however, was not seen in the top surface. Therefore, it is reasonable to conclude that the alignment of the myotubes in the additional layers may not be a direct consequence of the topography of the linear microstructures. Instead, the interlayer cell communication may dominate. This is further confirmed by the muscle cultured on 2 μm high PDMS microstructures, where the cells in first muscle layer were not well aligned. In this case, it was hard to observe myotubes alignment after a few layers. These results indicated that in the future design of engineering skeletal muscles, the morphologic difference between layers needs to be accounted since it is not uniform, which may strongly affects the contractile performance. Randomly oriented myotubes were observed in the control group where the cells were cultured on the plain PDMS surfaces for comparison.

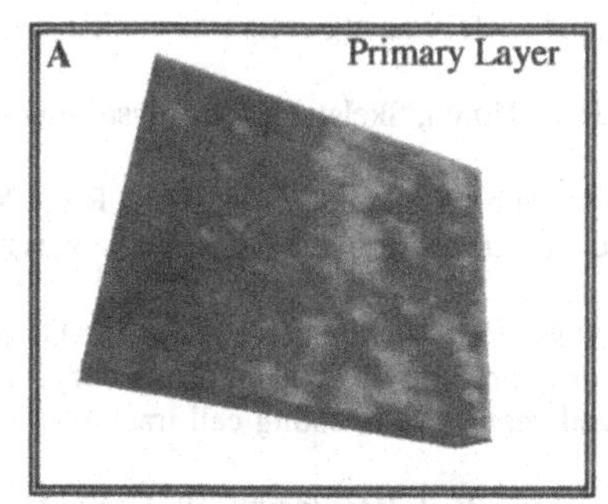

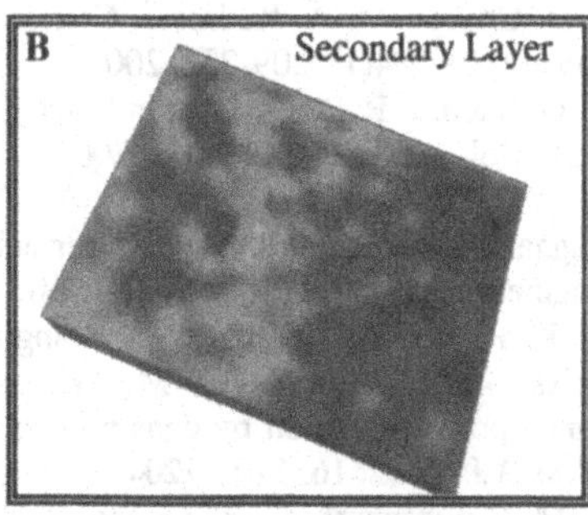

Figure 4: The microlines array directed the accommodated myoblasts into highly organized myotubes. Z-axis scanning of nuclei from the primary layer of cells (A) demonstrating regular intervals of bright and dark fields. Relative fluorescence of nuclei in the secondary layers (B) demonstrating lack of fluorescence distribution at regular intervals.

CONCLUSION

This work demonstrates an approach to fabricate skeletal muscle tissues with highly organized myotubes. The medium aspect ratio of the polymer microstructures creates a guidance to coax the immediate contacting myoblasts to proliferate and differentiate into desired directions, and further direct the alignment of additional muscle layers. Varying the heights of the microstructures could alter the degree of alignment of the cells in thick muscle layers so that the contractile and physiological performance of the fabricated muscles could be controllable. This work expected to foster the development of muscular tissue engineering and add a new dimension of knowledge in understanding the microtopography-regulated cell adaptation.

ACKNOWLDGEMENTS

The authors would like to thank Dr. Sudha Agarwal and her laboratory members, especially Dr. Mirela Anghelina and Dr. Ravi Chandran for their assistance in cell culture and immunofluorescence assay.

REFERENCES

[1] N. J. Fuller, M.A. Laskey, and M. Elia, Assessment of the composition of major body regions by dual-energy X-ray absorptiometry (DEXA), with special reference to limb muscle mass. Clin. Physiol. 12, 253-266, 1992

[2] P.K. Law, T.G. Goodwin, Q. Fang, M. B. Deering, V. Duggirala, C. Larkin, J. A. Florendo, D. S. Kirby, H. J. Li, M. Chen, Cell transplantation as an experimental treatment for Duchenne muscular dystrophy, Cell Transplant., 2, 485-505, 1993

[3] S. B. Chargé, M. A. Rudnicki, Cellular and molecular regulation of muscle regeneration, Physiol Rev. 84(1), 209-238, 2004.

[4] A. D. Bach, J. P. Beier, J. Stern-Staeter, R. E. Horch, Skeletal muscle tissue engineering, J. Cell. Mol. Med. 8(4), 413-422, 2004.

[5] W. Yan, S. George, U. Fotadar, N. Tyhovych, A. Kamer, M.J. Yost, R.L. Price, C.R. Haggart, J.W. Holmes, and L. Terracio. Tissue engineering of skeletal muscle, Tissue Engineering, 2007, 13(11), 2781-2790.

[6] K. K. Parker, A. L. Brock, C. Brangwynne, R. J. Mannix, N. Wand, E. Ostuni, N. A. Geisse, J. C. Adams, G. M. Whitesides, and D. E. Ingber, Directional control of lamellipodia extension by constraining cell shape and orienting cell tractional forces, The FASEB J., 2002, 16, 1195-1204.

[7] A. M. Collinsworth, C. E. Torgan, S. N. Nagda, R. J. Rajalingam, W. E. Kraus, and G. A. Truskey, Orientation and length of mammalian skeletal myocytes in response to a unidirectional stretch, Cell Tissue Res., 302(2), 243-251.

[8] T. Neumann, S. D. Hauschka, J. E. Sanders, Tissue engineering of skeletal muscle using polymer fiber arrays, Tissue Eng., 2003, 9(5), 995-1003.

[9] N. F. Huang, R. G. Thakar, M. Wong, D. Kim, R. J. Lee, S. Li, Tissue engineering of muscle on micropatterned polymer films, Conf Proc IEEE Eng Med Biol Soc. 2004, 7, 4966-4969.

[10] J. L. Charest, A. J. Garcia, W. P. King, Myoblast alignment and differentiation on cell culture substrates with microscale topography and model chemistries. Biomaterials, 2007, 28(13), 2202-2210.

[11] M. Das, K. Wilson, P. Molnar, and J.J. Hickman, Differentiation of skeletal muscle and integration of myotubes with silicon microstructures using serum-free medium and a synthetic silane substrate, Nature Protocols 2007, 2, 1795-1801.

[12] M. T. Lam, S. Sim, X. Zhu, and S. Takayama, The effect of continuous wavy micropatterns on silicone substrates on the alignment of skeletal muscle myoblasts and myotubes, Biomaterials, 2006, 27(24), 4340-7.

Mater. Res. Soc. Symp. Proc. Vol. 1052 © 2008 Materials Research Society 1052-DD03-28

RF MEMS Behavior, Surface Roughness and Asperity Contact

O. Rezvanian[1], M. A. Zikry[1], C. Brown[2], and J. Krim[2]
[1]Department of Mechanical and Aerospace Engineering, North Carolina State University, Raleigh, NC, 27695
[2]Department of Physics, North Carolina State University, Raleigh, NC, 27695

ABSTRACT

Modeling predictions and experimental measurements were obtained to characterize the electro-mechanical response of radio frequency micro-electro mechanical system (RF-MEMS) switches due to variations in surface roughness and finite asperity deformations. A Weierstrass-Mandelbrot fractal representation was used to generate three-dimensional surface roughness profiles. Contact asperity deformations due to applied contact pressures, were then obtained by a creep constitutive formulation. The contact pressure is derived from the interrelated effects of roughness characteristics, material hardening and softening, temperature increases due to Joule heating, and contact forces. The numerical predictions were qualitatively consistent with the experimental measurements and observations of how contact resistance evolves as a function of deformation time history. This study provides a framework that is based on an integrated modeling and experimental measurements, which can be used in design of reliable RF MEMS devices with extended life cycles.

INTRODUCTION

Surface roughness and asperity behavior are critical factors that affect contact behavior at scales ranging from the nano to the micro in microelectromechanical, electronic, and photonic devices. Specifically, in MEMS devices, large surface to volume ratios underscores that it is essential to understand how asperities behave in contact devices. MEMS switches, particularly those with radio frequency (RF) applications have demonstrated significantly better performance over current electromechanical and solid-state technologies [1].

The complex physical interactions between thermo-mechanical deformation, current flow, and heating at the contact, has made it extremely difficult to obtain accurate predictions of RF MEMS behavior, such that reliable devices can be designed for significantly improved life-cycles (see, for example, [2]). Various analytical and numerical methods have been employed to study the contact mechanics of ideally smooth surfaces (see, for example, [3]). Since surface topographies are critical in MEMS devices, some probabilistic models have been proposed to account for asperity height variations (for example, see [4]). The random and multiscale nature of the surface roughness can be better described by fractal geometry [5-6].

In this study a three dimensional fractal representation of surface roughness is used with a numerical framework to obtain predictions of thermo-mechanical asperity deformations of contacting surfaces as a function of time. Contact resistance behavior is then investigated and categorized for two surface roughness models with different roughness characteristics. The resistivity of the contact material is assumed to vary by strain hardening, and also by softening effects due to Joule heating at the asperity micro-contacts. The contact material used in this investigation is gold, which is one of the widely used contact materials for low-current MEMS switches [7]. To validate our approach, we also compared our predicted results with a set of experiments that were undertaken to characterize the contact resistance of RF MEMS switches.

CONTACT MECHANICS AND TOPOGRAPHY

Accurate modeling of the normal contact at the interface of the contact bump and the drain electrode in a RF MEMS switch requires that the roughness profiles of the two surfaces to be known. However, to simplify the contact problem, it can be assumed that the contact is between a rough and an infinitely smooth surface. This assumption is based on that the drain electrode is generally significantly smoother than the contact bumps [5]. Following the asperity-based model of Greenwood and Williamson [4], the asperities are dealt with individually. However, the deformation behavior of a contact asperity is influenced by other contact asperities, in that the share of the total applied load for each individual contact asperity will be determined by the set of all asperities that are in contact.

A realistic multi-scale three-dimensional fractal surface topography can be generated using a Weierstrass-Mandelbrot function [6], and can be expressed as

$$z(x,y) = L_0 (G/L_0)^{D-2} (\ln(\gamma)/M)^{1/2} \sum_{m=1}^{M} \sum_{n=0}^{n_{max}} \gamma^{(D-3)n}$$

$$\left\{ \cos\phi_{m,n} - \cos\left[(2\pi/L_0)\gamma^n (x^2+y^2)^{1/2} \cos\left(\tan^{-1}(y/x) - \pi m/M\right) + \phi_{m,n}\right]\right\}, \quad (1)$$

where L_0 is the sample length, G is the fractal roughness, D is the fractal dimension ($2 < D < 3$), γ is a scaling parameter, which is based on surface flatness and frequency distribution density, M is the number of superposed ridges used to construct the surface, $\phi_{m,n}$ is a random phase, n is a frequency index, where its maximum is a function of L_0, γ, and the cut-off length.

During the first few contacts, the applied pressure is normally higher than the yield stress of the contact material. During the period after this initial asperity deformation, contacting asperities would be susceptible to creep under compressive strain. Furthermore, creep deformation has been reported at micro-Newton level contact forces and low current levels [7]. The rate of creep deformation is assumed to have a power-law dependence on the stress as

$$\dot{\varepsilon} = A\sigma^p \exp\left(-\frac{Q_c}{kT}\right), \quad (2)$$

where $\dot{\varepsilon}$ is the strain-rate, A is a parameter relating to the material properties and the creep mechanism, σ is the stress, Q_c is the activation energy for creep, T is the absolute temperature, and k is the Boltzmann constant (k =1.38E-23 J/K). The stress exponent p in Eq. (2) is usually between 3 and 10, and is determined by the material composition.

PHYSICS OF CONTACT RESISTANCE AND SURFACE ROUGHNESS

Due to the surface roughness, when two surfaces are in contact, the contact is made at a finite number of points. In addition to the role that the roughness has in limiting the contact area and increasing the regular ohmic resistance, for an asperity contact radius on the order of the electron mean free path, more electrons will be ballistically transported, and the contribution of the boundary scattering of electrons to the total constriction resistance can increase. Based on the range of the electron mean free path, which is approximately 50 nm in Au [8], and considering that MEMS contact spots have been observed to be on the order of the electron mean free path [8], the boundary scattering effect can be critical for MEMS applications.

For a contact spot of radius a, and considering both ohmic and boundary scattering effects, the constriction resistance R_c can be given as [9,10]

$$R_c = f(\frac{\lambda}{a})\frac{\rho}{2a} + \frac{4\rho\lambda}{3\pi a^2}, \tag{3}$$

where ρ is the electrical resistivity, λ is the electron mean free path, and $f(\lambda/a)$ is an interpolation function, which has the limiting values of 1 as the Knudsen number λ/a approaches zero for $a>>\lambda$, and 0.624 as λ/a approaches infinity for $a<<\lambda$. A commonly used expression for $f(\lambda/a)$ is $(1+0.83(\lambda/a))/(1+1.33(\lambda/a))$. In this study it is assumed that the contact spots do not interact with each other, and are in parallel.

In the present model, strain-hardening and softening effects on resistivity are accounted for through a power-law formulation [11] as

$$\rho = \bar{\rho}(1+\frac{\varepsilon_p}{\varepsilon_{ref}})^q(1-\exp(-\frac{Q}{kT_c})), \tag{4}$$

where $\bar{\rho}$ is the average resistivity of a contact spot at temperature T_c, ε_p is the plastic strain, ε_{ref} is a reference strain, q is a material dependent parameter, Q is the activation energy for the mechanism by which stored dislocations are recovered or annihilated, and k is Boltzmann constant. The contact spot temperature can be stated as [8]

$$T_c = \sqrt{\frac{3f\pi a}{32L\lambda}V_c^2 + T_0^2}, \tag{5}$$

where L is Lorenz number (L = 2.45E-8 W. Ω /K^2), T_0 is the ambient temperature, and V_c is the voltage drop across the microcontacts.

EXPERIMENTAL SETUP

A number of experiments were carried out at the Nanoscale Tribology Laboratory at North Carolina State University, with the goal that a comparison can be made between the experimental and numerical modeling results. The RF MEMS devices used for these experiments are commercially available single pole double throw switches by wiSpry Inc.. The beam length is 135.5 μm with a width of 251 μm. The contact bumps are 6 μm in diameter and are separated by 55 μm. The travel distance of closure is 2 μm. The devices are mounted on gold sidebrazed ceramic packages, and the package is mounted inside of a vacuum system. The vacuum system is pumped to 1-5 mTorr, and then backfilled with Helium. This helium environment was also used to control stiction events. The vacuum environment created before backfilling the helium sufficiently removes moisture; however a thin layer of hydrocarbon is most likely present during testing. For the test device, 33 Volts generated a stable contact resistance for testing. Before each test, the switch was cold-switched 250 times. After this, the switch remained in the closed position, and a current of 100 μA was applied.

CONTACT MODELS

Two surface roughness models were obtained using Eq. (1), over an area of 4 by 4 microns, which is close to the area of the contact bumps in our fabricated RF MEMS switches. The topography was varied, so that we can understand how roughness can affect contact behavior. In Model 1 (Fig. 1(a)) a surface roughness profile with a peak-to-valley of 12.8 nm, and an RMS of 7.2 nm was generated. Model 2 (Fig. 1(b)) has a peak-to-valley of 27.1 nm and an RMS of 14.9 nm. The mating surface is assumed to be flat and infinitely smooth. The ambient temperature is 293 K. The applied contact force was chosen as 50 μN, pertaining to a gold cantilever beam with dimensions of 150×250×0.5 μm, and a gap of 2 μm. The switch is assumed to remain in close position for the duration of the simulation, and it is also assumed that there is no insulating film effect. This is done so that we can better understand the direct effects of the surface roughness. In RF MEMS switches with normal contact, the lateral deformations of asperities can be assumed to be negligible. An 80 by 80 grid of x and y coordinates has been used to take sampling points on the surface of the asperities. The far-field approach is determined from the collective creep response of the contact asperities. The asperity tips that establish contact are identified by comparing the far-field approach with the asperity heights. A detailed presentation of the computational approach is given in [11].

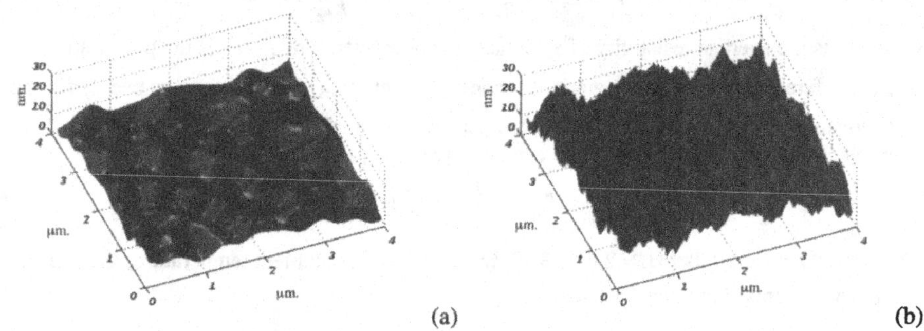

Figure 1. Three-dimensional fractal surface topographies generated using a Weierstrass-Mandelbrot function (a) Model 1 (b) Model 2.

RESULTS AND DISCUSSION

The ratio of the real contact area to the total area for the two roughness models is shown in Fig. 2(a). It is seen that the contact area increases with a steep slope at the beginning, which is due to the initial applied pressure. The initial applied pressure results in large deformations in the contact asperities. The asperity deformations, in turn, lower the applied pressure in two different ways. Firstly, as the asperities deform, more asperities come into contact. Secondly, cross sectional area of asperities generally increases as they are pressed down. The decreased applied pressure, then, along with the strain hardening of the material reduces the rate of increase of the contact area. The number of micro-contact spots for the two roughness models is shown in Fig. 2(b). Model 1 has very few asperities in contact, with their number being almost constant over time. In contrast, Model 2 has considerably more asperities in contact with their numbers increasing initially, and then fluctuating over time. The number of micro-contacts decreases when some micro-contacts coalesce, and it increases when new micro-contacts are established.

For the two roughness models, the constriction resistance has been shown in Fig. 3(a). Two stages of decrease can be seen in the curves of Fig. 3(a). The contact resistance initially decreases sharply until it reaches a rather stable level, during which it continues to gradually decrease. The initial stage can be associated with the initial deformation rate of asperities.

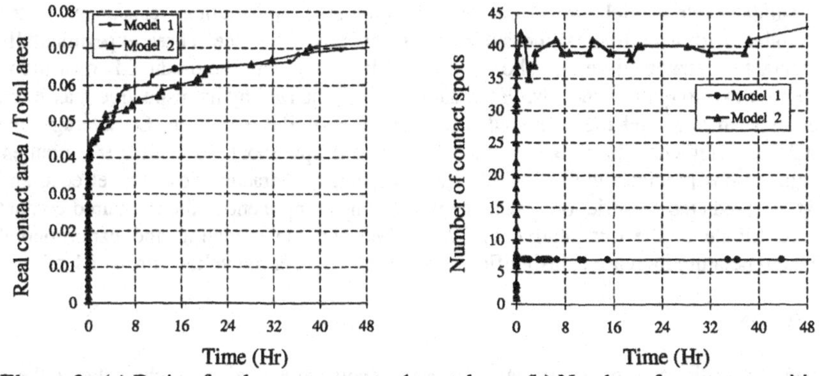

Figure 2. (a) Ratio of real contact area to the total area (b) Number of contact asperities.

Consequently, the increased contact area reduces the applied pressure, which in turn decreases the deformation rate of asperities. The results show that for the selected roughness models, the overall effect of increased resistivity due to the strain hardening is dominated by the contact area effect. In Model 1, maximum resistivity increases to about 4.1E-08 Ωm, which is 1.64 times the initial resistivity. In Model 2, the maximum resistivity reaches to about 4.4E-08 Ωm, which is 1.76 times the initial resistivity. Higher resistivity in Model 2 is an indication of asperities with higher curvatures. Model 1 has an initial contact resistance of approximately 44 mΩ. After 10 minutes the contact resistance decreases to about 32 mΩ, and after 48 hours to 20 mΩ (Fig. 3(a)). Model 2 has an initial contact resistance of 40 mΩ, which after 10 minutes decreases to 20 mΩ, and after 48 hours to about 11 mΩ (Fig. 3(a)). With having almost equal contact areas, the difference in contact resistance can be attributed to the number of the micro-contacts in each model. Majumdar et al. [5] reported a contact resistance between 60 and 130 mΩ for a population of 50 asperities, and a contact resistance between 80 and 110 mΩ for a population of five asperities, both pressed by a 50 μN contact force. Hence, it is seen that the predicted contact resistance values are within the same range of these cited studies.

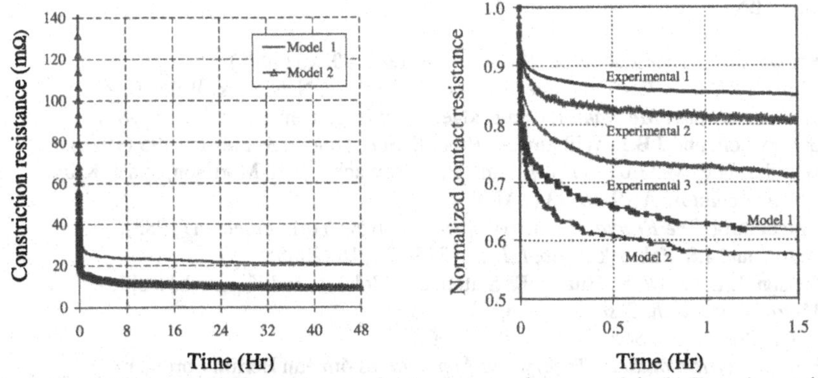

Figure 3. (a) The constriction resistance for the two roughness models (b) Experimentally measured contact resistance and predicted constriction resistance.

To validate our model, we compared the predicted constriction resistance of the roughness models with experimental measurements of contact resistance. Three experimentally measured contact resistance curves are shown in Fig. 3(b), along with the predicted constriction resistance for the two roughness models. Each data set is normalized by its respective maximum value. The experimental and the predicted results show a similar evolution. On average, the experimental measurements show that contact resistance drops about 20% after 90 minutes, while the simulations predict a 40% drop. Surface roughness characteristics, film effects, and the fact that the predicted constriction resistance is only one component of the measured contact resistance, contribute to the quantitative difference between the modeling and experimental results. The temperature increase is insignificant in both models for the voltage drop of 1 mV.

CONCLUSIONS

Modeling predictions and experimental measurements were obtained to characterize the electro-mechanical response of RF MEMS switches due to variations in surface roughness and finite asperity deformations. The interrelated effects of roughness, material hardening, and softening due to Joule heating were investigated on contact resistance. Changes in the constriction resistance of individual asperities were shown to depend on two competing events of increase in the contact area and the electrical resistivity. It is shown that there are two main stages related to the variation of the contact resistance. The initial decrease with a steep slope is attributed to the initial applied pressure, and the slightly strain-hardened contact material. During the subsequent gradual decrease in the second stage, the applied pressure had considerably decreased, and the contact material had extensively strain hardened. The results indicate that a very small percentage of the apparent contact area, conducts the electrical current. They also indicate that asperity coalescence limits the number of micro-contacts.

ACKNOWLEDGEMENTS

This work was funded by the AFOSR Extreme Friction MURI program, grant FA9550-04-1-0381.

REFERENCES

1. E.R. Brown, *IEEE Trans. Microw. Theory Tech.* **1868-80,** 46 (1998).
2. G.M. Rebeiz *RF MEMS, Theory, Design, and Technology*, New York: Wiley (2003).
3. K.L. Johnson *Contact Mechanics*, Cambridge: Cambridge University Press (1985).
4. J.A. Greenwood, and J.B.P. Williamson, *Proc. R. Soc. London A* **300-19,** 295 (1966).
5. S. Majumdar, N.E. McGruer, G.G. Adams, P.M. Zavracky, R.H. Morrison, and J. Krim, *Sensor Actuators A* **19-26,** 93 (2001).
6. B.B. Mandelbrot, *The Fractal Geometry of Nature*, New York: Freeman (1983).
7. S.T. Patton, and J.S. Zabinski, *Tribol. Lett.,* **215-230,** 18 (2005).
8. B.D. Jensen, L.L. Chow, K. Huang, K. Saitou, J.L. Volakis, and K. Kurabayashi, *J. Microelectromech. Syst.* **935-946,** 14 (2005).
9. G. Wexler, *Proc. Phys. Soc.,* **927-41,** 89 (1966).
10. R. Holm, *Electric Contacts: Theory and Applications* 4th edn Berlin: Springer (1967).
11. O. Rezvanian, M.A. Zikry, C. Brown, J. Krim, *J. Micromech. Microeng.,* **2006-15,** 17(2007).

Mater. Res. Soc. Symp. Proc. Vol. 1052 © 2008 Materials Research Society 1052-DD03-32

The Effect of Hydrophobic Patterning on Micromolding of Aqueous-Derived Silk Structures

Konstantinos Tsioris[1], Robert D White[1], David L Kaplan[2], and Peter Y Wong[1]
[1]Mechanical Engineering, Tufts University, Medford, MA, 02155
[2]Biomedical Engineering, Tufts University, Medford, MA, 02155

ABSTRACT

A novel micromolding approach was developed to process liquid biopolymers with high aqueous solvent contents (>90% water). Specifically silk fibroin was cast into a well-defined scaffold-like structures for potential tissue engineering applications. A method was developed to pattern the hydrophilicity and hydrophobicity of the polydimethylsiloxane (PDMS) mold surfaces. The water based biopolymer solution could then be directly applied to the desired regions on the cast surface. The variations in degree of hydrophilicity and hydrophobicity on the PDMS surfaces were quantified through contact angle measurements and compared to the outcome of the molded silk structures. Through this method free-standing structures (vs. relief surface-patterning) could be fabricated.

INTRODUCTION

Biopolymers, polymers synthesized by organisms, have many advantages including excellent biocompatibility and adjustable biodegradability [1]. Silk, for example, offers a wide spectrum of outstanding material properties such as good fracture toughness and excellent optical properties. Fabricating with biopolymers is also environmentally sound, since they can be processed in aqueous solutions. Due to these advantages biopolymers are often used as substrates in cell and tissue culture [1]. However they are rarely used in MEMS applications [2].

There is enormous potential for biopolymers in MEMS applications. In MEMS devices biopolymers could function as membranes or optical components. Devices which demand outstanding biocompatibility, such as implantable sensors, could be packed in or fully manufactured from biopolymers. The challenge today exists in understanding critical processing parameters in manufacturing structures with micron and submicron level features from biopolymers. In this research, the development of a micromolding technology, to produce microstructures from aqueous derived silk solutions was studied. In particular, well-defined cellular and tissue culture substrate (scaffold) fabrication was used as a model to study manufacturing methods.

The most important aspect of the proposed technology is the ability to produce freestanding structures vs. relief surface patterns through micromolding. In particular, the manufacturing challenges consist of processing materials with a high solvent content (> 90% water), producing well-defined structures and demolding the delicate structures. In this study we are addressing the molding process with the innovative solution of controlling the hydrophilicity and hydrophobicity of the cast surface to control the deposition of the biopolymer only in the cast cavities and not on cast surface.

EXPERIMENT

Soft lithography

Soft lithography is a well established method to create elastomer rubber stamps with micron size features [3, 4]. Briefly, standard UV lithography was performed with a transparency mask (Photoplotstore), SU-8 100 photoresist (Microchem®) of ~ 100 µm thickness on 4" silicon wafers to create a mold for the PDMS. The resists was exposed with an OAI 204 Aligner with a dose of 350 mJ/cm^2. We produced mold test pattern comprised a segmented band pattern of microchannels approximately 100 µm wide, 100 µm deep and with 200 µm interchannel spacing over an area of 10 mm x 10 mm. Furthermore, a variety of channels with wider dimensions was manufactured. PDMS was prepared with a ratio of 10:1 base to curing agent and degassed in a desiccator for approximately 1 hour. Subsequently, the liquid PDMS was poured over the microchannels. A pressurized nitrogen gun and a desiccator were used to remove air bubbles which where trapped during the molding process in the microchannels. The PDMS mold was cured at 60°C for ~ 5 hours. The PDMS stamp was released after curing by cutting along the wafer edge with a razor blade and peeling it off. No SU-8/ PDMS demolding agents were used.

Surface treatment

After the PDMS molds were released, they were treated with oxygen plasma to change the surface properties from hydrophobic to hydrophilic [5]. The PDMS stamps were placed with the patterned surface side up in the plasma chamber (Tegal Plasmod). A vacuum of ~200 mTorr was achieved before the valve which controls the oxygen flow was opened and the chamber was flushed with oxygen for 1 min at atmospheric pressure. Subsequently the valve was closed and the chamber was evacuated until the pressure reached ~200 mTorr. The procedure was repeated three times. Finally the valve was fully closed and the chamber pressure was allowed to drop to ~100mTorr. The AC power was adjusted to 50 Watts and fine tuned to allow uniform, bright and stable plasma in the chamber. The exposure times were varied from 10s, 20s to 30s depending on the experiment. After oxygen plasma treatment the PDMS molds were immediately further processed. Three test groups were established: 1) The PDMS surface was not further modified after plasma treatment. 2) The hydrophilic top most PDMS surface was carefully swiped with a common stainless-steel razor blade to expose the underlying hydrophobic material (see figure 1) and 3) a thin layer of silicone-based vacuum grease (Dow Corning® High Vacuum Grease) was deposited at the very top surface of the mold to cover the hydrophilic surface with a hydrophobic film. To obtain the configuration of group 3, a thin grease film was deposited on a microscope slide and the PMDS stamp was placed on the slide and removed. Through this the hydrophobic grease was placed only on the very top surface of the PDMS master.

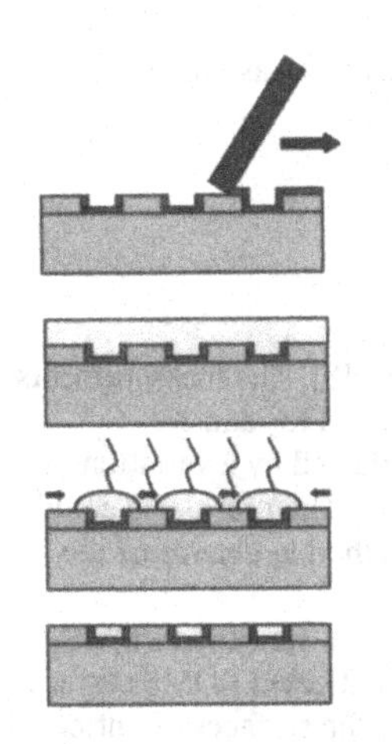

Figure 1: Cross-sectional view of schematic description of casting silk solution on surface modified PDMS channels (Group 2). From top to bottom: Stripping of hydrophilic top most layer with razor blade. Silk solution is placed on hydrophilic patterned PDMS channels. Silk solution is allowed to dry; simultaneously the surface tension of the solution forces the liquid away from the hydrophobic top most surface into the hydrophilic channels. Water has fully evaporated and silk fibroin is deposited in the channels.

Quantitative surface characterization

The degree of hydrophilicity or hydrophobicity was determined through contact angle measurement. Drops of water or silk solution (5µl each) were placed on the following surfaces: Glass, PDMS untreated, PDMS treated with oxygen plasma for 10s, 20s and 30s, PDMS treated with oxygen plasma for 20s and subsequently treated with a razor blade or vacuum grease. This led to 14 different surface/ liquid pairs (full data not shown). The images were recorded with a CCD camera and saved onto a PC. All images were recorded 10 seconds after the droplet was placed on the substrate. Further image analysis and automated contact angle measurement were performed with the ImageJ plug in LBADSA [6]. Three measurements, one droplet each, were conducted for each liquid/substrate pair. Statistical analysis (mean, standard deviation and t-test) was performed in Microsoft Excel®.

Silk fibroin extraction

A ~7.7% w/v silk fibroin solution was prepared from *B. mori* silk cocoons. The procedure of extracting silk fibroin from cocoons is described in detail by Sofia et al [7]. Briefly, fibroin is extracted in .02 M Na_2CO_3 solution. Subsequently the dried silk fibers are dissolved in a 9.3 M LiBr solution and dialyzed against DI water until the desired concentration is achieved.

Micromolding process and characterization of structure

A volume of 0.2 ml of the silk solution was placed on the PDMS negative with a 1ml syringe. The tip of the syringe was used to spread the solution evenly across the patterned surface. The silk was allowed to dry; simultaneously the surface tension of the solution forced the liquid away from the hydrophobic top most surface into the hydrophilic channels. The silk solution was allowed to dry for 2-5 days at room temperature in order for the silk fibroin to deposit into the channels (figure 1). Subsequently the silk structures were released by manually peeling them from the PDMS mold. The scanning electron microscopy (SEM) images where taken on a JEOL

JSM 840A SEM. All manufacturing and testing procedures were kept as clean as possible by being performed in a class 1000 clean room.

DISCUSSION

Surface characterization

Lawton et al. reported a contact angle of water on PDMS of 110° [8]. The measurements in our study showed a mean contact angle and standard deviation (n=3) of 113° and 1° respectively. The variation from the value in the literature could be explained by a variation of the PDMS processing method, the drop size, air temperature or humidity. Therefore, one could argue that this result is in good agreement with the literature and the method to determine the contact angle is satisfactory.

Figure 2 shows the results of the contact angles of silk solution on a variety of PDMS surfaces. The most significant difference ($p = 0.00$) exists between the untreated PDMS surface and the surface treated for 20 s with oxygen plasma. Next we compared the surfaces of untreated PDMS and PDMS treated with oxygen plasma for 20 seconds and subsequently stripped or coated. No significant difference ($p = 0.94$) between untreated and stripped surface was found. The coated surface shows a significant difference ($p = 0.06$) in the contact angle. Table 1 shows the numerical values for mean and standard deviation associated with the bar graph of figure 2. The high standard deviation of the group *PDMS 20s stripped* and *–PDMS 20s coated* indicates the difficulty associated with the method of manually stripping and coating the surface. Though, one could argue that even if those methods are inconsistent it can still produce useful results with the stripping method for the micromolding technique (see figure 3). This could be explained by the stripping method returning the surface back close to the original surface configuration. PDMS consists of repeating $–O\text{-}Si(CH_3)_2$ units. Upon plasma treatment, polar groups (silanol groups (Si-OH)) are introduced on the surface and these groups replace the methyl groups ($Si\text{-}CH_3$)[5]. The silanol groups are subsequently removed or modified by the razor blade so that the surface becomes hydrophobic again. Figure 3 shows an association of the different surface treatments with the final outcome of the molded silk structures, details of the images will be discussed in the next section.

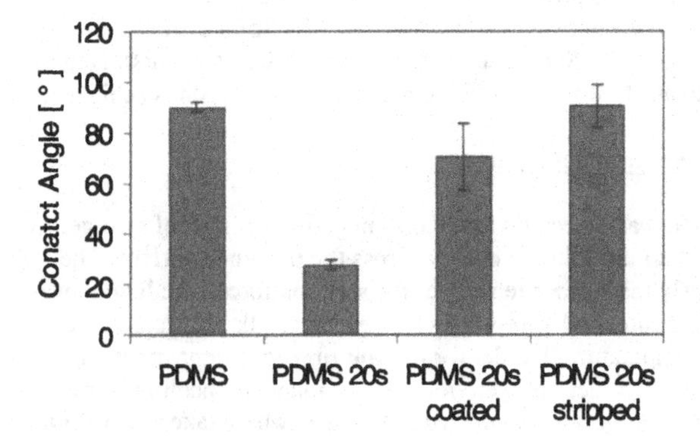

Figure 2: The bar graphs indicate contact angle measurements of silk solution droplets on treated and untreated PDMS surfaces (mean value n=3). Black bars indicate standard deviation. *20s* refers to oxygen plasma treatment for duration of 20 seconds. *Coated* refers to a deposited a thin silicone grease film on the surface. *Stripped* refers to removal of the top most layer of PDMS surface with a razor blade.

Table 1: Numerical values of mean and standard deviation (n=3) of contact angles from figure 2.

PDMS	PDMS 20s	PDMS 20s coated	PDMS 20s stripped
90	27	70	91
2	2	13	8

Micromolding

The structure seen in figure 4B was obtained by using the surface stripping approach to generate the differences in hydrophillicity and hydrophobicity. Figure 4A shows a micrograph of an approximately 100 μm x 100 μm PDMS microchannel. The channel sidewalls are vertical and good replication of the corners was achieved from the PDMS on SU-8 molding process. The bottom is perfectly smooth and shows a replica of the top SU-8 surface. The sidewalls show some roughness which can be contributed to the low quality of the low cost lithography mask. Figure 4B shows the casted silk structure with similar dimensions to the PDMS microchannel. The corners where replicated with high accuracy with a radius of approximately a few microns.

Figure 1 shows a schematic description of the casting process and mechanisms that lead to the separation of the silk solution and the deposition into the individual channels. Figure 3 shows an association of the different surface treatments with the final outcome of the casted

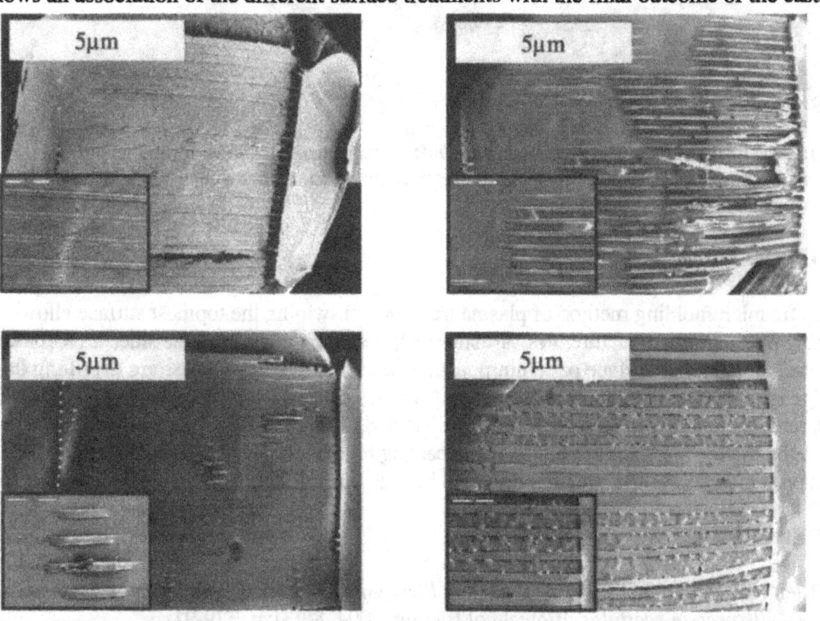

Figure 3: Decasted silk films, clockwise from top left, PDMS untreated, PDMS treated for 20 seconds with oxygen plasma, PDMS treated for 20 seconds with oxygen plasma and surface subsequently stripped with a razor blade, and PDMS treated for 20 seconds with oxygen plasma and surface subsequently coated with vacuum grease. Bottom left corner in all images shows a magnified region of a section of the image.

structures. The method using a razor blade led to the desired free standing segmented band patterns as seen in the micrograph in the bottom right corner of figure 3. The other three cast preparation methods led to only surface patterning with insufficient imprints of the features.

Limitations

The major difficulty with micromolding is the decasting process of the delicate structures [9, 10]. In this study the minimum feature size was mainly determined by the decasting process. The smallest features that could be replicated were rectangular fiber with a ~ 100 µm x 100 µm cross section (Figure 4). Consistently reproducible, the smallest structures were ~ 200 µm wide and 100 µm thick (bottom right corner Figure 3). The major reason for this limitation relate to the attachment of the silk structures to the PDMS side walls. The inconsistency in producing the smallest features is due to the limits of the process parameters of the plasma tool and the inconsistency caused by manually modifying the top surface of the cast substrate with a razor blade.

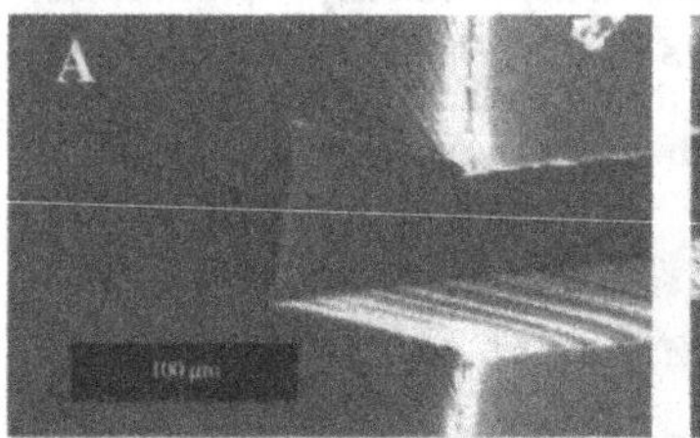

Figure 4: SEM image A on the left is showing a PDMS casting channel with width and height dimensions of approximately 100 µm x 100 µm. Image B on the right is showing a single casted silk fiber with the same dimensions.

CONCLUSION

The specific micromolding method of plasma treating and swiping the topmost surface allows fabricating free standing structures (vs. structures supported by a layer on one side, what could also be considered relief surface patterning). The two critical parts to success are (1) filling the channel features without having liquid on the top surface of the PDMS stamp and (2) decasting the structures without severe deformation. The method is able to fill the channels well; however, the feature sizes are limited by the decasting process in which the geometries and material properties of the silk are important and have to be consider.

REFERENCES

1. Meinel, L., et al., *Engineering cartilage-like tissue using human mesenchymal stem cells and silk protein scaffolds.* Biotechnol Bioeng, 2004. **88**(3): p. 379-91.
2. Bettinger, C.J., Cyr, K. M., Matsumoto, A., LangerR., Borenstein J. T., Kaplan D. L., *Silk Fibroin Microfluidic Devices.* Biomaterials, 2007 (in press).
3. Kim, E., Y.N. Xia, and G.M. Whitesides, *Micromolding in capillaries: Applications in materials science.* Journal of the American Chemical Society, 1996. **118**(24): p. 5722-5731.

4. Xia, Y.N. and G.M. Whitesides, *Soft lithography.* Annual Review of Materials Science, 1998. **28**: p. 153-184.
5. Rangel, E.C., G.Z. Gadioli, and N.C. Cruz, *Investigations on the stability of plasma modified silicone surfaces.* Plasmas and Polymers, 2004. **9**(1): p. 35-48.
6. Stalder, A., et al., *A snake-based approach to accurate determination of both contact points and contact angles.* Colloids and Surfaces A: Physicochemical and Engineering Aspects, 2006. **286**(1): p. 92-103.
7. Sofia, S., et al., *Functionalized silk-based biomaterials for bone formation.* J Biomed Mater Res, 2001. **54**(1): p. 139-48.
8. Lawton, R., et al., *Air plasma treatment of submicron thick PDMS polymer films: effect of oxidation time and storage conditions.* Colloids and Surfaces A: Physicochemical and Engineering Aspects, 2005. **253**(1): p. 213-215.
9. Guo, Y.H., et al., *Analysis of the demolding forces during hot embossing.* Microsystem Technologies-Micro-and Nanosystems-Information Storage and Processing Systems, 2007. **13**(5-6): p. 411-415.
10. Heckele, M. and W.K. Schomburg, *Review on micro molding of thermoplastic polymers.* Journal of Micromechanics and Microengineering, 2004. **14**(3): p. R1-R14.

MEMS Devices I

Mater. Res. Soc. Symp. Proc. Vol. 1052 © 2008 Materials Research Society 1052-DD04-07

Fabrication and Characterization of Normal and Shear Stresses Sensitive Tactile Sensors by Using Inclined Micro-cantilevers Covered with Elastomer

Masayuki Sohgawa[1], Yu-Ming Huang[1], Minoru Noda[1,2], Takeshi Kanashima[1], Kaoru Yamashita[1,2], Masanori Okuyama[1], Masaaki Ikeda[3], and Haruo Noma[4]

[1]Graduate School of Engineering Science, Osaka University, 1-3 Machikaneyama-cho, Toyonaka, Osaka, 560-8531, Japan

[2]Graduate School of Science and Technology, Kyoto Institute of Technology, Matsugasaki, Sakyo-ku, Kyoto, 606-8585, Japan

[3]Omron Corporation, Kizugawa, Kyoto, 619-0283, Japan

[4]Advanced Telecommunications Research Institute International, Seika, Souraku, Kyoto, 619-0288, Japan

ABSTRACT

Tactile sensors of Si microcantilevers with a piezoresistive layer have been proposed for detection of both normal and shear stresses. Micro-cantilevers were fabricated by the surface micromachining of SOI wafers and were adequately inclined by controlling deflection with a Cr layer. The cantilevers were embedded in the PDMS elastomer to create a human-friendly surface. When a stress is applied to the surface of elastomer, the deformation of the cantilevers along with the elastomer is detected as a resistance change in the piezoresistive layer of the cantilevers. The piezoresistive response of the cantilever was analyzed by FEM calculations. The fabricated tactile sensor is sensitive to both normal and longitudinal shear stresses and its responses agree closely with the calculated value. Moreover, it has little sensitivity to shear stress in the transverse direction to the cantilever, which means that the tactile sensor can distinguish the direction of shear stress. This sensor can be utilized for tactile sensing in human support robots.

INTRODUCTION

In recent years, human support robots for nursing-care have attracted much attention because of acceleration of the average age of the population [1]. If human support robots do not have the ability to recognize allowable forces for contact with the human body, there is possibility that the human is hurt by excess force. To assure safety for the human body, human support robots require human-like tactile sensing. However, tactile sensors with high resolution and sensitivity to both normal and shear stresses like the human tactile sense have not yet been realized. It is believed that a tactile sensor for human support robots requires high sensor density on the robot's skin, ability to measure normal and shear stresses and a human-friendly surface.

We have proposed tactile sensors for human support robots which can detect both normal stress and shear stress and have a human-friendly surface [2]. In previous work, we have examined Si LSI process compatibility on the premise of integration with signal processing units and controllability of deflection of the cantilevers [3]; it has been found that out-of-plane deflection control by the addition of a Cr thin film layer is suitable for fabrication of tactile sensitive cantilevers. In this work, tactile sensor devices were fabricated by using cantilevers of Cr/Si and its performance was characterized in response towards normal stress and shear stress.

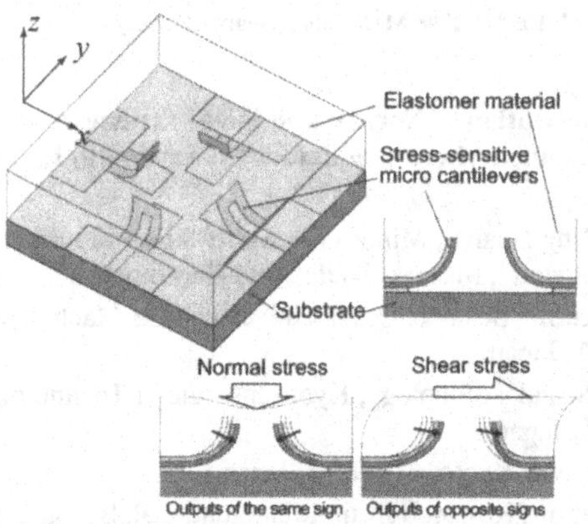

Figure 1. Schematic illustration and operation of proposed tactile sensor.

EXPERIMENT

Figure 2 shows a cross-sectional view of the fabricated cantilever. Cantilevers are adequately inclined by Cr layer as a deflection control layer. To form p-Si layer as piezoresistors, boron ions were implanted in the surface of the active Si layer of SOI wafer (active Si layer: 2.5 μm, buried oxide (BOX) layer: 1.0 μm), and the wafer was annealed in N_2 atmosphere at 900°C for 30 min. The density of impurities is about 10^{18} cm^{-3} and depth of p-layer is about 700 nm. Then a SiN layer to insulate between Cr and p-Si layers was deposited by LPCVD with a thickness of about 200 nm. Cr thin film (the thickness of about 100 nm) was deposited as the deflection control layer by vacuum evaporation and patterned by conventional photolithography process. The cantilever pattern was formed by etching SiN and Si layer of the area other than the cantilever. A Au/Cr layer (thickness: ~500 nm) to form an electrode was deposited and patterned by a lift-off process. Finally, etch holes for sacrificial etching were made by photolithography and the BOX layer was etched in buffered HF (NH_4HF_2: 20%) for a few hours and then the cantilever structure was released from the substrate. After the cantilever was released, a mixture of base resin and curing agent for PDMS (SILPOT 184, Dow Corning Corp.) was coated on the substrate and dried in air at room temperature for 1 day.

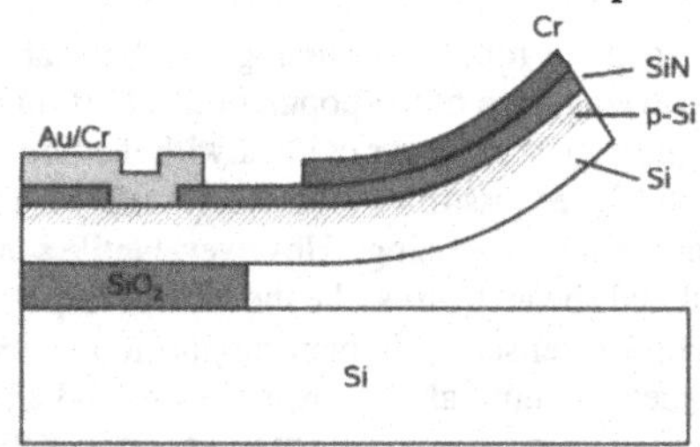

Figure 2. A cross-sectional view of the cantilever of the tactile sensor.

RESULTS AND DISCUSSION

Configuration of fabricated tactile sensor

Figure 3 (a) shows a photograph of the cantilever before embedding in the elastomer. The distance between the unetched line and the tip of the cantilever is 500 μm. It is observed that the cantilever structure has curved away from the substrate and has upward deflection as designed. The longitudinal direction of the cantilever is directed to <110> direction in Si (100) plane. After curing the elastomer, the tactile sensor chips were die-bonded on a printed board and connected electrically to the board by Au-bonding wire. A photograph of the completed sensor device is shown in figure 3(b). The chip size is 1 cm square; 16 cantilevers can be included in one chip.

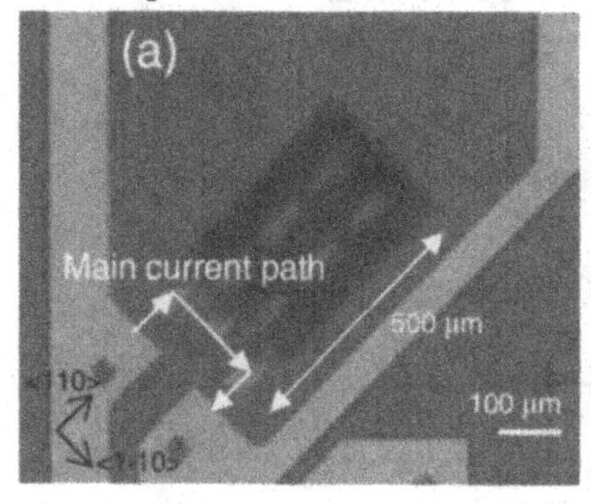

Figure 3. Photographs of (a) fabricated cantilever without elastomer, (b) tactile sensor chip (and a dime).

Analysis of deformation of cantilever and sensor output

Deformation of the cantilever induced by internal stresses of the multilayers has been evaluated by finite element method (FEM) using ANSYS (Rel. 11.0) software. The cantilever models were meshed by using SHELL elements which can only take account of bending and in-plane deformation without antiplane shear, because thickness of cantilever is much smaller than the lateral size [3]. First, the initial inclined shape of the cantilever is calculated as deflection induced by a difference in thermal expansion among the layers. The material parameters used for FEM calculation are shown in Table I [5, 6]. The element size for the mesh is several micrometers. The calculated initial tip height of the cantilever is 106 μm.

Table I. Material parameter for FEM calculation.

	Young's modulus (GPa)	Poisson's ratio	CTE ($10^{-6}K^{-1}$)
Si	150	0.25	2.6
SiN	300	0.24	3.2
Cr	180	0.35	8.4

In addition, deformation of the elastomer is calculated by using rubber elasticity theory [7]. Deformation due to a normal stress, σ_z, is calculated by the following equations:

$$s_z = \frac{E_e}{3}\left(\lambda^2 - \frac{1}{\lambda}\right) \quad (1)$$

$$\lambda = 1 - \frac{\Delta d}{d}, \quad (2)$$

where E_e, λ and d are the Young's modulus, the extension ratio and the initial thickness of the elastomer, respectively. On the other hand, deformation in the case of shear stress along the longitudinal direction of the cantilever, s_l, is calculated by the following equation:

$$s_l = \frac{E_e}{3}\gamma, \quad (3)$$

where γ is shear strain. Figure 4 shows the deformation ratio (strain) for an applied normal stress and shear stress. The Young's modulus of PDMS is about 1 MPa.

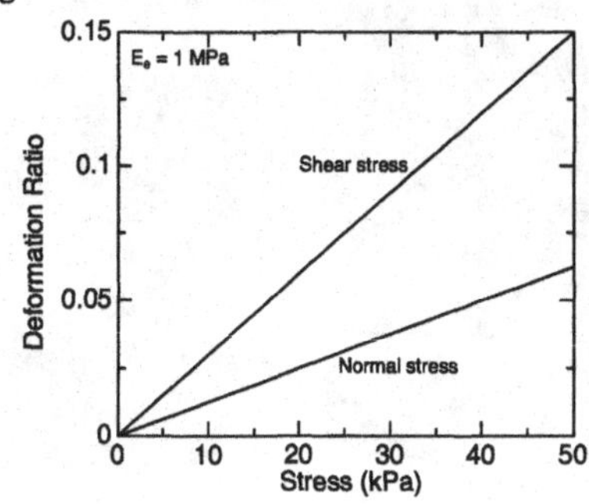

Figure 4. Calculated deformation ratio of elastomer for normal and shear stresses.

Assuming that the tip of the cantilever moves perfectly together with the elastomer, the change in tip height of the cantilever, Δh, is expressed as follows:

$$\Delta h = \frac{h\Delta d}{d}, \quad (4)$$

where h is the initial tip height of the cantilever (h = 106 μm from the calculation described above). Similarly, lateral displacement of the tip of the cantilever, Δl, is expressed as:

$$\Delta l = h\gamma. \quad (5)$$

The internal stress distribution in the cantilever was obtained by FEM calculation with the base of the cantilever fixed and constant displacement applied from calculations using by eqs. (4) or (5) to the tip of the cantilever. The change rate of piezoresistance in the cantilever is expressed as the product of the piezoresistive coefficient and the integral of the stress, σ, obtained by FEM calculation along the path of current flow:

$$\frac{\Delta R}{R} = \frac{\pi_l}{r}\int_r \sigma_l dr + \frac{\pi_t}{r}\int_r \sigma_t dr, \quad (6)$$

where subscript l and t represent the longitudinal and transverse direction, respectively, and r is the current path length. If current flows in the <110> direction in (100) plane of p-Si, piezoresistive coefficients π_l and π_t are 7.2×10^{-10} Pa^{-1} and -6.6×10^{-10} Pa^{-1}, respectively [8]. The calculated resistance change rate to a normal stress of 5 kPa is 3.5×10^{-5}. On the other hand, the calculated value for a shear stress of 5 kPa is 2.1×10^{-4}. Therefore, the response of this sensor to shear stress is more sensitive than the response to normal stress.

Measurement of piezoresistive response under normal and shear stresses

The surface of the fabricated tactile sensor was depressed by a flat plate to measure the response to normal stress. Reference force and moment were measured by a 6-axis force sensor (UFS 2A-05 Nitta Corp.). The normal stress was calculated by dividing the reference force by the contact area. The resistance due to piezoresistance was measured by a high-precision digital multimeter (R6581 Advantest). Figure 5 (a) shows the resistance change rate of a fabricated sensor as a function of applied normal stress. The open diamond in figure 6 shows the resistance change calculated by using FEM. This value agrees well with the measured one. To measure the response against shear stress, the tactile sensor was pressed and moved laterally. Shear stress was calculated by dividing the reference moment by contact area and distance between the center of the reference sensor and the contact point to sensor surface. Figure 5 (b) shows the resistance change rate as a function of shear stress. The resistance change rate for 5 kPa of shear stress in the longitudinal direction to the cantilever is about 1×10^{-4}, which is smaller than calculated value (2.1×10^{-4}). It is believed that this error is due to neglecting other forces besides simple shear in the PDMS deformation calculations. On the other hand, the sensitivity to shear stress in the transverse direction in relation to the cantilever is smaller than that for the longitudinal direction. It means that the tactile sensor can distinguish the direction of shear stress.

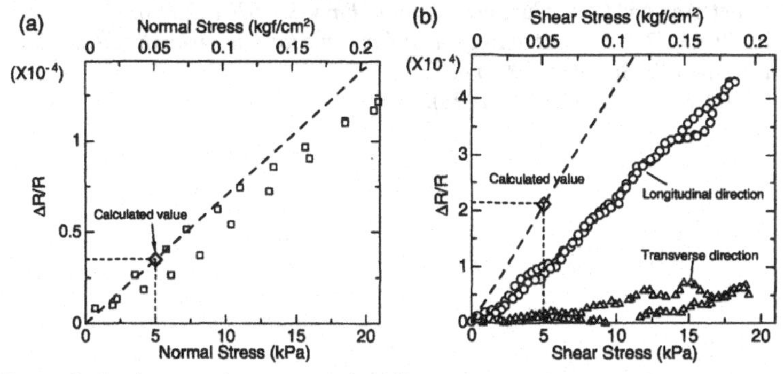

Figure 5. Resistance change rate of tactile sensor as a function of normal stress.

CONCLUSIONS

Tactile sensors for human support robot application having Cr/Si cantilever structure and a human-friendly surface have been fabricated by using surface micromachining and PDMS coating, and characterized. Cr/Si cantilevers have upward initial deflection as designed. The fabricated tactile sensor is sensitive to both normal and shear stresses in the longitudinal direction to the cantilever. On the other hand, the sensitivity to shear stress in the transverse direction in relation to the cantilever is much smaller than the sensitivity in the longitudinal direction; thus, the tactile sensor can distinguish the direction of shear stress. The responses of the tactile sensor against normal and shear stresses were measured as piezoresistance change and the deformation and stress distributions were estimated by using FEM calculation. It is found that calculated sensitivity found by the FEM almost agrees with the measured sensitivity. The proposed tactile sensor can be utilized for sensing both normal and shear stresses in the human support robot and the sensitivity of tactile sensor can be improved much further by designing with FEM estimation.

ACKNOWLEDGMENTS

This study was carried as a part of "Research and Development of Nanodevices for Practical Utilization of Nanotechnology" from New Energy and Industrial Technology Development Organization, Japan.

REFERENCES

1. 1. J. Forlizzi, *Interactions* **12**, 16 (2005).
2. S. Yoshida, T. Mizota and H. Noma, *IEEE-Virtual Reality*, Charlotte, NC, USA (March 2007).
3. Y. -M. Huang, M. Sohgawa, M. Noda, K. Yamashita, M. Okuyama and H. Noma, *Mater. Res. Soc. Symp. Proc.* **969**, 0969-W05-01 (2007).
4. K. Noda, K. Hoshino, K. Matsumoto and I. Shimoyama, *Sens. Act. A* **127**, 295 (2006).
5. "Rika Nenpyo", edited by National Astronomical Observatory of Japan (Maruzen, 2000).
6. K. E. Petersen and C. R. Guarnieri, *J. Appl. Phys.* **50**, 6761 (2006).
7. G. R. Strobl, "*The Physics of Polymers: Concepts for Understanding Their Structures and Behavior*", (Springer, 1997) p. 315.
8. Y. Kanda, *Sens. Act. A* **28**, 83 (1991).

Mater. Res. Soc. Symp. Proc. Vol. 1052 © 2008 Materials Research Society 1052-DD04-08

Design and Fabrication of an Optical-MEMS Sensor

Vaibhav Mathur, Jin Li, and William D. Goodhue
Photonics Center, Department of Physics and Applied Physics, University of Massachusetts, Lowell, 720, Suffolk St., Lowell, MA, 01854

ABSTRACT

A novel optical-MEMS sensor based on the AlGaAs material system is designed and fabricated. The device consists of micro-beam waveguides butt-coupled with their ends separated by approximately 2 to 4 μm. The device works on the principle that when acoustically driven by an external source, the waveguides misalign, leading to coupling loss. The device design parameters were determined using FEM (Finite Element Method) modeling. The dielectric waveguide beams were designed for single mode propagation at 785 nm and longer wavelengths. A combination of dry and wet etching process followed by precision laser cutting was used to fabricate the suspended beams. Beams ranging from 100 μm to 400 μm with fundamental frequencies of 50 KHz to 200 KHz were successfully fabricated. Initial uncut waveguide test results will be discussed along with the plan for characterizing the devices using an acoustically coupled piezoelectric driver. These devices may be utilized for vibration sensing, or optical intensity modulation.

INTRODUCTION

Micro-electro-mechanical System (MEMS) based sensors have been used in the past decade for a variety of applications ranging from pressure sensing, bio-sensing and accelerometers. Most vibration sensors in use today are accelerometers in which the suspended structure oscillates harmonically. For sinusoidal vibrations, the displacement, velocity and acceleration amplitudes are related directly to the frequency. For example, the iMEMS family of accelerometers developed by Analog Devices is widely used in the automobile industry [1]. Most of these devices work on the principle of capacitance changes induced in a parallel plate capacitor type configuration due to the movement of a proof mass with respect to the substrate. The device presented here employs a novel technique based on the change in the optical signal which could potentially have faster response and higher sensitivity. In addition, conventional MEMS devices have relied heavily on the use of silicon as the substrate material. Developing MEMS based on III-V semiconductor materials offers the great potential of integration with opto-electronic devices such as lasers and photodetectors.

The AlGaAs/GaAs based MEMS devices reported here work on the principle of change in the optical coupling due to micro-cantilever waveguide misalignment when driven acoustically with a piezo chip. The device design parameters were determined using FEM (Finite Element Method) modeling. The dielectric waveguide beams were designed for single mode propagation at 785 nm and longer wavelengths. A combination of dry and wet etching process followed by precision laser cutting was used to fabricate the suspended beams. Beams ranging from 100 μm to 400 μm with fundamental frequencies of 50 KHz to 200 KHz were successfully fabricated. Progress is reported on designing, fabricating and testing the devices, with the emphasis on device design and fabrication.

DEVICE DESIGN

The schematic of the optical-MEMS device under investigation is depicted in figure 1. Each sensor consists of a pair of micro-cantilever beams with free ends butt coupled. Light coupling from one to the other is altered when the micro-cantilever is driven acoustically close to its resonance frequency by a piezo chip. An array of sensors with varying cantilever beam length was fabricated to characterize the frequency response of the device.

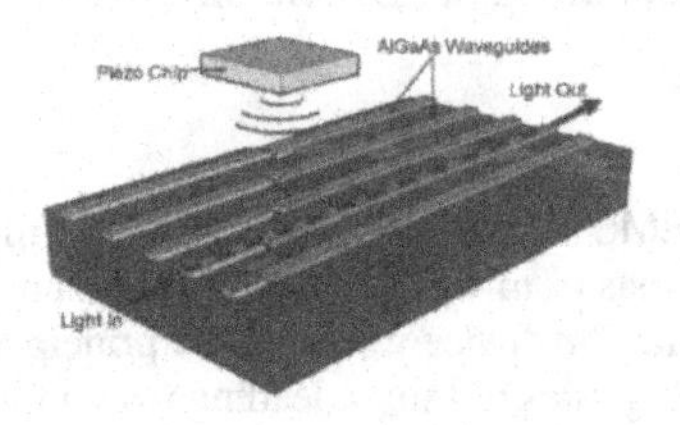

Figure 1. Schematic of the MEMS device.

The micro-cantilever beams are essentially dielectric symmetric rib waveguides. The AlGaAs material system is selected here as the refractive index varies with the AlAs mole fraction [2]. The AlAs mole fraction should be above 0.12 for both cladding and core layers for low loss propagation at 785 nm or longer wavelength. Also in order to obtain the single-mode operation at this or longer wavelength, the refractive index difference between the core and cladding layers must satisfy the condition $\Delta n \approx 0.0239$ for a waveguide core layer in the range of 0.55 µm in thickness [3]. The $Al_{0.3}Ga_{0.7}As$ (n = 3.405) and $Al_{0.24}Ga_{0.76}As$ (n = 3.429) satisfy both requirements and were selected for the upper/lower cladding layers and guiding layer, respectively. The guiding condition is $n_2 < n_{eff} < n_1$, i.e. in our case $3.405 < n_{eff} < 3.429$. The thickness of the upper and lower cladding $Al_{0.3}Ga_{0.7}As$ layers were 1.2 µm to prevent the light from coupling into the GaAs substrate that has high absorption loss at 785 nm wavelength.

A rib width of 5 µm was chosen for the FEM simulation studies here. The cross section of the final waveguide design arrived from the analysis is illustrated in figure 2(a). The COMSOL package was employed for a 2-D waveguide analysis to determine the effective refractive index of propagation in the device and the single mode profile distribution. Figure 2(b) shows the Magnetic field, z component plot on the cross section of the waveguide assuming TE (Transverse Electric field) mode propagation.

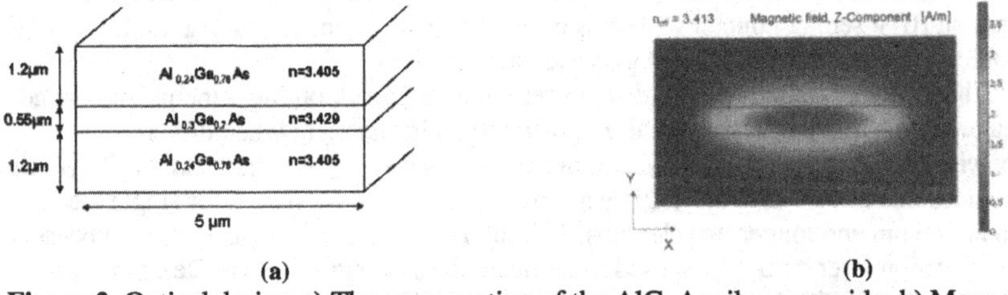

(a) **(b)**

Figure 2. Optical design a) The cross section of the AlGaAs rib waveguide; b) Magnetic field, z component surface plot for the fundamental mode, $n_{eff} = 3.413$

A micro-cantilever beam reaches maximum vibration amplitude at its fundamental resonance frequency. The fundamental frequency is inversely proportional to its length [4]. For a beam of 200 μm in length, the resonance frequency is calculated to be approximately 50 KHz, assuming the bulk material properties of $Al_{0.3}Ga_{0.7}As$ [3]. The 2-D FEM Eigen-frequency analysis results for a pair of cantilevers of 200 μm and 100 μm in length, respectively, are as shown in figure 2(a). A plot of fundamental frequency as a function of the length is shown in Figure 2(b). Based on these studies a range of 100-400 μm was chosen for the beam lengths for the device driven with a frequency range from 0-200KHz. A detailed FEM analysis and light propagation simulation of the cantilever sensors can be found in our earlier work [5].

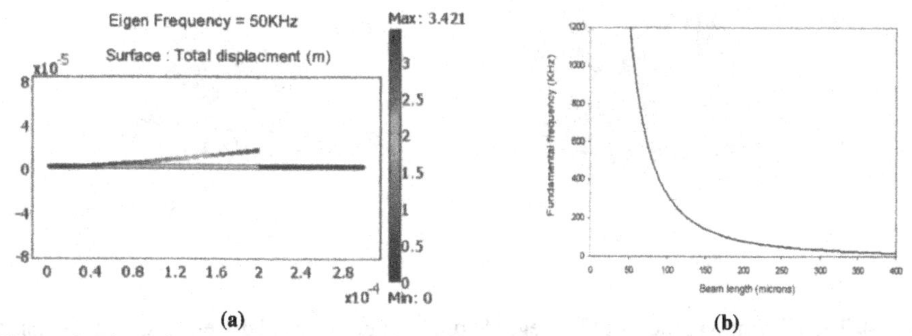

(a) (b)

Figure 3. Eigen-frequency analysis (a) A cantilever beam of length 200μm at resonance; (b) Plot of fundamental frequency vs the beam length.

DEVICE FABRICATION

The AlGaAs waveguide epitaxial layers with total thickness approximately 3 μm were grown on a GaAs substrate using a Veeco Gen-II Molecular Beam Epitaxy (MBE) system. A thin cap layer of GaAs was grown on top to protect the AlGaAs surface from oxidation. First, a ridge pattern with periodic beam widths of 6, 8 and 10 μm was lithographically transferred on to the sample. The sample was then dry etched using a Br-IBAE (Br-Ion Beam Assisted Etching) system [6] to form AlGaAs ridges, as illustrated in figure 4(a). The sample was etched at 100°C for 70 mins at an etch rate of 100 nm/min. Straight smooth sidewalls were observed, as shown below in figure 7(a) for a cross section of the device. The total etch depth around 7 μm ensures the exposure of GaAs for the subsequent wet etch based undercutting below the AlGaAs. After second photolithography, pattern as shown in figure 4(b) was transferred on the sample. A Peroxide/Alkaline system with volume ratio 30:1 of 30% H_2O_2 and 29% NH_4OH that exhibits high etching selectivity towards GaAs [7] was used as the wet etchant. The sample was immersed in the etchant for approximately 4 mins and suspended beams were formed as illustrated in figure 4(c). The ammonium hydroxide leads to oxide formation on the AlGaAs surface, which was easily removed by further treatment with HCl and DI Water.

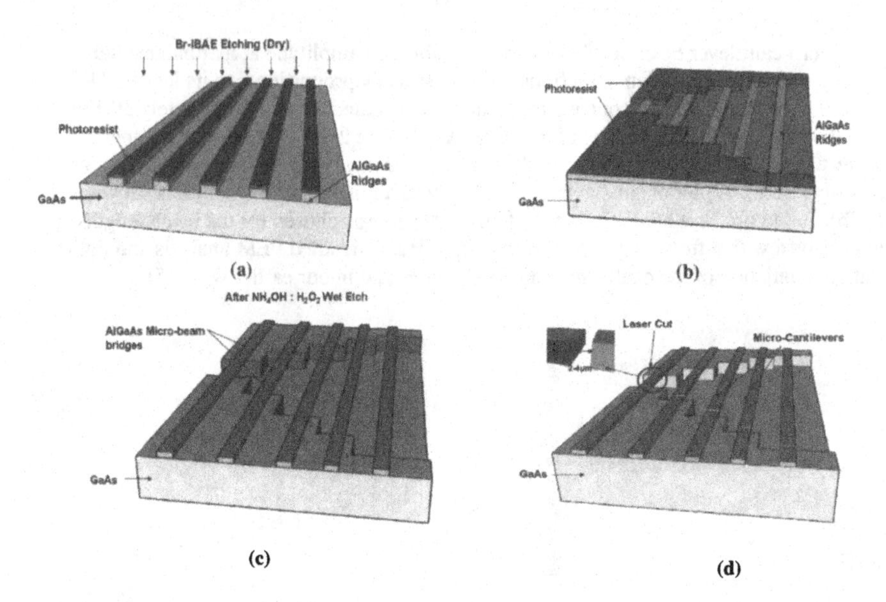

Figure 4. The fabrication process a) Waveguide ridges after Br-IBAE dry etching for 60 mins; b) Second photoresist pattern transferred covering the ridges; c) Waveguide ridges selectively undercut to form bridges of varying lengths; d) Free moving cantilevers formed by laser ablation at the edges.

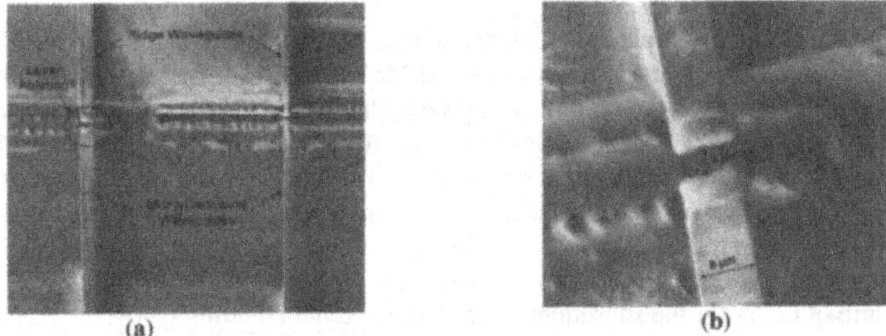

Figure 5. SEM images of the device a) Micro-bridges after laser ablation; b) Close up of an 8μm width beam showing a laser ablated spacing of 4μm.

Finally, to form free ends of cantilevers from the suspended beams, a high energy 193nm Excimer pulsed laser was employed. About 20 pulses with energy densities of the order of 1.5-1.8 J/cm^2 are impinged at the edge of a micro-bridge structure to ablate the material and form micro-cantilevers. A spacing of approximately 4 μm was observed between the cantilevers. Figure 5 shows an SEM image of the micro-beams after laser ablation.

DEVICE CHARACTERIZATION

The schematic of the experimental setup to characterize the waveguides is as shown in figure 6. Light emitting from a 785nm fiber pigtailed laser was coupled into the sample using a microscope objective. A spot size diameter of about 300 microns was estimated here using gaussian beam propagation calculations for the given wavelength. The sample was mounted on a submicron resolution XYZ stage for optical alignment. In the preliminary wave guiding test, a CCD camera was positioned at the output end to calibrate the optical coupling and the device alignment. Note,these tests were conducted on uncut, micro-bridge structures.

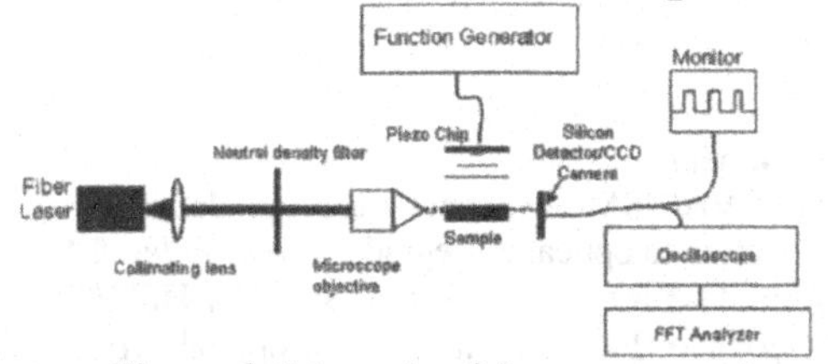

Figure 6. Schematic of the experimental setup

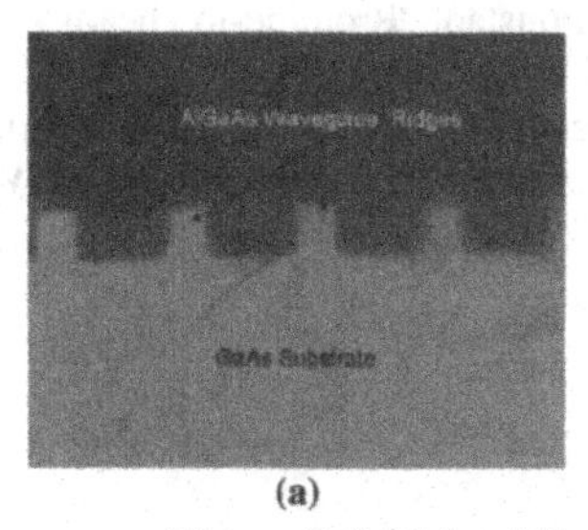

(a)

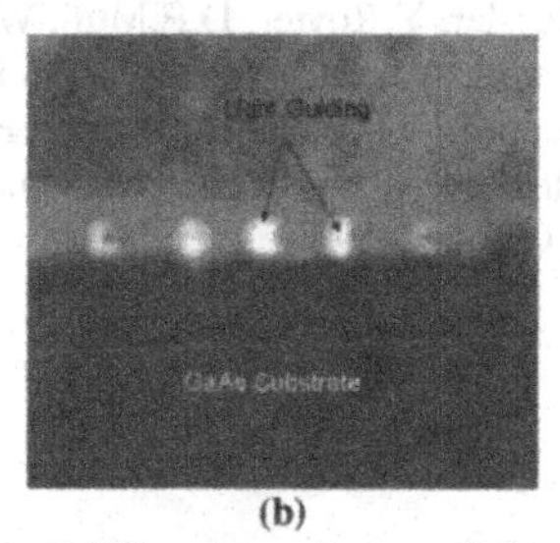

(b)

Figure 7. Initial guiding tests on the device a) Microscope image of the sample cross-section; b) Wave guiding observed using a CCD camera at the output end

Figure 7 shows the images of the sample output plane before (a) and after (b) switching on the laser source. The sample was carefully aligned by tuning the XYZ stage till the maximal output power was observed at the ouput beam. Note that the spacing between the beams was 50 μm and the light is seen to couple into at least 4 ridges. From this the diameter of the focused beam at the input can be estimated to be 200-300μm, consistent with our calculation.

Currently work is progressing to fully characterize the laser-cut devices using an acoustically coupled piezo-electric driver as depicted in Fig. 1. In this case the slow CCD camera used in the initial work is being replaced by a high speed silicon detector, and the optical signal modulation intensity from the output waveguide studied as a function of the piezo-actuation frequency.

SUMMARY

A novel AlGaAs/GaAs based MEMS vibration sensor has been designed and an initial process developed for fabricating the devices. FEM studies were conducted to determine the operation range of the device both structurally and optically. The uncut suspended waveguides were characterized and an experimental setup was described for future full vibration testing of the devices.

ACKNOWLEDGMENTS

The authors would like to thank Jie Fu and Douglas Pulfer of JPS Associates Inc. for laser cutting the samples. We would also like to thank Abhishek Kumar of Center For Advanced Materials, UMass Lowell for taking the SEM images of the samples.

REFERENCES

1. http://www.analog.com/imems/
2. http://www.ioffe.rssi.ru/SVA/NSM/Semicond/ AlGaAs/index.htm
3. M.J.Adams "An introduction to optical waveguides" *John Wiley & Sons*
4. Y.Warren "Roark's formulas for stress and strain" McGraw-Hill Education
5. V.Mathur, J.Li,W.D.Goodhue "FEM simulation of an optical-MEMS sensor" in proceedings of *COMSOL users conference(2006),Boston*
6. J.M.Rossler, Y.Royter, D.E.Mull, W.D.Goodhue, C.G.Fonstad "Bromine-Ion beam assisted etching of InP and GaAs" *J. Vac. Sci. Technol.* **B 16N30**,May/June (1998)
7. Yuji Uenishi, Hidenao Tanaka and Hiroo Ukita "Characterization of AlGaAs microstructure fabricated by AlGaAs/GaAs micromachining" *IEEE transactions on electron devices*, **Vol.41**, No.**10**,October (1994)

MEMS Devices II

Mater. Res. Soc. Symp. Proc. Vol. 1052 © 2008 Materials Research Society 1052-DD05-03

Relative Resistance Chemical Sensors Built on Microhotplate Platforms

Joshua L. Hertz, Christopher B. Montgomery, David L. Lahr, and Steve Semancik
Chemical Science and Technology Laboratory, National Institute of Standards and Technology,
100 Bureau Dr., MS 8362, Gaithersburg, MD, 20899

ABSTRACT

The selectivity, sensitivity, and speed of metal oxide conductometric chemical sensors can be improved by integrating them onto micromachined, thermally-controlled platforms (i.e., microhotplates). The improvements largely arise from the richness of signal inherent in arrays of multiple sensing materials and the ability to rapidly pulse and collect data at multiple temperatures. Unfortunately, like their macroscopic counterparts, these sensors can suffer from a lack of repeatability from sample-to-sample and even run-to-run. Here we report on a method to reduce signal drift and increase repeatability that is easily integrated with microhotplate chemical sensors. The method involves passivating one of a pair of identically-formed sensors by coating it with a highly electrically resistive and chemically impermeable film. Relative resistance measurements between the active and passive members of a pair then provide a signal that is reasonably constant over time despite electrical, thermal and gas flow rate fluctuations. Common modes of signal drift, such as microstructural changes within the sensing film, are also removed. The method is demonstrated using SnO_2 and TiO_2 microhotplate gas sensors, with a thin Al_2O_3 film forming the passivation layer. It is shown that methanol and acetone at concentrations of 1 μmol/mol, and possibly lower, are sensed with high reproducibility.

INTRODUCTION

Chemical sensor arrays are desired as inexpensive means to detect a wide range of gaseous analytes. They have the potential to fill vital roles in homeland security, industrial process control, personal safety, and other application areas. At NIST, we have been researching sensors built on micromachined, thermally controlled platform ("microhotplate") arrays [1,2]. These platforms consist of a silicon oxide membrane that has been grown on a silicon substrate and then, to a large extent, thermally isolated by micromachining. Buried within the membrane is a polysilicon line used for both heating and temperature measurement. On top of the membrane are electrodes for contact to a chemiresistive sensor film. Typically, the film is a metal oxide semiconductor. The purpose of the heater and the high thermal isolation are to enable very rapid, low-power temperature control with good temperature uniformity. Rapidly pulsing the temperature has been shown to create a rich data stream and thus allow differentiation amongst a wide range of analytes, despite a sensor's broad sensitivity [3].

Generally, sensor arrays must first be "trained" by recording the sensor output upon exposure to a number of relevant analytes and, possibly, interferences. In use, the sensor output is then compared against the recorded training measurements to determine the presence or absence of various analytes. The ability to match the measurements taken during use with those taken during training thus relies on the sensor output being relatively invariant over time, given the same chemical environment. Temporal variations, or drift, in the sensor output can come from a few sources, namely: irreversible adsorption of chemical species onto the sensor ("chemical drift"), structural evolution in the active element in the sensor ("primary physical drift"), evolution in electrical contacts and other support elements in the sensor ("secondary physical drift"), variations in flow rates and other environmental factors ("environmental drift"), and measurement noise. Manufacturing variations are an additional component of drift if trying to use training data collected on one sensor with other, nominally identically produced sensors.

A number of different techniques have been reported to correct for sensor drift [4-10]. Typically, these rely on modification of data based on previously measured drift; essentially, drift is included as a parameter during the training. Often, the methods include periodic recalibration using exposure to a known test gas. The weakness in all of these methods is that they rely on extrapolation from previous values and thus cannot adjust in real time.

Here, we describe a method for real-time correction of drift in chemical sensors using a relative resistance measurement between a sensor and an identical but chemically insensitive replica. The components of drift which may be addressed by this approach include all but the chemical drift. The sensors were built on microhotplate arrays using SnO_2 and TiO_2 active elements. For each sensor, a replica was made on the device and desensitized by coating with an Al_2O_3 film. Combining the high signal-to-noise of relative resistance sensing with the analytical richness of data that is achieved with rapid thermal programmed sensing yields a high performance chemical sensing system.

EXPERIMENT

SnO_2 and TiO_2 films were grown in a vacuum chamber using the thermal self-lithographic technique. This technique uses the thermal decomposition of chemical vapor precursors on specifically activated microhotplates to provide local control of film growth [1]. Multiple, nominally identical films of SnO_2 were grown at 375 °C using a $Sn(NO_3)_4$ precursor and growth time of 20 s. Similarly, films of TiO_2 were grown at 500 °C using a $Ti(OCH(CH_3)_2)_4$ precursor and growth time of 120 s. The films were all grown on a single 16-element microhotplate array. A roughly 100 nm thick Al_2O_3 film was then sputter deposited on eight of the sensors to form a passivation layer. A shadow mask prevented deposition on the other eight sensors. X-ray photoelectron spectroscopy (XPS) verified that the Al_2O_3 completely coated the desired sensors and was not present on the active elements.

Relative resistance sensor measurements were made using a simple voltage divider, with a constant voltage placed across the active sensor and the coated sensor ("compensator") in series. A diagram of the measurement setup is given in Figure 1. The recorded sensor output was the voltage drop measured across the compensator, V_c, divided by the total source voltage, V_{tot}, (which was highly stable, yet measured periodically). Thus, the sensor output, V', was

$$V' = \frac{V_c}{V_{tot}} = \frac{R_c}{R_c + R_s} \tag{1}$$

where R_c and R_s are the resistances of the compensator and the sensor, respectively. In a first order model for drift, $R^{drift} = \alpha \cdot R^0$, where R^0 is the original resistance, α is a drift coefficient, and R^{drift} is the resistance after drift. If R_c and R_s drift as in this model, and α is the same for the compensator and sensor, then according to equation (1), V' is unchanged. Thus, components of drift—to the extent that they equally and linearly affect the compensator and sensor—are canceled using this approach. This may include components of primary physical drift, such as recrystallization of the sensor film; secondary physical drift, such as aging of the embedded heater; manufacturing drift, such as film thickness variations; and environmental drift, such as variations in absolute pressure. Note that chemical drift cannot be canceled by this technique, since the Al_2O_3 covering blocks the irreversible adsorption of species onto the compensator.

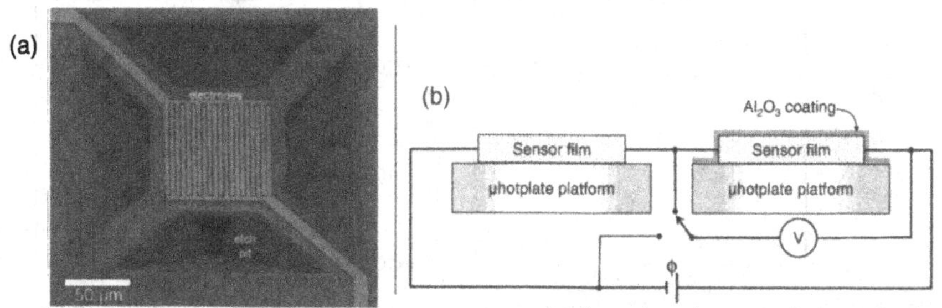

Figure 1. The relative resistance measurement system. (a) Scanning electron micrograph of a microhotplate sensor (before sensor film deposition). (b) Schematic diagram of measurement.

The device was then placed into a flow system with a constant total flow rate of 1 standard L/min for sensing measurements. Lab generated zero-grade dry air was the background ambient. Methanol or acetone was added to this background using commercially supplied gas cylinders containing a dilution of one of the analytes in dry air. Gas from the cylinder was further diluted to achieve concentrations between 1 μmol/mol – 20 μmol/mol. The temperature of a sensor and corresponding compensator were matched at all times. During measurement in a particular gas environment, the temperature profile was either a ramp (e.g., 50°C → 450 °C) or a cycle (e.g., 50°C → 450 °C → 50 °C) with a data point collected at every 1 °C. Heating and cooling rates during measurement were 10 °C/s (platform thermal time constants ≈ 3 ms).

RESULTS AND DISCUSSION

The resistances of active sensors and passivated compensators were first measured in a few gas environments, using linear temperature ramps between 50 °C and 450 °C, in order to test the chemical impermeability of the Al_2O_3 film. Measurements in dry air, 5 μmol/mol methanol and 20 μmol/mol methanol for sensors and compensators of SnO_2 and TiO_2 are shown in Figure 2. It can be seen that, whereas the resistances of the sensors are dependent on the measurement environment, the compensators are insensitive to the methanol presence. Thus, the sputtered Al_2O_3 films were suitable for passivation of sensors.

Figure 2 also shows that the resistances of the compensators were less than those of the sensors (this was the case for all 8 compensator/sensor combinations on the device). This is believed to have occurred due to a slight oxygen deficiency in the Al_2O_3 film. Thus, the SnO_2 and TiO_2 films buried underneath the Al_2O_3 films in the compensators are slightly reduced on the surface and are, in effect, sensing the Al_2O_3. Nevertheless, because the temperature coefficients of resistance are similar for each sensor and its corresponding compensator, V' should remain fairly flat across all temperatures. A flat response in dry air would indicate that any subsequent shape to the curve arises from a particular analyte species, with minimal contribution from background thermally generated carriers. Another disadvantage of traditional sensing is demonstrated in Figure 2d, namely that the resistance of the TiO_2 sensor is extremely high and thus difficult to measure below 300 °C. This problem may be circumvented in relative resistance mode, since the measurement is sensitive not to the absolute magnitude of the resistance, but rather to the ratio of the sensor and compensator resistances.

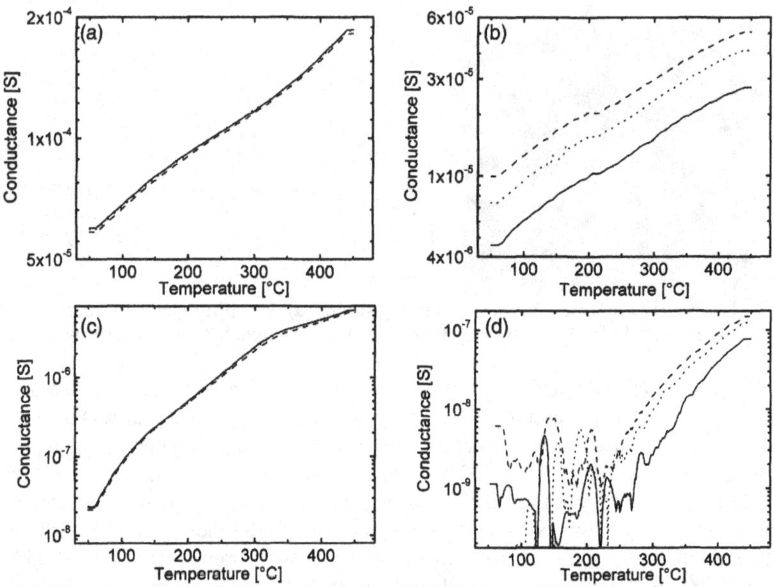

Figure 2. Resistance of a (a) SnO_2 compensator, (b) SnO_2 sensor, (c) TiO_2 compensator, and (d) TiO_2 sensor, measured during a 50°C → 450 °C ramp. Measurement ambient is: dry air (solid line), 5 μmol/mol methanol (dotted line) or 20 μmol/mol methanol (dashed line). In subplots (a) and (c), the three lines are overlapping and difficult to distinguish.

Relative resistance measurements made using SnO_2 and TiO_2 in a few gas environments at temperatures cycled between 50 °C and 450 °C are shown in Figure 3. For SnO_2, the separation between the measurements in dry air, 20 μmol/mol methanol and 20 μmol/mol acetone exceed 3 standard deviations. TiO_2 has good performance, though seemingly less sensitive then the SnO_2. The observed increase in V' due to the presence of methanol or acetone is due to the fact that SnO_2 and TiO_2 are n-type semiconductors and thus have reduced resistance in a chemically reducing environment. As seen in equation (1), reduced R_s leads to increased V'.

Clearly visible in figure 3 is that V', even in dry air, has a slight temperature-dependence. The reason for the non-flat response likely arises from surface oxygen stoichiometry changes of the sensor films in dry air vs. those coated with the Al_2O_3 film. Ways to mitigate this unwanted effect will be examined in the future. Nevertheless, there is a change in the shape of the V' curve that depends on the analyte present. For the TiO_2, the difference is subtle: the 'light-off temperature' occurs earlier for acetone compared to methanol. On the other hand, with SnO_2, there is a noticeable bump in the signal during measurements in methanol near 350 °C on the decreasing temperature part of the cycle. This feature is not found during measurements in dry air or acetone, and so is likely due to desorption or ionization of a methanol-related species.

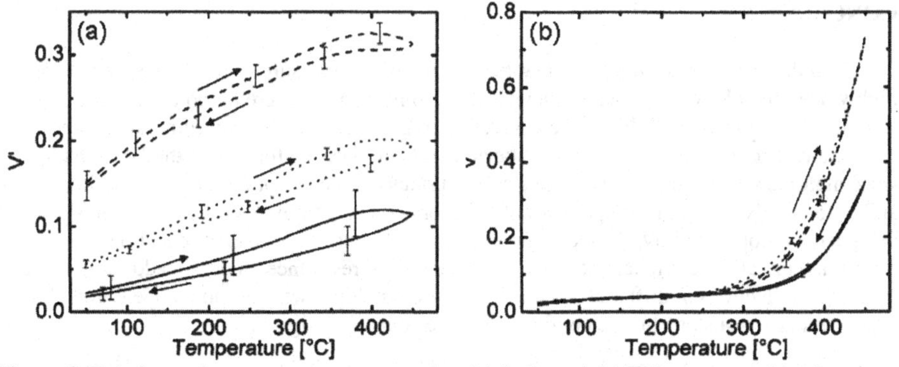

Figure 3. Relative resistance measurements using (a) SnO_2 and (b) TiO_2, measured in: dry air (solid line), 20 µmol/mol acetone (dotted line) and 20 µmol/mol methanol (dashed line). Arrows show direction of temperature. Lines represent an average of ~5 runs; error bars indicate ± 3σ.

In order to examine this feature more closely, repeated measurements were made using a SnO_2 sensor/compensator in dry air and methanol, with temperature cycled between 300 °C and 500 °C. These measurements are shown in Figure 4. The measurements were repeated over the course of a few hours, including multiple cycles of methanol presentation and removal. The shape of the curve changes from nearly linear in dry air to noticeably concave-down in methanol. Since this curved feature was not found previously for acetone, this is a promising method for differentiation between analytes. It can also be seen that the measurements are repeatable, with very small standard deviations and with the final measurement in dry air very close to the dry air measurements from the beginning of testing.

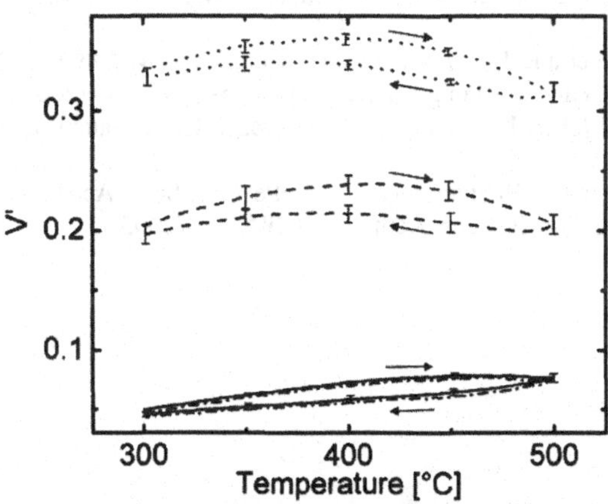

Figure 4. Averaged raw sensor data, measured using relative resistance mode, of SnO_2, measured in: dry air (solid line), 1 µmol/mol methanol (dashed line) and 10 µmol/mol methanol (dotted line). The final measurement, after returning the ambient to dry air, is the dash-dot line. Measurement temperature was cycled between 300 °C and 500 °C; arrows indicate direction of measurement. Lines represent average of ~8 runs; error bars indicate ± 3σ.

CONCLUSIONS

In this work, we have shown that relative resistance measurements can be used to achieve highly reproducible sensing measurements. Thin films of Al_2O_3 were found to have the requisite chemical impermeability to block the interaction of gas analytes with thin film gas sensors. Microfabricated hotplates were found to be an excellent platform for this method of sensing, since multiple sensors and compensators can be produced in close proximity. In addition, the ability to closely match the temperature of a sensor and compensator is easily accomplished using a microhotplate array. Work is ongoing to directly measure the drift reduction when operating in relative resistance mode vs. the more typical resistance mode. In addition, methods for further reductions in drift and noise by employing multiple sensors and compensators, for example in a Wheatstone bridge configuration, are being investigated.

ACKNOWLEDGMENTS

The authors gratefully acknowledge financial support from the National Research Council postdoctoral program.

REFERENCES

1. R.E. Cavicchi, J.S. Suehle, K.G. Kreider, B.L. Shomaker, J.A. Small, M. Gaitan and P. Chaparala, Appl. Phys. Lett. **66**, 812 (1995).
2. S. Semancik and R. Cavicchi, Acc. Chem. Res. **31**, 279 (1998).
3. T.A. Kunt, T.J. McAvoy, R.E. Cavicchi and S. Semancik, Sens. Actuators B **53**, 24 (1998).
4. T. Artursson, T. Eklöv, I. Lundström, P. Mårtensson, M. Sjöström and M. Holmberg, J. Chemom. **14**, 711 (2000).
5. S. Holmin, C. Krantz-Rülcker, I. Lundström and F. Winquist, Meas. Sci. Technol. **12**, 1348 (2001).
6. O. Tomic, H. Ulmer and J.-E. Haugen, Anal. Chim. Acta **472**, 99 (2002).
7. D. Hui, L. Jun-hua and S. Zhong-ru, Sens. Actuators B **96**, 354 (2003).
8. M. Paniagua, E. Llobet, J. Brezmes, X. Vilanova, X. Correig and E.L. Hines, Electron. Lett. **39**, 40 (2003).
9. M. Zuppa, C. Distante, P. Siciliano and K.C. Persaud, Sens. Actuators B **98**, 305 (2004).
10. B.C. Sisk and N.S. Lewis, Sens. Actuators B **104**, 249 (2005).

Mater. Res. Soc. Symp. Proc. Vol. 1052 © 2008 Materials Research Society 1052-DD05-10

Novel Differential Surface Stress Sensor for Detection of Chemical and Biological Species

Kyungho Kang, and Pranav Shrotriya
Mechanical Engineering Department, Iowa State University, 2025 Black Engineering Building, Ames, IA, 50011

ABSTRACT

A miniature sensor consisting of two adjacent micromachined cantilevers (a sensing/reference pair) is developed for detection of chemical and biological species. A novel interferometric technique is utilized to measure the differential bending of sensing cantilever with respect to reference. Presence of species is detected by measuring the differential surface stress associated with adsorption/absorption of chemical species on sensing cantilever. Surface stress associated with formation of alkanethiol self-assembled monolayers (SAMs) on the sensing cantilever is measured to characterize the sensor performance.

INTRODUCTION

Microcantilever based sensors are increasingly being investigated to detect the presence of chemical and biological species in both gas and liquid environments. Thundat et. al. [1] reported the static deflection of microcantilevers due to changes in relative humidity and thermal heating and thus opened a myriad of possibilities for the use of AFM cantilever deflection technique for chemical and biological sensing. Cantilever based sensors have been successfully demonstrated for DNA sequence recognition and as electrical noses to detect chemical mixtures [2]. In majority of the current state of art sensors, molecule absorption induced surface stress change is inferred from the deflection of a single or multiple laser beams reflected from the sensing surface. A large optical path is required between sensitized surface and position sensitive detectors to achieve high sensitivity in surface stress measurement. As a result, it is difficult to implement the sensing scheme into a single micro-fabricated device. In the current paper, we report a novel differential surface stress sensor that utilizes a single-mode fiber based Mach-Zehnder interferometer for measuring cantilever deflection and consequently, the detection of chemical and biological species. The interferometric technique is amenable to miniaturization and may facilitate the integration of all components of sensors into a single microfabricated chip.

Measurement of surface stress associated with formation of alkanethiol self assembled monolayer (SAM) on a gold surface is utilized to characterize the performance of differential surface stress sensor. Berger et. al. [3] reported the generation of compressive stresses on the order of 0.1-0.5 N/m during the formation of alkanethiol self-assembled monolayer on the cantilever's surface and also reported that the magnitude of surface stress increased linearly with the carbon chain backbone of the monolayer. Since the first report by Berger et. al. [3], SAMs have been used as test system for almost all cantilever based sensing techniques [4-6]. This is because they are relatively easy to prepare, form well-ordered close packed films and offers limitless possibilities of variations in chain length, end group and ligand attachments [7]. One of the commonly studied SAMs are alkanethiol SAMs ($HS-(CH_2)_{n-1}CH_3$) in which n is the number of carbon atoms in the alkyl chain. Godin et. al. [4] have shown that the kinetics of formation of self-assembled monolayers on gold-coated cantilevers and the resulting structure are dependent on the microstructure of the gold film and also the rate at which the SAM reaches the surface.

SENSING PRINCIPLE

The differential surface stress sensor consists of two adjacent cantilevers, a sensing/reference pair, where only the sensor surface is activated for adsorption of chemical or biological molecules (Figure 1 A). Absorption/adsorption of analyte species on the sensitized surface is expected to induce differential bending and deflection between the sensing and reference cantilevers. The microcantilevers and a pair of microlens arrays are arranged in the optical arrangement shown schematically in Figure 1B to measure the differential displacement between sensing and reference cantilevers. In this optical configuration, incident laser beam at points A and C always arrives to points B and D, respectively, regardless of their incident angle and differential bending produces a change in path length difference between the beams reflected from the two cantilevers. .

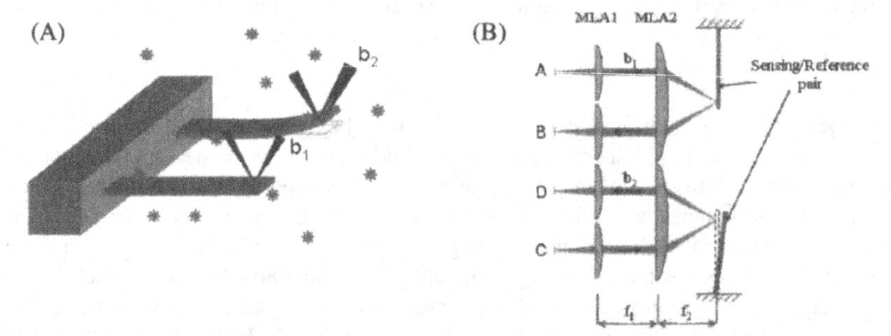

Figure 1. Schematic illustration of the sensing principle. (A) Conceptual view of alkanethiol molecule absorption on the gold surface of sensing/reference cantilevers and the consequent differential bending. (B) Principle of differential surface stress measurement.

After reflecting from the sensing and reference surfaces, the two beams accumulate a path length difference, l, equal to twice the differential displacement between sensing and reference surface. The beams are interfered to measure the path length difference and differential surface stress ($\Delta\sigma$) between the two cantilevers is determined using Stoney's formula [8].

$$\Delta\sigma = \left(\frac{E}{3(1-\nu)}\right)\left(\frac{t}{L}\right)^2 l$$

where, E, is the elastic modulus and ν is the Poisson's ratio, L is the length and t is thickness of the cantilevers. Measurement of differential surface stress ensures that detected signal is proportional to specific absorption of analyte species on the sensing cantilever and eliminates the influence of environmental disturbances such as nonspecific adsorption, changes in pH, ionic strength, and especially the temperature

EXPERIMENT

Differential surface stress sensor realization

An optical circuit shown in Figure 2(A) is utilized for assembling the surface stress sensor. In the system, two adjacent rectangular-tipless AFM cantilevers were used as a sensing/reference pair. A pair of MLAs (microlens arrays) with lens of 240 μm and 900 μm diameters and pitches of 250 μm and 1mm respectively were used to direct the beams towards the sensing/reference pair. Motorized and manual actuators were used to assist in aligning of MLAs with respect to the sensing/reference cantilevers. A bi-directional coupler was applied to split the beam from a 635nm fiber coupled laser source and delivered to MLA1 at 50/50 ratio. Other two reflected beams were interfered using the second bi-directional coupler and intensity of interfered beam was monitored using photodetectors.

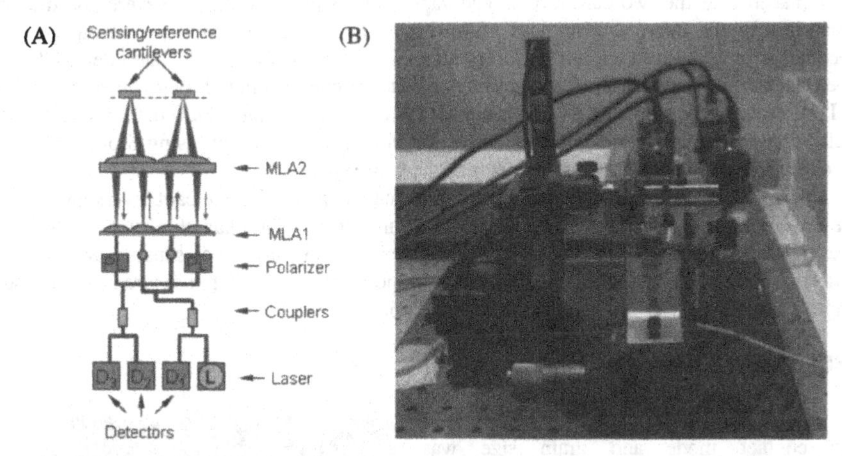

Figure 2. Optical circuit of differential surface stress sensor (A) and Photograph of experimental setup (B).

The polarization plane of the reflected beams was matched and common mode rejection was utilized to ensure maximum fringe visibility in the inferred beams. An isolation box covered all fiber couplers as well as sensor components to eliminate acoustic and vibrational noise from the system.

Cantilever

Silicon cantilevers used in the sensor realization are 480 μm long, 80 μm wide, and 1 μm thick with a top side coating of 5nm titanium and 30nm gold film. (Nanoworld, Switzerland). AFM cantilevers are batch produced with large variation of dimensions and mechanical properties from the manufacture's quote [9, 10]. In order to accurately measure surface stress development, the thickness of each cantilever is calculated based on the experimentally measured spring constant with the material constants [9]. Microstructure and surface roughness of the gold film were determined using contact mode atomic force microscope imaging.

Experimental procedure

Liquid octanethiol [$CH_3(CH_2)_7SH$] was selected as alkanethiol solution and purchased from Sigma-Aldrich. All the AFM cantilevers were cleaned by immersing for 30 minutes in piranha solution (70% H_2SO_4, and 30% H_2O_2 by volume) and were then rinsed in deionized water and dried in the gentle N2 flow. Only the reference cantilevers were incubated in pure octanethiol solution for 12 hours to ensure the formation of a self assembled monolayer (SAM) on the gold film. Formation of a stable SAM on the reference cantilever ensures that alkanethiol molecules are only absorbed on the sensing cantilever during subsequent experiments.

Surface stress development associated with alkanethiol SAM formation was measured in three steps. In the first step, reference and sensing cantilever were mounted in the sensor and stability of the interferometer was first checked to ensure that measured signal is not affected by drift and ambient noise. In the second step, 20 mL of pure liquid octanethiol was injected into a beaker placed near the two cantilevers. The vapors of alkanethiol solutions were confined near the cantilevers and interferometer was utilized to measure the deflection of sensing cantilever associated with deposition and formation of alkanethiol SAMs. Intensity of the interfered beams as well as back reflection from the first coupler were monitored through photodetectors (D2, D3 and D1, respectively in Figure 2(A)) and a data acquisition system. Differential surface stress which is proportional to the cantilever deflection is then calculated by using Stoney's Formula with obtained spring constant and geometry of the cantilever.

After the exposure to alkanethiol, both the sensing and reference cantilevers are expected to be covered with alkanethiol SAM; therefore, reintroduction of alkanethiol vapors should not cause further differential bending of the cantilevers. In the last step, sensing and reference cantilevers were again exposed to alkanethiol vapors to ensure that measured surface stress change is associated with only alkanethiol formation.

RESULTS AND DISCUSSION

Gold film on the cantilever was imaged using contact mode and grain size was determined to be 40 ± 10 nm (Figure 3). The mean square roughness of the gold surface was 2.07 ±0.23 nm for the 750 nm scan size. The stiffness of the cantilever were found to be in the range of 0.16-0.18 N/m resulting in a calculated thickness of approximately 1.7- 1.8 μm.

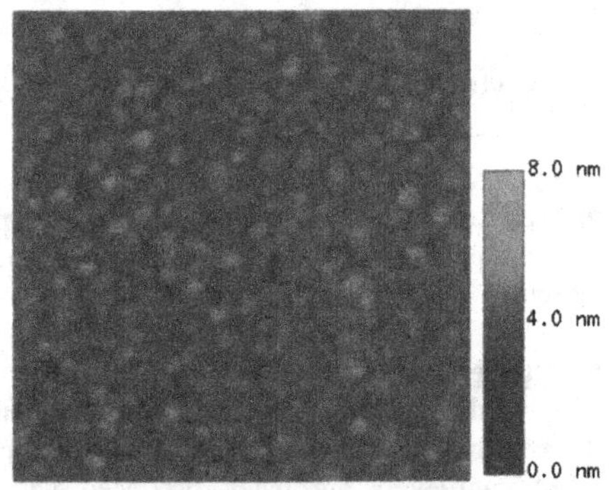

Figure 3. AFM image (750nm×750nm) of gold film microstructure on AFM cantilevers

Experimental measurements of surface stress induced due to vapor phase deposition of alkanethiol during a typical run are plotted in Figures 5(a), (b) and (c). Intensity of interfered beams monitored before introduction of alkanethiol vapor are plotted in Figure 5(a). Measured intensities are nearly constant and do not drift with time. The differential bending and the corresponding surface stress change during formation of alkanethiol SAMs on the sensing cantilever respect to reference cantilever are shown in Figure 5(B). As soon as alkanethiol solution is injected, the sensing cantilever undergoes tensile surface stress change first and subsequently, as expected compressive surface stress develops on the sensing cantilever.

According to Figure 5(B), alkanethiol SAMs rapidly form in the early stages, 10 minutes after injection, but it took about 50 minutes to complete SAM formation (final saturation). Surface condition of cantilevers such as cleanliness, roughness, and the condition of gold deposition on the cantilever's surface influence on the pattern as well as magnitude of surface stress change. In addition, the distance of cantilever to the location where alkanethiol droplets are introduced was 10 cm away. The second development of surface stress change in Figure 5(B) is due to slow arrival of alkanethiol molecules or slow saturation of closely packed SAM on sensing cantilever. As a result, final surface stress change was 0.28 ± 0.02N/m and the corresponding differential banding was 180 ± 10 nm at grain size of gold surface was 40 ± 10 nm.

After the SAM formation on the sensing cantilever, sensor was again exposed to alkanethiol vapors. Intensity of the interfered beams measured during second exposure of alkanethiol is plotted in Figure 5(c). As shown in the Figure 5 (c), variations of the interfered beam intensities were within the system's normal noise range. A minimal surface stress change during re-introduction of the alkanethiol vapors indicates that both sensing and reference cantilever are covered with alkanethiol SAM. Furthermore, it indicates surface stress change observed during the first introduction is unambiguously associated with SAM formation on sensing cantilever. Previous reports [4] have indicated that distance of cantilever to the location where alkanethiol droplets are introduced, condition of gold surface like cleanliness and roughness, and grain structure of the gold on the cantilever's surface affects the kinetics and magnitude of surface stress development. Among those conditions, the microstructure of gold film significantly influences the development of the surface stress during the formation of Alkanethiol SAMs [4].

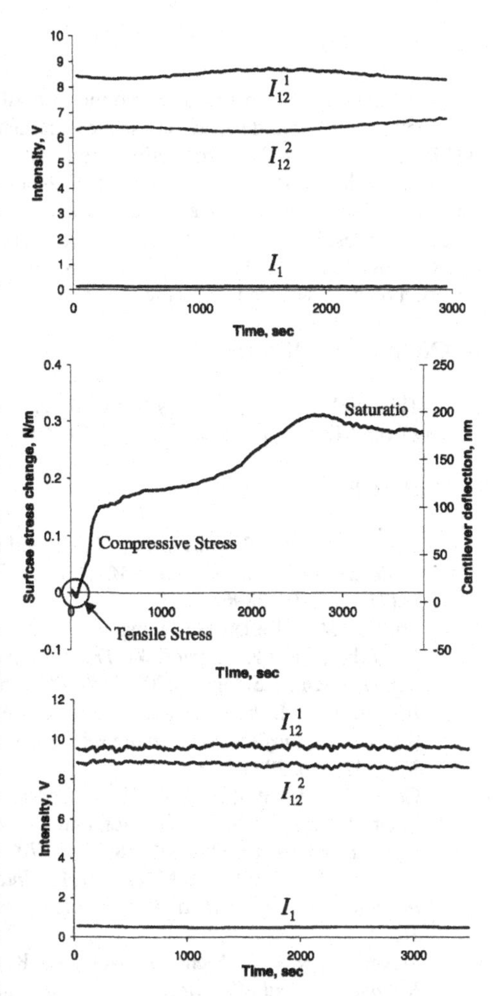

Figure 5. Surface stress change and the corresponding sensing cantilever deflection respect to reference cantilever (A) Intensity of interfered beams before deposition, (B) Differential surface stress during deposition; (C) Inensity of interfered beams due to alkanethiol exposure after deposition.

CONCLUSIONS

A miniature sensor based on two microcantilevers – a sensing and reference pair – is developed for differential surface stress measurement and is explored for detection of chemical and biological species. High resolution interferometry is utilized to measure the differential surface stress developed due to absorption of chemical species on the sensing cantilever. Sensitivity of sensor measurement is not dependent on distance between the sensing surface and detector; as a result, surface stress sensor is amenable for miniaturization and array of sensors would be easily fabricated on a single MEMS device. Surface stress associated with alkanethiol formation on gold surface is measured to characterize the response of the sensor.

ACKNOWLEDGMENTS

This work was supported by National Science Foundation Career Award and Seed Funding from Midwest Forensic Research Center

REFERENCES

1. Thundat, T., R.J. Warmack, G.Y. Chen, and D.P. Allison, *Thermal and Ambient-Induced Deflections of Scanning Force Microscope Cantilevers.* Applied Physics Letters, 1994. **64**(21): p. 2894-2896.
2. Fritz, J., M.K. Baller, H.P. Lang, H. Rothuizen, P. Vettiger, E. Meyer, H.J. Guntherodt, C. Gerber, and J.K. Gimzewski, *Translating biomolecular recognition into nanomechanics.* Science, 2000. **288**(5464): p. 316-318.
3. Berger, R., E. Delamarche, H.P. Lang, C. Gerber, J.K. Gimzewski, E. Meyer, and H.J. Guntherodt, *Surface stress in the self-assembly of alkanethiols on gold.* Science, 1997. **276**(5321): p. 2021-2024.
4. Godin, M., P.J. Williams, V. Tabard-Cossa, O. Laroche, L.Y. Beaulieu, R.B. Lennox, and P. Grutter, *Surface stress, kinetics, and structure of alkanethiol self-assembled monolayers.* Langmuir, 2004. **20**(17): p. 7090-7096.
5. Raiteri, R., H.J. Butt, and M. Grattarola, *Changes in surface stress at the liquid/solid interface measured with a microcantilever.* Electrochimica Acta, 2000. **46**(2-3): p. 157-163.
6. Stevenson, K.A., A. Mehta, P. Sachenko, K.M. Hansen, and T. Thundat, *Nanomechanical effect of enzymatic manipulation of DNA on microcantilever surfaces.* Langmuir, 2002. **18**(23): p. 8732-8736.
7. Ulman, A., *An Introduction to Ultrathin Organic Films: From Langmuir--Blodgett to Self--Assembly* 1991, New York: Academic Press. 442.
8. Stoney, G.G., *The tension of metallic films deposited by electrolysis.* Proceedings of the Royal Society of London: A, 1909. **82**: p. 172-175.
9. Sader, J.E., J.W.M. Chon, and P. Mulvaney, *Calibration of rectangular atomic force microscope cantilevers.* Review of Scientific Instruments, 1999. **70**(10): p. 3967-3969.
10. Sader, J.E. and L. White, *Theoretical-Analysis of the Static Deflection of Plates for Atomic-Force Microscope Applications.* Journal of Applied Physics, 1993. **74**(1): p. 1-9.

Poster Session: MEMS

Mater. Res. Soc. Symp. Proc. Vol. 1052 © 2008 Materials Research Society 1052-DD06-03

Excimer Laser Induced Patterning of PSZT and PLZT Films

Patrick William Leech[1], and Anthony S. Holland[2]
[1]CSIRO, Molecular and Health Technologies, Clayton, Victoria, 3168, Australia
[2]School of Electrical and Computer Engineering, RMIT, Melbourne, Australia

ABSTRACT

The patterning of strontium-doped lead zirconium titanate (PSZT) and lanthanum-doped lead zirconium titanate (PLZT) films has been examined using excimer laser radiation. Both types of film were deposited by rf magnetron sputtering using in-situ heating and a controlled cooling rate in order to obtain the perovskite-oriented phase. The depth of laser ablation in both PSZT and PLZT films showed a logarithmic dependence on fluence. The threshold fluence required to initiate ablation was ~1.25 mJ/cm^2 for PLZT and ~1.87 mJ/cm^2 for PSZT films. The ablation rate of PLZT films was slightly higher than that of PSZT films over the range of fluence (10-150 J/cm^2) and increased linearly with number of pulses. The higher ablation rate of PLZT films has been attributed to the finer grain size (160-200 nm) than in the PSZT films (1.0-1.2 μm).

INTRODUCTION

Lead zirconium titanate (PZT) has become an important material in the fabrication of electromechanical devices such as bio-sensing cantilevers and sensors/ actuators. In these applications, the exceptional piezoelectric properties of PZT have been coupled with structural elements of silicon. A further significant enhancement in the piezoelectric properties of PZT has been reported by doping with strontium [1]. In particular, the growth of PSZT films in the perovskite orientation to obtain these enhanced piezoelectric properties has recently been demonstrated using magnetron sputter deposition on Si [2]. A similar sputter technique has also been used to deposit the perovskite phase in films of lanthanum doped PZT [3]. PLZT in a perovskite orientation has been characterized by a high electro-optic coefficient with the potential for application in a range of optoelectronic and optical devices [4].

The realization of device structures in PZT has required the development of techniques of lithographic patterning. The etching of PZT has been reported by several methods which have included wet chemical etching, reactive ion etching (RIE) and ion beam etching (IBE) as summarized in ref [5]. The wet chemical etching of PZT has been restricted to coarser patterns while the definition of fine structures using IBE and RIE has been characterised by a low etch rates. An alternative technique of micromachining bulk PZT by excimer laser radiation was initially reported by Eyett et.al. [6] and further demonstrated by Konishi [7] and Desbiens [8]. These studies of laser ablation were confined to the patterning of bulk PZT. However, the use of KrF laser applied to the direct patterning of PZT-based films offers a simple process compared with the multi-step methods associated with RIE and IBE. In this paper, we examine for the first time the patterning by excimer laser of Sr and La-doped PZT films comprising a perovskite structure. The use of single and multiple pulse experiments has enabled an examination of the progressive etching of the layers. The processing of the films has been determined as a function of laser fluence.

EXPERIMENTAL DETAILS

Films of $(Pb_{0.92}Sr_{0.08})(Zr_{0.65}Ti_{0.35})O_3$ and $(Pb_{0.91}La_{0.09})(Zr_{0.60}Ti_{0.40})O_3$ were deposited by rf magnetron sputtering on substrates of Pt/Ti/(100) silicon. For sputtering of each film, a single target of the oxide composite was used with an atmosphere of either Ar for PLZT or $Ar:O_2$ (9:1) gas for PSZT. In the case of PSZT films, the temperature of deposition (600 °C) and the cooling rate (5 °C/ min) were selected to give a dominant (110) Perovskite-orientation in the as-grown structure [2]. The PLZT films were also obtained in a Perovskite phase by in-situ heating at 700 °C during sputter deposition followed by rapid cooling [3]. The films were ~1 µm thick with an average grain size of 160-200 nm (PLZT) or 1.0-1.2 µm (PSZT). Experiments on ablation of the films were performed using a KrF excimer laser (Optec). The laser was operated at a wavelength of 248 nm with 5 ns pulse width. The beam was focused to dimensions of 30 x 30 µm with examination of fluence in a range from 10 to 150 J/cm^2. A series of experiments was performed as arrays of both single and multiple (1-5) pulses using irradiation in air. An upper limit of 5 pulses was sufficient to ablate through each of the films. The characteristics of the ablated region were measured by scanning probe microscopy (SPM) and scanning electron microscopy.

RESULTS AND DISCUSSION

Figure 1 shows SPM images of an ablated region of a single 30 µm square in the PLZT and PSZT films. The partial ablation through the film thickness has created surfaces with a globular morphology. The average diameter of the globules was ~250-330 nm in PLZT and ~1.4-1.6 µm in PSZT films, a similar dimension to the original grain size. Measurements of root mean square roughness, R_{ms}, on the ablated surface have been plotted versus fluence in Figure 2.

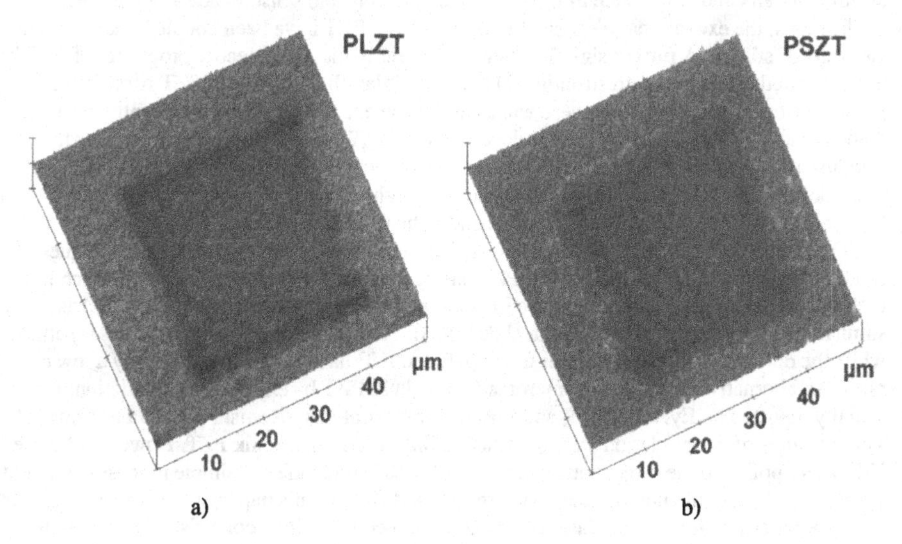

Figure 1. Laser ablated region in a) PLZT and b) PSZT films for single pulse at 100 J/cm^2.

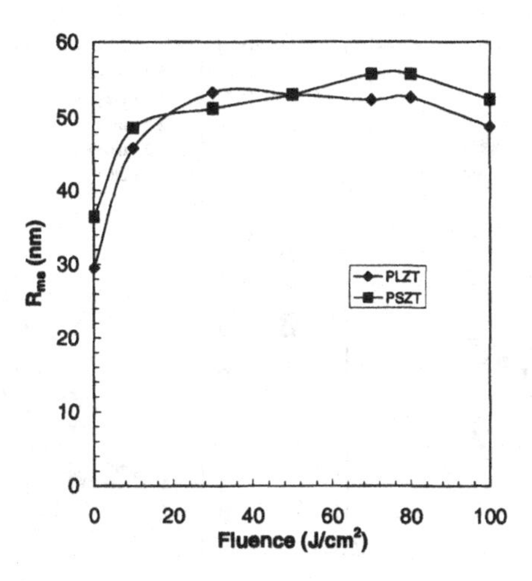

Figure 2. R_{ms} of laser ablated surface as a function of fluence for PLZT and PSZT films.

For the as-deposited surfaces which were plotted as zero fluence, the measured value of R_{ms} was lower for PLZT (~30 nm) than for PSZT (~36 nm) films. Figure 2 shows that after laser ablation, the surface roughness of both films was increased to R_{ms} = 50-55 nm at fluences of $\geq$ 10 J/cm^2. Both the PLZT and PSZT films have shown an approximately constant value of R_{ms} (50-55 nm) in the range of fluence 10-100 J/cm^2.

Figure 3 (a) and (b) shows the average depth of ablation as a function of laser fluence for PLZT and PSZT films, respectively. These plots comprise data from both single and multiple pulse (2-5) experiments with each point representing an average of ~5 measurements. In Figure 3(a) for PLZT films, the etch depth in the single pulse experiments has shown a logarithmic dependence on the incident laser fluence. For single pulse experiments, the threshold fluence for initiation of etching in Figure 3(a) was determined as ~1.25 J/cm^2. The depth of etch for a single pulse in PLZT at a fluence of 125 J/cm^2 was measured as ~175 nm. The series of experiments using multiple pulses has produced the same trend in etch depth versus incident fluence but with a cumulative effect on depth of each additional pulse. The slope of depth versus fluence increased with the number of pulses. Comparison with Figure 3(b) has shown that a similar trend was evident in PSZT films, but with a threshold fluence for etching of ~1.87 J/cm^2. The ablation depth for a single pulse in PSZT at 125 J/cm^2 was measured as ~158 nm.

The plots in Figure 3 have allowed an evaluation of the results in terms of the Beer-Lambert law. According to this law, the optical penetration depth due to light absorption in a material is inversely proportional to the absorption coefficient, α, thereby enabling an estimate of the depth of efficient absorption at a given wavelength. The Beer-Lambert law can be expressed in terms of etch rate, R_e, as

$$R_e = (1/\alpha)\ln(I_{in}) - (1/\alpha)\ln(I_{th}) \quad (1)$$

where I_{in} is the incident laser fluence and I_{th} is the threshold laser fluence or intercept with the horizontal axis.

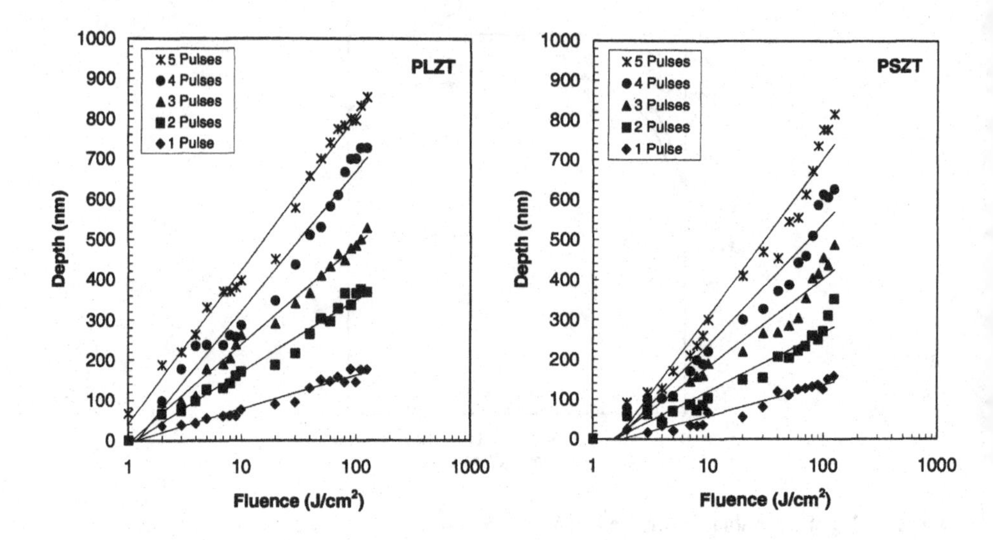

Figure 3. Measurements of etch depth as a function of fluence in a) PLZT and b) PSZT films.

With the absorption coefficient, $\alpha = 30\ \mu m^{-1}$ for PZT at 248 nm [10], the equation has given an etch rate of 154-160 nm/pulse for PLZT and ~140-147 nm/pulse for PSZT films at 125 J/cm^2. These estimates of depth were slightly lower than the experimental measurements at the same fluence. A least squares fit of equation 1 to each of the series of data in Figure 3 has shown a logarithmic dependence on I_{in}. As an indication of the quality of fit of the data, a value of r^2 (square of the correlation coefficient) was calculated for each plot. A summary of r^2 for each plot has been given in Table I. The value of r^2 was calculated in the range 0.93-0.96 for PSZT and 0.96-0.99 for PLZT films. These two ranges of r^2 indicate good and very good fits of the data with equation 1, respectively. The results show that etching has occurs more uniformly in PLZT than PSZT films. The values of I_{th} and etch depth are also included in Table I. Throughout the experiments, the PLZT films have exhibited a higher etch depth and a lower threshold fluence, I_{th} than the PSZT films. The magnitude of I_{th} was essentially constant with the number of pulses.

Table I. r^2, I_{th}, and Etch Depth at 125 J/cm^2 for PLZT and PSZT films

Film	PLZT			PSZT		
	r^2	I_{th} (J/cm^2)	Etch depth (nm)	r^2	I_{th} (J/cm^2)	Etch depth (nm)
1 pulse	0.99	1.25	176	0.96	1.87	158
2 pulses	0.97	1.25	368	0.95	1.62	352
3 pulses	0.99	1.07	528	0.95	1.50	488
4 pulses	0.98	1.25	727	0.92	1.62	627
5 pulses	0.96	0.87	853	0.93	1.62	815

Fig. 4 shows a plot of etch rate of the PLZT and PSZT films versus number of pulses at a fluence of 100 J/cm^2. The etch rate increased near linearly with the number of pulses as indicative of a cumulative process of ablation. In these experiments, a slightly higher etch rate was measured for PLZT than PSZT films, a trend which was also evident at other fluences.

Since both types of film were similar in composition (except for La or Sr dopant elements), the results indicate that the difference in ablation rate was due to the variation in grainsize. The addition of >1 mole % La in PZT has been reported to reduce grain size by inhibiting grain growth [9]. The presence of La and the rapid cooling in PLZT films has evidently resulted in a finer grain size (160-200 nm) than in PSZT films (1.0-1.2 μm). At the high temperatures of laser ablation, PZT has been reported to undergo a thermally induced decomposition into solid ZrO_2 and a liquid phase of PbO and $PbTiO_3$ [10]. The low boiling point of PbO has resulted in the preferential evaporation and depletion of this component in the liquid [10]. Previous studies have also shown morphological changes occurring at the surface during laser irradiation of PZT. The early stage nucleation of globules at the surface was followed by an evolution into elongated cone shapes [11]. The raised regions at the centre of the cones were shown to be deficient in Pb compared with surrounding recessed regions [11]. In the present work, the globular shapes at the erosion surface were typical of the early stage of development of morphology in laser ablated PZT. The larger grain boundary area at the surface of PLZT films has thereby potentially enhanced the loss of PbO and the erosion rate of the surface compared with PLST films.

In Figure 5, the width of ablated squares has been plotted as a function of fluence at the exposed beam width of 30 μm. For both PLZT and PSZT films, the ablation width increased slightly with fluence attained at a measured width of 30 μm over the range ~20-50 J/cm^2. The spread in the measurements of feature width with number of pulses has been attributed to the granular nature of the erosion process. At a given fluence, the width of the ablated squares was essentially independent of the number of laser pulses.

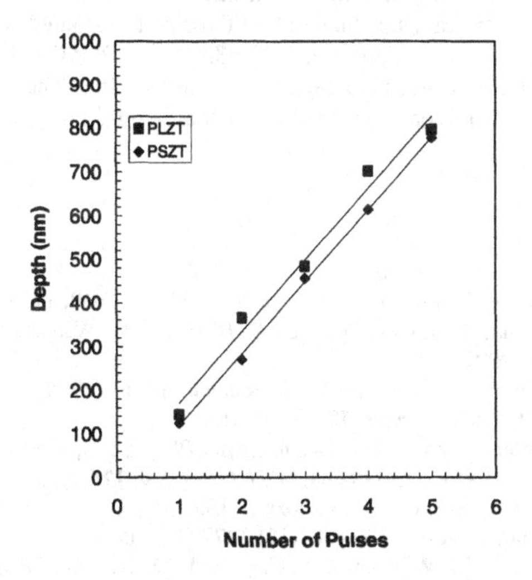

Figure 4. Etch rate of PLZT and PSZT versus number of pulses at a fluence of 100 J/cm^2.

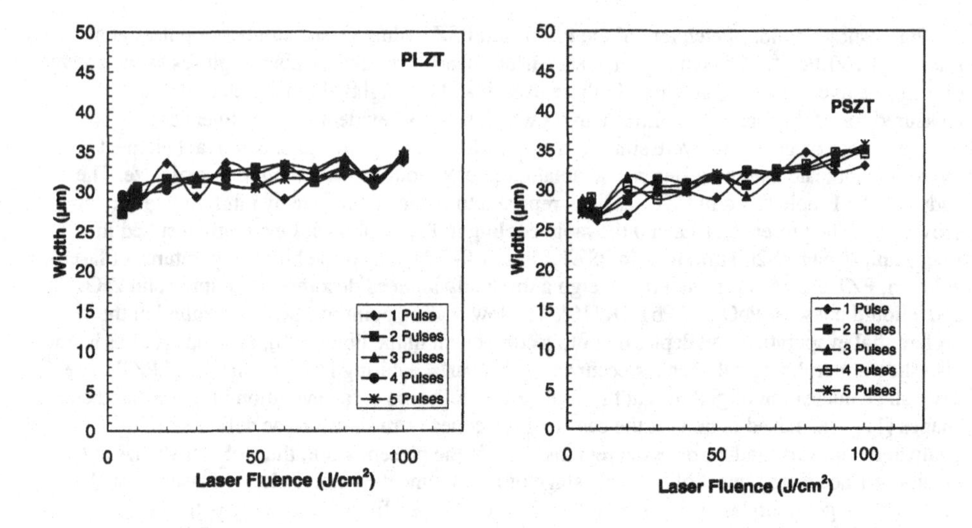

Figure 5. Width of ablation as a function of fluence for a) PLZT and b) PSZT films.

CONCLUSIONS

The ablation depth of PLZT and PSZT films using a KrF laser has shown a logarithimic dependence on fluence. The slightly higher ablation rate and lower threshold fluence in PLZT films has been attributed to a finer grain size than in PSZT films. The ablated surfaces comprised a globular morphology with an average diameter of ~250-330 nm in PLZT and 1.4-1.6 μm in PSZT films. These were similar dimensions to the grain size in the films. The measured R_{ms} of the ablated surfaces increased with fluence until leveling-off at $\geq$30 mJ/cm^2.

REFERENCES

1. E.B. Araujo and J.A. Eiras, *J.Phys.D:Appl.Phys.* **36**, 2010 (2003).
2. S. Sriram, M.Bhaskaran and A.S. Holland, *Semicond. Sci.Technol.* **21**(9), 1236 (2006).
3. S. Kandasamy, M. Ghantasala, A. Holland, Y.X. Li, V. Bliznyuk, W. Wlodarski and A. Mitchell, *Materials Letters*, In Press (2007).
4. B. Tunaboylu, P. Harvey and S.C. Esener, *Integr.Ferroelectr.* **19**, 11 (1998).
5. J. Baborowski, *Journal of Electroceramics,* **12** 23 (2004).
6. M. Eyett, D. Bäuerle, M. Wersing and H. Thomann, *J.Appl.Phys.* **62**(4), 1511 (1987).
7. T. Konoshi, M. Ide, K. Tanaka and S. Sugiyama, *IEEJ Trans. SM*, **126**(7), 302 (2006).
8. J-P. Desbiens and P. Masson, *Sensors and Actuators,* **A 136** 554 (2007).
9. M. Hammer and M. Hoffmann, *J.Am.Cer.Soc.* **81**(12) 3277 (1998),
10. D.W. Zeng, K. Li, K.C. Yung, H.L.W. Chan, C.L. Choy and C.S. Xie, *Appl.Phys.*, **A78** 415 (2004).
11. X.Y. Chen and Z.G. Liu, *Appl.Phys.*, **A 69** S523 (1999).

Mater. Res. Soc. Symp. Proc. Vol. 1052 © 2008 Materials Research Society 1052-DD06-14

Optically Actuated Deformable Micro-Mirrors for Adaptive Optics

Troy Ribaudo[1], Jin Li[1], Bahareh Haji-saeed[2], Jed Khoury[3], and William Goodhue[1]
[1]Physics, University of Massachusetts at Lowell, 1 University Ave., Lowell, MA, 01854
[2]Solid State Scientific Corporation, Hollis, NH, 03049
[3]Air Force Research Laboratory, Hanscom AFB, MA, 01731

ABSTRACT

In this paper we report on our latest efforts to fabricate and characterize optically actuated deformable micro mirrors for wavefront correction in an adaptive optics system. The optically actuated DMM device consists of a Silicon Nitride (Si_3N_4) thin film patterned into a spring plate array, an SU-8 photoresist supporting structure that provides the space through which the mirror is allowed to deform, and a PIN photodiode which allows an optical control signal to actuate the DMM.

INTRODUCTION

MEMS based deformable micro mirrors (DMM's) have been employed in a variety of optical processing systems. They are used for phase correction in adaptive optics, as switches in fiber optics communications and as image production mechanisms in DLP technology.

Adaptive optics technology has found applications in a multitude of fields including astronomy, retinal imaging, and free space optical communications. Wavefront correction is essential to enhance the performance of optical systems and to approach the diffraction limit of geometrical optics. As light propagates through a turbulent medium, wavefront distortion may be introduced. This aberration gives rise to a blurring of the image that can be restored with an adaptive optics system. One of the key components in such a system is a deformable mirror that can deflect under appropriate control and correct the distortions. Currently, the main technologies for addressing membrane mirror light modulation are electron beam addressing, optical addressing, and electrical addressing with integrated circuit technology.

In the group's previous work, optically actuated deformable micro mirrors were fabricated to serve as wavefront correction mechanisms. The use of aluminized Mylar films demonstrated the concept, but large applied voltages were required due to high spring constants [1]. To overcome this problem, spring plate arrays that were patterned in a Si_3N_4 thin film were employed which successfully reduced these voltages [2]. However, the fabrication procedure for this device included a difficult manual alignment step that hindered its repeatability.

In this work, we demonstrate the improved fabrication of a deformable micro mirror that can be optically actuated with low voltage requirements. The layer supporting the Si_3N_4 micro-mirror arrays is replaced with a web of SU-8 photoresist (Microchem) and is deposited directly around the mirrors. SU-8 is a versatile negative photosresist that has been employed in MEMS fabrication procedures as both mechanically actuated [3], and passive components. This change allows for precise alignment, and a stronger, more durable device. In addition, a GaAs PIN photodiode is used for optical actuation as compared with the initial configuration using electron hole pair generation in a semi insulating GaAs substrate with the expectation of improvement in device responsivity. The proof of concept for this design was initiated at reference [4].

DEVICE DESIGN

The deformable micro mirrors consist of a spring plate patterned Si_3N_4 layer (Fig. 1a) that has been metalized for electrical contact and suspended by a dielectric supporting web of SU-8 photoresist over a PIN photodiode. The SU-8 web provides the free space through which the spring plate can deform as well as the insulation that creates the actuating capacitance. The spring plate arrays are aligned on top of a GaAs PIN photodiode and the cross section of the device is illustrated in Fig. 1b.

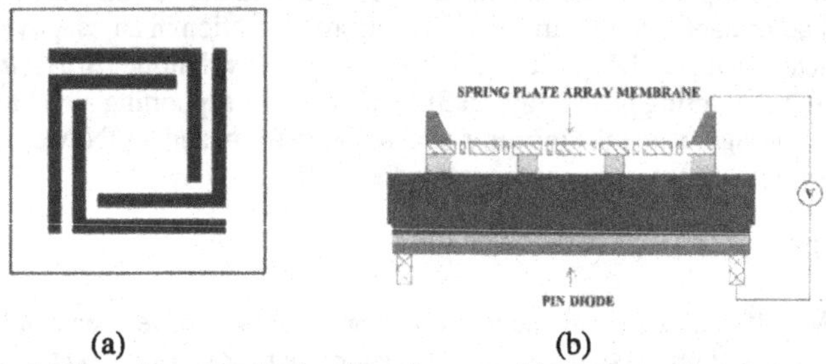

(a) (b)

Figure 1. (a) Top view of a single plate element; (b). Cross sectional view of an optically actuated spring plate array.

The spring plate array and the PIN photodiode work as two impedances in series when biased [5]. One associated with the spring plate array membrane and GaAs substrate interface and the other with the PIN diode. Currently, the diode size is larger than an individual spring plate element, but the goal in the near future is to develop a design where each element has its own diode.

In the device shown in Fig. 1, the application of a voltage across the capacitor like structure results in an electrostatic force that pulls the spring plate downward. The resulting deformation is counter acted by a restoring force from the cantilever arms of the spring plate until equilibrium is achieved. The application of light illumination on the diode causes an impedance change in a photosensitive substrate, and in turn alters the voltage drop across the membrane-substrate, thereby resulting in the change in the membrane deformation.

FABRICATION

The spring plate arrays and PIN diode were fabricated separately and then bonded together with appropriate alignment. Figure 2 illustrates the fabrication process of the spring plate arrays using an Indium Phosphide (InP) sacrificial substrate. First, a 1 μm thick Si_3N_4 film was deposited on the InP substrate by low pressure chemical vapor deposition (CVD) (2a). Standard photolithography was used to generate the spring plate array pattern in a layer of photoresist (2b). The pattern was transferred to Si_3N_4 film with Reactive Ion Etching (RIE) (2c). After photoresist removal (2d), SU-8 negative photoresist was spin-coated onto the patterned Si_3N_4 surface (2e). A sensitive pre-exposure baking of this photoresist was then performed by ramping the temperature up to 95 ° C over 30 minutes, a five minute bake at this temperature, and then immediate removal from the hotplate. Next, a contact mask aligner was utilized to expose the web structure around the spring plate arrays, thereby assuring excellent alignment (2f). After development and hard baking, the wafer was mounted with SU-8 side down on a glass slide using black wax (2g). The perimeter of the InP substrate was also coated with the wax. The sample was then etched in Hydrochloric acid (HCl) at a temperature of 50° C until all the exposed InP had been removed. Trichloroethylene was then used to remove the black wax, leaving a freestanding spring plate array membrane suspended by an InP window frame. A thin film of Ti/Au (350/2000 Å) was deposited as a metal contact layer by electron beam evaporation (2h). Figure 2i. shows the excellent alignment between a single spring plate element and the SU-8 web.

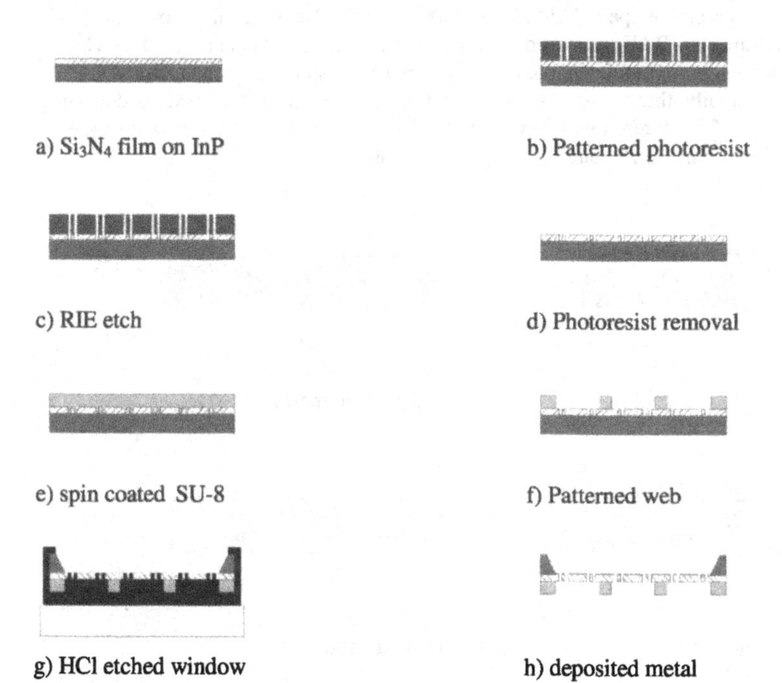

i) Single spring plate element.

Figure 2. (a-h) DMM fabrication procedure and (i) Top view of a suspended spring plate element well aligned with the SU-8 web underneath.

The fabrication process of the PIN photodiode is shown in Fig. 3. The PIN structure (1 μm each of N+, N-, and P+ GaAs epitaxial layers) was grown on an n-type GaAs substrate by Molecular Beam Epitaxy (MBE) (3a). Using conventional lithography and lift-off processes, a pattern of ohmic ring contacts was generated on the p-type surface using a metal recipe of Ti/Au (350/2000 Å) (3b). Next, Bromine ion beam assisted etching (Br-IBAE) was used to create mesas aligned with the rings (3c). The sample was then lithographically patterned with SU-8 for sidewall passivation of the mesas (3d). Finally, the spring plate arrays and the diode were aligned and bonded using small droplets of SU-8 resist which were thermally cured at elevated temperatures (3e). Figure 3f shows a metalized and suspended device array.

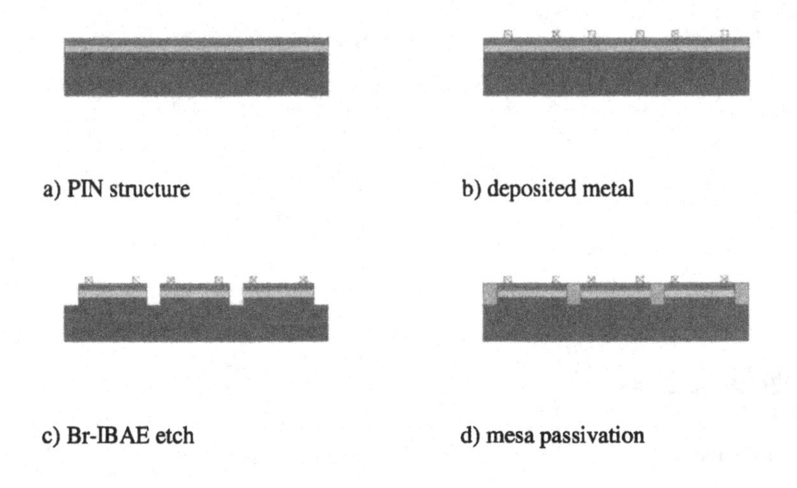

a) PIN structure

b) deposited metal

c) Br-IBAE etch

d) mesa passivation

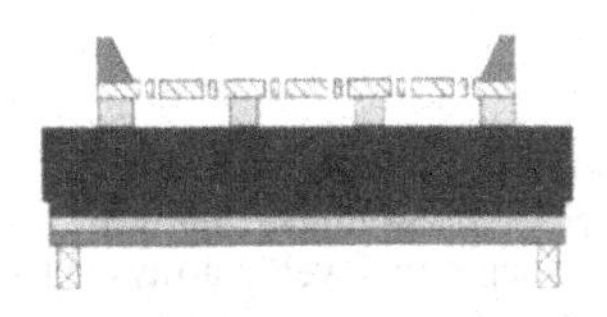

e) bonded device

f) top view spring plate array

Figure 3. (a-e) PIN diode fabrication process, and (f) Top view of the final device.

RESULTS AND DISCUSSION

Testing of the SU-8 suspended spring plates was performed using the Michelson interferometer setup shown schematically in Fig. 4. In this initial test, a DC voltage was applied between the diode n-type substrate and the metalized plate. Fully testing the device with a control optical beam is planned in the near future as soon as several packaging issues are resolved. The test bed used a 632 nm Helium Neon laser and control voltages were varied from 0 to 10 volts. Figure 4 shows interference patterns generated as the device bias was changed form 0 V to 7.7 V. This change in the pattern indicates a height change of 158 nm corresponding to a π phase shift. By assuming that the electrical force, F_e, is in balance with the mechanical force, F_m, one finds that

$$F_e = \varepsilon_0 A V^2/2(g-x)^2 = kx = F_m$$

where ε_0 is the permittivity of free space, A is the spring plate area, V is the applied voltage, g is the initial gap height, x is the displacement, and k is the effective spring constant. The expected value of k calculated from the data with this simplified model is 1.28 N/m.

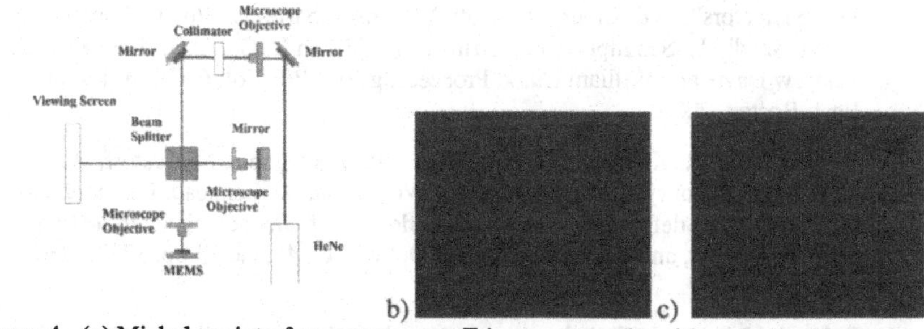

Figure 4. (a) Michelson interferometer setup. Fringe pattern (b) with and (c) without bias

SUMMARY

The new device fabrication method using SU-8 photoresist appears to be robust and allows more precise alignment of the spring plate devices to the supporting web structure. With this fabrication process in place, work will begin to develop arrays with individual spring plates coupled to individual photodiodes.

ACKNOWLEDGMENTS

This work was supported by the United States Air force under contract# FA8718-05-C-0081. The opinions, interpretations, conclusions, and recommendations are those of the authors and not necessarily those of the United States Air Force.

REFERENCES

1 B. Haji-saeed, R. Kolluru, D. Pyburn, R. Leon, S. Sengupta, M. Testorf, W. Goodhue, J. Khoury, A. Drehman, C. Woods, and J. Kierstead, Photoconductive optically driven deformable membrane for spatial light modulator applications utilizing GaAs substrates, Applied Optics Vol. 45, No.12 pp. 2615-2622 2006.

2 G. Griffith, B. Haji-saeed, S. K. Sengupta, W. Goodhue, J. Khoury, C. L. Woods, and J. Kierstead, Patterned Multipixel Membrane Mirror MEMS Optically Addressed Spatial Light Modulator With Megahertz Response IEEE Photonics Technology Letters, Vol. 19, No.3, pp. 173- 175, 2007.

3 T. Fujita, K. Maenaka and Y. Takayama, Dual-axis MEMS mirror for large deflection-angle using SU-8 soft torsion beam, Sensors and Actuators A: Physical, Vol. 121, Issue 1, pp. 16-21, 2005.

4 "Optically Addressed MEMS Deformable Mirrors Driven via an Array of Photodetectors", Jed Khoury, Kenneth Vaccaro, Charles L. Woods, Bahareh Haji-saeed, Sandip K. Sengupta, Craig Armiento, William D. Goodhue, John Kierstead, Andrew Davis and William Clark, Proceedings of SPIE Vol. 6368, SPIE Optics East, Boston, 2006.

5 B. Haji-saeed, R. Kolluru, D. Pyburn, R. Leon, S. Sengupta, M. Testorf, W. Goodhue, J. Khoury, A. Drehman, C. L. Woods, and J. Kierstead, Photoconductive optically driven deformable membrane under high-frequency bias: fabrication, characterization, and modeling, Applied Optics Vol. 45, No.14 pp. 3226-3236, 2006.

6 Bahareh Haji-saeed, "Development of Novel Device Assemblies and Techniques for Improving Adaptive Optics Imaging Systems", PhD. Thesis Dissertation, University of Massachusetts at Lowell (2006)

Mater. Res. Soc. Symp. Proc. Vol. 1052 © 2008 Materials Research Society 1052-DD06-18

High Temperature Annealing Studies on the Piezoelectric Properties of Thin Aluminum Nitride Films

R. Farrell[1], V. R. Pagán[1], A. Kabulski[1], Sridhar Kuchibhatla[1], J. Harman[1], K. R. Kasarla[1], L. E. Rodak[1], J. Peter Hensel[2], P. Famouri[1], and D. Korakakis[1]
[1]Lane Department of Computer Science and Electrical Engineering, West Virginia University, PO BOX 6109, Morgantown, WV, 26506
[2]USDOE/NETL, Morgantown, WV, 26505

ABSTRACT

A Rapid Thermal Annealing (RTA) system was used to anneal sputtered and Metal Organic Vapor Phase Epitaxy (MOVPE)-grown Aluminum Nitride (AlN) thin films at temperatures up to 1000°C in ambient and controlled environments. According to Energy Dispersive X-Ray Analysis (EDAX), the films annealed in an ambient environment rapidly oxidize after five minutes at 1000°C. Below 1000°C the films oxidized linearly as a function of annealing temperature which is consistent with what has been reported in literature [1]. Laser Doppler Vibrometry (LDV) was used to measure the piezoelectric coefficient, d_{33}, of these films. Films annealed in an ambient environment had a weak piezoelectric response indicating that oxidation on the surface of the film reduces the value of d_{33}. A high temperature furnace has been built that is capable of taking in-situ measurements of the piezoelectric response of AlN films. In-situ d_{33} measurements are recorded up to 300°C for both sputtered and MOVPE-grown AlN thin films. The measured piezoelectric response appears to increase with temperature up to 300°C possibly due to stress in the film.

INTRODUCTION

Aluminum nitride is an attractive material for the fabrication of MicroElectroMechanical Systems (MEMS) due in part to its piezoelectricity, inertness and tolerance to high temperatures. AlN remains piezoelectric after annealing at temperatures as high as 1000°C, while other materials with a higher piezoelectric response (e.g. PZT) lose their properties after annealing above their Curie temperature well below 1000°C. Annealing AlN at high temperatures in an ambient environment causes the surface of the film to oxidize. AlN begins to oxidize in an ambient environment at 800°C and the oxidation abruptly increases at an annealing temperature of 1000°C [1]. This surface oxidation drastically reduces the displacement of piezoelectric films.

In-situ measurements of AlN films using Laser Doppler Vibrometry are not widely reported. Kano *et al* used rf-sputtered 1200 nm thick AlN films to reveal that AlN can be used as an actuator with a constant output velocity up to 300°C [2]. They reported a constant d_{33} value of 1.38 pC/N at temperatures ranging from 20°C to 300°C. The results of high temperature measurements of sputtered and Metal Organic Vapor Phase Epitaxy (MOVPE)-grown AlN films as actuators will be discussed in this paper.

EXPERIMENT

This study reports the effects of high temperature annealing on sputtered and MOVPE-grown AlN thin films. Sputtered films were deposited onto a 350 μm thick (1 0 0)-oriented *p*-

type silicon substrate using a DC magnetron sputtering technique in an Ar/N_2 environment at room temperature. The sputtering parameters are outlined in Table 1. MOVPE-grown AlN was also grown in an AIXTRON 200/4 RF-S horizontal reactor on 2 inch (1 0 0) *p*-type silicon substrates. The growth temperature was approximately 1100 °C and the reactor pressure was 50 mbar. Trimethylaluminum (TMAl) and ammonia (NH_3) were used as Al and N source precursors with flows of 12 μmol/min and 1.5 slm respectively. Hydrogen was used as the carrier gas.

Table 1 - Sputtering conditions of the AlN films

Parameter	Value
Al Target (%)	99.999
DC Power (W)	500
Substrate Temperature	Room Temperature
Sputtering Pressure (mTorr)	30
Sputtering time (minutes)	35
N_2 concentration (%)	99.999
Ar/N_2	3/27

The AlN thin films were annealed in an Annealsys AS-Micro Rapid Thermal Annealing (RTA) furnace at temperatures ranging from room temperature to 1000°C for 5 minutes. Annealing was done in an ambient environment to allow the films to oxidize. The atomic percentage of oxygen on the surface of these films is measured using Energy Dispersive X-Ray Analysis (EDAX). The EDAX results displaying the oxidation trends in sputtered and MOVPE-grown AlN thin films are shown in Figure 1.

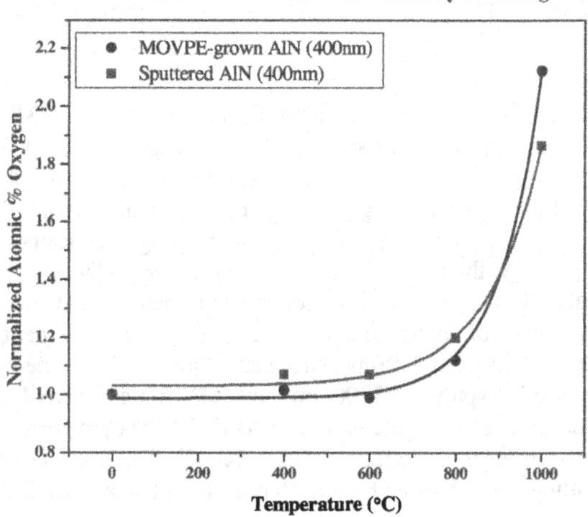

Figure 1 – Oxidation trends in sputtered and MOVPE-grown AlN

Solid circular platinum contacts were deposited on the top of the AlN films after annealing using a standard photolithography technique and an aluminum backside contact was sputtered onto the back of each wafer. An AC signal was used to actuate the films and a PSV-400 Scanning Head Laser Doppler Vibrometer (LDV) was employed to measure the picometer level displacement of the AlN films in order to determine how the piezoelectric properties of thin AlN films respond after exposure to high temperatures. The results are shown in Table 2.

Table 2 – 400 nm AlN films annealed in the RTA in an ambient environment

Sample	d_{33} (pm/V)
Sputtered Unannealed	3.31
Sputtered Annealed (1000°C)	1.03
MOVPE-grown Unannealed	15.38
MOVPE-grown Annealed (1000°C)	4.08

A high temperature furnace was built that is capable of measuring the d_{33} of sputtered and MOVPE-grown AlN films while annealing the films in a high temperature environment. The furnace consists of a stainless steel chamber with a 1 inch diameter Al_2O_3 button heater that is rated up to 1200°C and connected to a DC power supply. An electrical feed-through on the side of the chamber allows for electrically probing the sample while the chamber is taken to high temperatures and a gas feed-through allows for the flow of various gasses during high temperature testing. The gas flow rate can be increased up to 20 scfh (approximately 9500 sccm). The top of the furnace has a sapphire window (1 inch thick and 1 inch in diameter) located directly above the button heater. The LDV is mounted over the top of the furnace so that the laser travels through the window and onto the sample which is resting on top of the heater.

Prior to high temperature testing, sputtered AlN samples are annealed in the RTA in a nitrogen environment at 600°C for 1 hour. Annealing the samples prior to in-situ tests assures that no annealing effects of the crystal will be encountered during testing. This preparation is done in a N_2 environment so the films will not oxidize causing a lower d_{33} at the beginning of the measurements. MOVPE-grown films are grown at a much higher temperature than the in-situ temperature will reach, so the pre-test annealing preparation is unnecessary for these films.

A bondpad was fabricated for the AlN samples because the 400 μm platinum contacts cannot be probed directly from the electrical feed-through. Large Pt contacts were deposited onto a Si/SiO_2 wafer using a standard photolithography technique and a thin layer of titanium was used as an adhesion layer. The backside of the AlN test sample was bonded to one of the Pt contacts on the bondpad using a temperature resistant silver paste then the paste was cured in room temperature for 2 hours followed by a 2 hour 90°C curing bake. Gold wire-bonding was used to probe the 400 μm Pt contacts on the test sample from another Ti/Pt contact on the bondpad. By probing the contacts on the bondpad from the

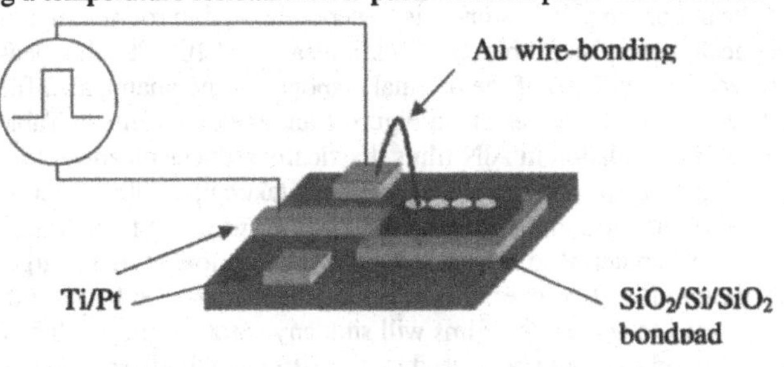

Figure 2 – AlN sample on bondpad with gold wire-bonding used to probe top contacts

electrical feed-through, the signal is relayed to the top and backside of the sample for actuation.

The furnace is heated by increasing the power to the button heater inside the furnace. During testing, the temperature of the furnace was increased at a rate of 10°C/minute while N_2 gas was flowing through the chamber at a rate of 2 scfh (approximately 950 sccm). The temperature inside the furnace was held to within ±5°C of the target temperature while d_{33} measurements were taken. A 3.5 KHz AC signal was supplied to the electrical feed-through at varying peak-peak voltages up to 20 V while the picometer level displacement was measured by the LDV. The d_{33} of sputtered and MOVPE-grown films were measured at temperatures up to 300°C. The results of these measurements are summarized in Figure 3.

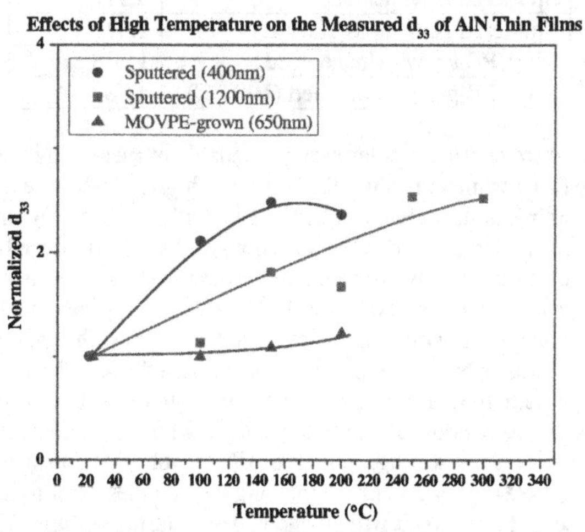

Figure 3 – In-situ measurements of the piezoelectric response of sputtered and MOVPE-grown AlN thin films

DISCUSSION

The results in Figure 1 show that sputtered films oxidize more than MOVPE-grown films at lower temperatures probably due to the different crystalline quality achieved by the two methods of deposition. At temperatures close to 1000°C both films oxidize rapidly as expected and the d_{33} of annealed samples is lower compared to room temperature measurements of unannealed samples (Table 2). After annealing at 1000°C, the measured d_{33} of both films reduced to about 25% of the original response of the unannealed films. There is a correlation between the oxidation results in Figure 1 and the d_{33} results in Table 2 leading to the conclusion that surface oxidation in AlN films drastically reduces piezoelectric response.

In-situ d_{33} measurements have been taken up to 300°C. In Figure 3, it is evident that there is an increase in the measured d_{33} as the annealing temperature increased. AlN is a pyroelectric material so this increase could be attributed to a charge density increase due to high temperature [3]. However, this effect can be disregarded because it is likely that the power supply used to actuate the films will sink any extra charge produced in the films. The increase in the measured d_{33} could be caused by the difference in thermal expansion coefficients between

AlN and Si. At 100°C, the thermal expansion coefficients of Si and a-plane AlN are roughly 3.1×10^{-6}/K and 4.2×10^{-6}/K, respectively [4,5]. At high temperatures, the AlN film expands more than the Si substrate, decreasing the lattice mismatch allowing the AlN to behave more like an unclamped film. However, a more detailed study of clamped and unclamped AlN films, including in-situ measurements of the d_{33}, will need to be conducted to verify these trends.

At temperatures higher than 300°C, the thin sputtered AlN films become conductive and when the temperature is decreased to 23°C, these previously insulating films still conduct. This was not seen in films that had been annealed in the RTA prior to d_{33} measurements. To find the relation between oxidation and d_{33}, AlN films were annealed at 1000°C prior to measurements and AlN films used for in-situ tests were pre-annealed 600°C. These films were not conducting when room temperature measurements were taken, so applying an electric field across the sample at high temperatures appears to damage the films. MOVPE-grown films became conductive at 400°C leading to the conclusion that a film with a poor crystalline quality may be more susceptible to damage from applying a voltage at those temperatures.

CONCLUSIONS

AlN is a robust material that remains piezoelectric above 1000°C but if the film is annealed in a harsh environment, oxidation could reduce the piezoelectric response. This paper has shown that AlN films oxidize drastically when annealed in an ambient environment causing a large reduction in the d_{33} of the material. If the films are protected from oxidation, the d_{33} of the material may be preserved up to 1000°C in any environment. Methods used to protect the films from oxidation while retaining the original piezoelectric response are currently under investigation.

In-situ d_{33} measurements of AlN films have been taken at temperatures up to 300°C. The d_{33} appears to increase as temperature increases. This trend is likely caused by stress at the AlN/Si interface increasing with high temperatures. The films become conductive at temperatures above 300°C which compromises d_{33} measurements. It is important to understand why the films conduct at this temperature in order to proceed with in-situ measurements.

ACKNOWLEDGMENTS

This work was supported in part by NETL, AIXTRON and NSF RII contract EPS 0554328 for which WV EPSCoR and WVU Research Corp matched funds.

REFERENCES

1. J.W. Lee, I. Radu, M. Alexe, J. Materials Science, 13, 131-137, 2002.
2. Kazuhiko Kano, Kazuki Arakawa, Yukihiro Takeuchi, Morito Akiyama, Naohiro Ueno, Nobuaki Kawahara, Sensors and Actuators, A, 130-131, 397-402, 2006
3. V. Fuflyigin, E. Salley, A. Osinsky, P. Norris, App. Phys. Letters, V. 77, 19, 2000
4. Yasumasa Okada, Yozo Tokumaru, J. Appl. Phys. 56 (2), 1984
5. W.M. Yim, R.J. Paff, J. Appl. Phys. 45(3), 1974

Mater. Res. Soc. Symp. Proc. Vol. 1052 © 2008 Materials Research Society 1052-DD06-21

A Miniature Silicon Condenser Microphone Improved with a Flexure Hinge Diaphragm and a Large Back Volume

H. J. Kim, S. Q Lee, J. W. Lee, S. K. Lee, and K. H. Park
Electronics and Telecommunications Research Institute, 161 Gajeong-dong, Yuseong-gu, Daejeon, 305-700, Korea, Republic of

ABSTRACT

We have developed a miniature silicon condenser microphone improved with a spring supported hinge diaphragm and a large back volume, which is designed in order to increase sensitivity of microphones. MEMS Technology has been successfully applied to miniature silicon capacitive microphones, and we fabricated the smallest condenser silicon microphone in the presented reports. We used the finite-element analysis (FEA) to evaluate mechanical and acoustic performances of the condenser microphone with a flexure hinge diaphragm. From the simulation results, we confirmed that the sensitivity of a flexure hinge diaphragm can be improved about 285 times higher than a flat diaphragm. The first and second modes occurred at 15,637Hz and 24,387Hz, respectively. The areas of the miniature condenser microphones with a hinge diaphragm are 1.5 mm x 1.5 mm. We measured the impedance characteristics and sensitivities of the fabricated condenser microphones. The sensitivities of microphones are around 12.87 μV/Pa (-60 dB ref. 12.5 mV/Pa) at 1 kHz under a low bias voltage of 1 V, and the frequency response is flat up to 13 kHz.

INTRODUCTION

Today, MEMS technology has been successfully applied to miniature silicon capacitive microphones [1-3], a lot of research has been published in order to achieve high performance miniature silicon microphones. Most miniature microphones are adopted by diaphragm-based capacitive type, i.e., condenser microphones and electret microphones because they have the flat frequency response in broad bandwidth, high SNR (signal-to-noise ratio), high sensitivity, low power consumption, and large quantities of manufacture, as compared with other types of microphones as like piezoelectric or piezoresistive microphones [4-6].

A typical silicon condenser microphone consists of a diaphragm, a rigid backplate, an air gap, electrodes and a back volume. And the sensitivity of the condenser microphone is determined by the electrical field strength exerting across the capacitor gap, the deformation (deflection) of diaphragm and a large air damping of back volume. Thus, to obtain high sensitivity in broad bandwidth, the diaphragm has to be designed to be flexible more and the backchamber has to have larger volume. So far, an important number of corrugated diaphragms and large backchamber designs have been introduced in order to achieve higher sensitive condenser microphone [7-9]. This paper presents a miniature condenser microphone with a spring supported hinge diaphragm and a large back volume, which contributes to high sensitivity and flat frequency response, even though the small diaphragm diameters of 500 μm.

EXPERIMENT

This paper presents a miniature silicon condenser microphone with a flexure hinge diaphragm and a large back volume, which is designed to obtain high sensitivity of microphones, compared with other condenser microphones. We fabricated a very small silicon condenser microphone through MEMS technology.
The schematic view of a condenser microphone with a flexure hinge diaphragm and a large back volume is shown in Figure 1, which means the cross-sectional structure of microphones, passing through the center of diaphragm. The diaphragm has a spring supported hinge design and the backchamber are designed to have a large back volume using 2 step DRIE process.

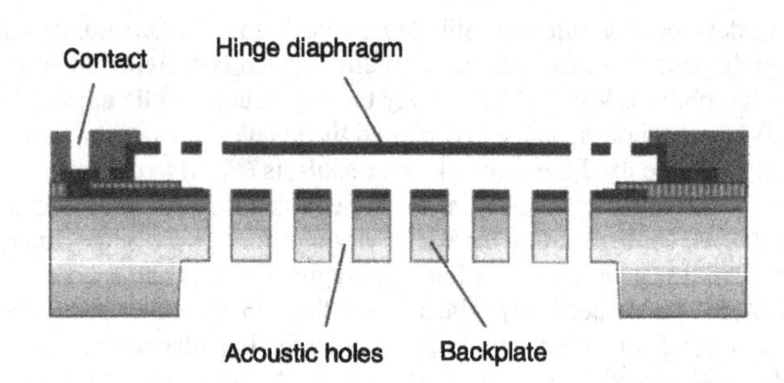

Figure 1. Schematic cross-sectional view of the miniature condenser microphone with a flexure hinge diaphragm.

The process sequence starts with a <100> oriented, double side polished 5 inch silicon wafer, as shown in Figure 2. First, a LTO layer with a thickness 0.5 μm is deposited on the double side of a wafer as an insulating layer by LPCVD (Figure 2a). And the backplate electrodes and highly insulated nitride of 0.3 μm are patterned well (Figure 2b). The backplate electrodes are well designed to form acoustic holes of the rigid backplate in order to obtain higher sensitivity.

Next, we use the sacrificial layer to realize the devices in one single chip. The PSG sacrificial layers of about 2.5 μm were defined, and then small corrugations are implemented by etching narrow trenches of a depth of 1.5 μm in the PSG layer (Figure 2c). These sacrificial layers determine the capacitive gaps of condenser microphones. And silicon nitride of 0.2 μm is deposited by PECVD equipment.

Next, the contacts for the backplate electrodes are patterned (Figure 2d) and Au membranes of 0.5 μm with a hinge pattern are formed by lift-off process (Figure 2e). And the polyimide was well patterned as posts of the hinge diaphragm (Figure 2f). Then, acoustic holes of the rigid backplate are etched by 1st Deep-RIE process and then backside cavities are etched by 2nd Deep-RIE backside etching technique to make a larger back volume (Figure 2g). Finally, the miniature condenser microphones are completed by sacrificial layer releasing in BHF solution (Figure 2h).

DISCUSSION

It is well known that the sensitivities of condenser microphones are very dependent on the stress distribution and deformation of diaphragms. In this paper, we used the finite-element analysis (FEA) using ANSYS to evaluate the mechanical and acoustic performance of the devices.

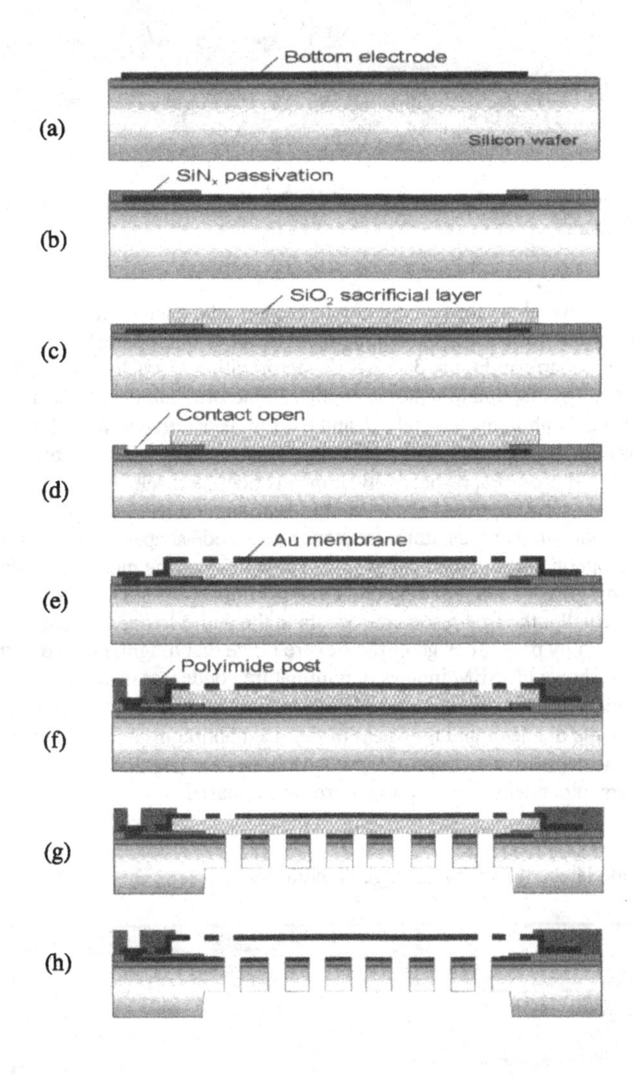

Figure 2. A schematic view of the microphone fabrication processes.

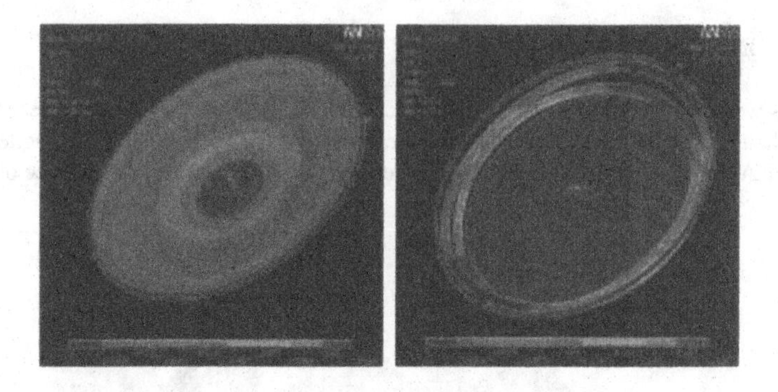

Figure 3. The deformation characteristics of the flat diaphragm (left) and the flexure hinge diaphragm (right).

Based on several parameters, we conducted a modal analysis to determine the vibration characteristics of the flexure hinge diaphragm. A reference microphone with a flat diaphragm was used for a comparison. Figure 3 shows the deformation characteristics of the flat diaphragm and the flexure hinge diaphragm with the applied sound pressure of 100 Pa. The diameters and thicknesses of the diaphragms are 500 μm and 0.6 μm, respectively. With the applied sound pressure of 100 Pa, the mechanical compliances of the flat/hinge diaphragms are 0.5818E-4 μm/Pa and 0.1657E-1 μm/Pa, respectively. It means that the mechanical sensitivity of a flexure hinge diaphragm can be improved about 285 times higher than a flat diaphragm.

It is also known that the natural frequency and mode shapes are very important factors for condenser microphones with broad bandwidth. Table I shows the modal analysis for the flat/hinge diaphragms. The first and second modes occurred at 15,637Hz and 24,387Hz, respectively. Actually, the first resonance occurs in the audio frequency range, but we can obtain broader bandwidth by proper design of the flexure hinge diaphragm, i.e. wider bridge patterns.

Figure 4 shows FE-SEM images of a miniature condenser microphone with a hinge diaphragm fabricated with silicon-based MEMS technology. The areas of the devices are 1.5 mm x 1.5 mm. And, the diameter and thickness of the Au diaphragms are 500 μm and 0.5 μm, respectively. The capacitive air gaps of the microphones are less than 2.5 μm, and the PSG sacrificial layers of condenser microphones are well released.

Table I. The modal analysis for flat/hinge diaphragms.

	1st resonance freq.	*2nd resonance freq.*
Flat diaphragm	*153,814 Hz*	*269,266 Hz*
Hinge diaphragm	*15,637 Hz*	*24,387 Hz*

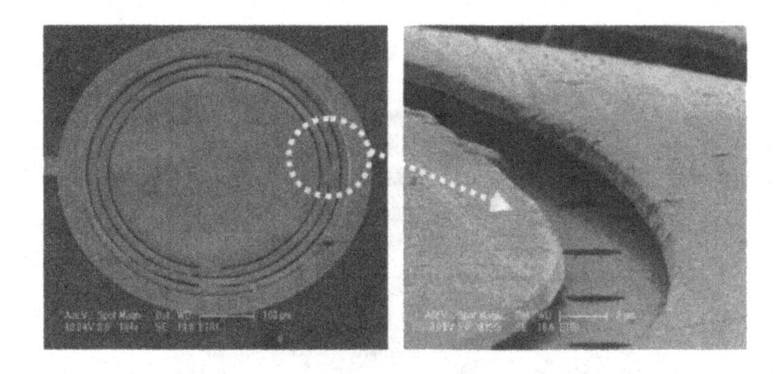

Figure 4. FE-SEM images of fabricated miniature condenser microphones.

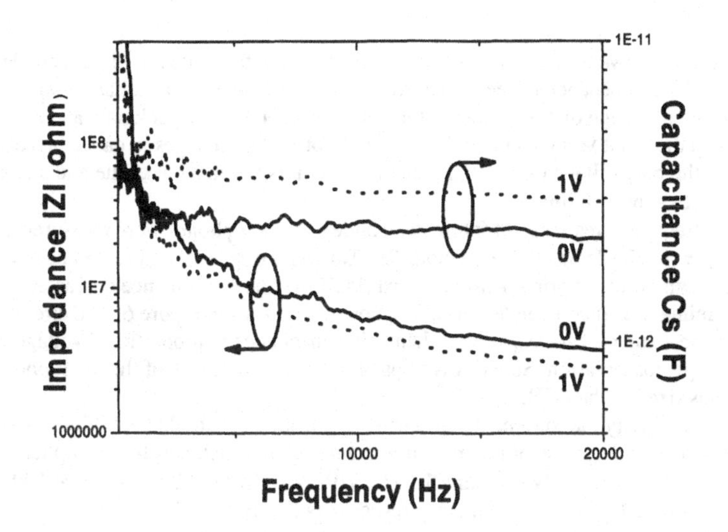

Figure 5. Impedance characteristics of the fabricated condenser microphone with a flexure hinge diaphragm.

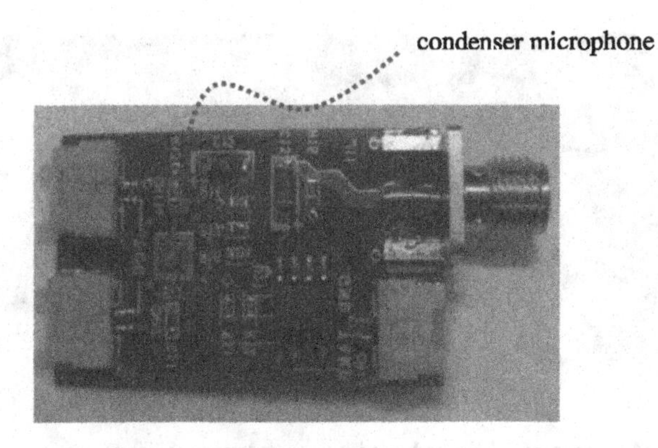

Figure 6. The printed circuit board (PCB) with a fabricated condenser microphone and a pre-amplifier for the performance characterization.

Figure 5 shows the impedance characteristics of the fabricated miniature condenser microphone. The impedance of the device was decreased while the capacitance was increased with the applied voltage of 1 V. It means that an interval between a backplate and a membrane is reduced with the input voltage due to the Coulomb force. The changes in capacitance are about 0.8 pF with the bias voltage of 1 V and then the interval between a backplate and a membrane was reduced to about 1.8 μm.

The frequency response of fabricated condenser microphones were measured using a Brüel & Kjær (B&K) Type 4232 Anechoic Test Box with a speaker and reference microphone. We designed an adequate printed circuit board (PCB) for the performance characterization of the fabricated miniature silicon condenser microphones, as shown in Figure 6. The circuit works by translating the condenser microphone's change in capacitance to proportional voltage change through charge conservation. Sound was transmitted to the front side of the microphone, whose backside was fixed to the PCB.

The sensitivity measurements are carried out from 20 Hz to 20 kHz. Figure 7 shows the measured frequency response of the miniature condenser microphones for an applied voltages of 1 V. The measured sensitivity is around 12.87 μV/Pa (-60 dB ref. 12.5 mV/Pa) at 1 kHz under a low bias voltage of 1 V, and the frequency response is flat up to 13 kHz.

CONCLUSIONS

This paper presented a miniature condenser microphone with a flexure hinge diaphragm and a large back volume. We fabricated a single-chip silicon condenser microphone using MEMS technology. The fabricated condenser microphones are among the smallest in the presented reports.

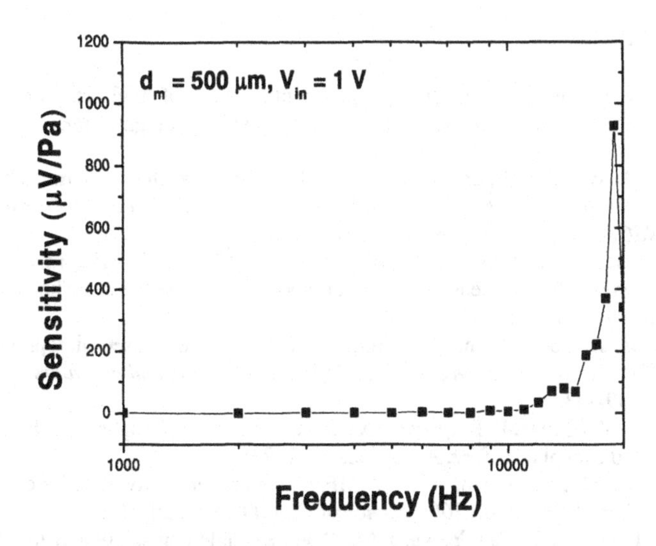

Figure 7. Sensitivity measurement of the miniature silicon condenser microphone with a flexure hinge diaphragm and a large back volume.

We used the finite-element analysis (FEA) using ANSYS to evaluate the mechanical and acoustic performance of the miniature condenser microphone with a flexure hinge diaphragm. From the simulation results, we confirmed that the mechanical sensitivity of a flexure hinge diaphragm can be improved about 285 times higher than a flat diaphragm.

The areas of fabricated condenser microphones are 1.5 mm x 1.5 mm, and the diameter and thickness of the membrane is 500 μm and 0.5 μm, respectively. The measured sensitivity of fabricated microphones is around 12.87 μV/Pa (-60 dB ref. 12.5 mV/Pa) at 1 kHz under a low bias voltage of 1 V. And the frequency response is flat up to 13 kHz, which can be improved by proper design revision of the flexure hinge diaphragms, i.e. wider bridge patterns.

ACKNOWLEDGMENTS

The authors would like to acknowledge the help of D. S. Kim and Prof. S. M. Wang of Gwangju Institute of Science and Technology (GIST) for simulation and measurement of this project.

REFERENCES

1. Q. Zou, A. Li, L. Liu, "Theoretical and experimental studies of single-chip-processed miniature silicon condenser microphone with corrugated diaphragm," *Sens. Act. A*, **63**, 209 (1997).
2. N. B. Ning, A. W. Mitchell, and R. N. Tait, "Fabrication of a silicon micromachined capacitive microphone using a dry-etch process," in *Digest Tech. Papers Transducers '95 & EUROSENSORS IX* (Sweden, 1995), pp.704-707.
3. P. R. Scheeper, B. Nordstrand, J. O. Gulloy, B. Liu, T. Clausen, L. Midjord, and T. Storgaard-Larsen, "A new measurement microphone based on MEMS technology," *J. Microelectromech. Sys.*, **12**, 880 (2003).
4. J. J. Bernstein, J. T. borenstein, "A micromachined silicon condenser microphone with on-chip amplifier," in *Technical Digest of the Solid-State Sensors and Actuators Workshop* (1996), pp. 239-243.
5. M. Pedersen, W. Olthuis, P. Bergveld, "A silicon condenser microphone with polyimide diaphragm and backplate," *Sens. Act. A*, **63**, 97 (1997).
6. J. W. Weigold, T. J. Brosnihan, et al., "A MEMS condenser microphone for consumer applications," in *IEEE MEMS 2006 Technical Digest* (2006), pp. 86-88.
7. J. Chen, L. Liu, Z. Li, Z. Tan, Y. Xu, J. Ma, "On the single-chip condenser miniature microphone using DRIE and backside etching techniques," *Sens. Act. A*, **103**, 42 (2003).
8. R. Kressmann, M. Klaiber, G. Hess, " Silicon condenser microphones with corrugated silicon oxide/nitride electret membranes," *Sens. Act. A*, **100**, 301 (2002).
9. Takashi Kasai, Yoshitaka Tsurukame, Toshiyuki Takahashi, Fumihiko Sato and Sumio Horiike, "Small silicon condenser microphone improved with a backchamber with concave lateral sides," in *Digest Tech Papers Transducers '07 & EUROSENSORS XXI* (France, 2007), pp. 2613-2616.

Mater. Res. Soc. Symp. Proc. Vol. 1052 © 2008 Materials Research Society 1052-DD06-22

Comparison of 1D and 2D Theories of Thermoelastic Damping in Flexural Microresonators

Sairam Prabhakar, and Srikar Vengallatore
Mechanical Engineering, McGill University, Montreal, H3A 2K6, Canada

ABSTRACT

Thermoelastic damping (TED) represents the lower limit of material damping in flexural mode micro- and nanoresonators. Current predictive models of TED calculate damping due to thermoelastic temperature gradients along the beam thickness only. In this work, we develop a two dimensional (2D) model by considering temperature gradients along the thickness and the length of the beam. The Green's function approach is used to solve the coupled heat conduction equation in one and two dimensions. In the 1D model, curvature information is lost and, hence, the effects of structural boundary conditions and mode shapes on TED are not captured. In contrast, the 2D model retains curvature information in the expression for TED and can account for beam end conditions and higher order modes. The differences between the 1D and 2D models are systematically explored over a range of beam aspect ratios, frequencies, boundary conditions, and flexural mode shapes.

INTRODUCTION

Flexural mode microresonators with low damping are the building blocks of micro-electromechanical systems (MEMS) used for sensing, communications and energy harvesting applications. Energy dissipation due to thermoelastic coupling represents the lower bound on damping in microscale flexural resonators [1]. In a beam undergoing flexural vibrations, the thermoelastic effect implies that an oscillating stress gradient will generate a corresponding temperature gradient such that the compressive regions are hotter, and the tensile regions colder, than the equilibrium temperature of the beam. This finite temperature gradient will inevitably lead to irreversible heat conduction, entropy generation, and energy dissipation. This mode of energy dissipation is called thermoelastic damping (TED).

In 1937, Zener provided a closed form expression for TED by considering irreversible heat conduction in one-dimension (through the thickness of a vibrating Euler-Bernoulli beam of thickness h). Zener's formula for TED is given by [2]:

$$Q_{Zener}^{-1} = \frac{E\alpha^2 T_0}{C}\frac{\omega_n \tau}{1+\omega_n^2\tau^2} = \frac{E\alpha^2 T_0}{C}\frac{\Omega}{1+\Omega^2}, \quad \tau = \frac{h^2 C}{\pi^2 k} \tag{1}$$

Here E is the Young's modulus, α is the linear coefficient of thermal expansion, T_0 is the equilibrium temperature of the beam, C is the specific heat per unit volume, ω_n is the undamped natural frequency of vibration of the beam, $\Omega = \omega_n \tau$ is a normalized frequency, and τ is a time constant. Recently, Lifshitz and Roukes [3] improved upon Zener's work by developing exact closed-form expressions for thermoelastic damping in a beam resonator within the context of a 1D theory. Their expression is given by:

$$Q_{LR}^{-1} = 6\frac{E\alpha^2 T_0}{C}\frac{1}{\xi^2}\left(1-\frac{1}{\xi}\left\{\frac{\sinh\xi+\sin\xi}{\cosh\xi+\cos\xi}\right\}\right), \quad \xi = h\sqrt{\frac{\omega_n C}{2k}} \tag{2}$$

These 1D models are now widely used to assess whether thermoelastic damping is dominant in microresonators by comparing theoretical predictions against experimental measurements [1], for guiding the design of resonators in MEMS, and as the basis of advanced models for TED [4,5]. However, all these models share the following set of underlying assumptions:

- *Assumption 1*: *The temperature gradient is one-dimensional through the thickness of the beam.* Therefore, thermoelastic temperature gradients along the length of vibrating beams are ignored.
- *Assumption 2*: *Thermoelastic damping is independent of the mode shape of flexural vibrations.* 1D analytical model predict the magnitude of thermoelastic damping as functions of frequency and beam thickness, but do not account for the effects of higher order mode shapes on TED.
- *Assumption 3*: *Thermoelastic damping is independent of structural boundary conditions.* Miniaturized resonators with either cantilevered or doubly-clamped boundaries are commonly employed in MEMS, but 1D models do not distinguish between them.

Higher dimensional models of TED developed using analytical techniques have not quantified the errors of the 1D models over a full range of parameters [6,7]. Here, we describe the development of exact 1D and 2D theories of TED using a common mathematical framework that utilizes the method of Green's function [8] to solve the governing equation for thermoelastic heat conduction. The results of 1D and 2D models of TED are then compared for different structural boundary conditions, mode shapes and beam aspect ratios.

THEORY

Consider a slender, isotropic and homogenous beam of length L, width b and thickness h with geometry defined by: $0 \le x \le L;\ \ 0 \le y \le h;\ \ 0 \le z \le b$, being subjected to flexural vibrations about the z-axis. The transverse deformation is given by:

$$Y(x,t) = Y_0\,\varphi_n(x)\,exp(i\omega_n t). \tag{3}$$

Here, Y_0 is the amplitude in the $y-$direction, $\varphi_n(x)$ is the natural mode shape, and ω_n is the natural frequency of the nth mode.

The temperature field may be represented by the difference between the temperature of the beam, T, and the surrounding temperature T_0, $\theta = T - T_0$, as:

$$\theta(x,y,t) = \theta_0(x,y)\,exp(i\omega_n t), \tag{4}$$

where θ_0 is, in general, a complex valued quantity.

The temperature field can be calculated from the one-way coupled heat equation [9]:

$$C\frac{\partial\theta}{\partial t} = k\,\nabla^2\theta - \frac{E\alpha\,T_0}{(1-2\nu)}\frac{\partial}{\partial t}\left(\varepsilon_{xx}+\varepsilon_{yy}+\varepsilon_{zz}\right). \tag{5}$$

Here, ν is Poisson's ratio. For an Euler-Bernoulli beam (considering only the longitudinal strain and ignoring Poisson deformation), we have $\varepsilon_{xx} = -(y-h/2)\dfrac{\partial^2 Y}{\partial x^2}$. Hence,

$$C\frac{\partial \theta}{\partial t} = k\,\nabla^2\theta + i\omega_n (y-h/2) E\alpha T_0 \frac{\partial}{\partial t}\frac{\partial^2 Y}{\partial x^2}. \tag{6}$$

The solution of this boundary-value problem will lead to the temperature field, θ. Subsequently, the thermal strain can be calculated as:

$$\varepsilon_{xx}^{thermal} = \alpha\theta. \tag{7}$$

The work lost per cycle of vibrations is [10]:

$$\Delta W = -\pi \iiint_V \sigma_{xx}\, Im\left(\varepsilon_{xx}^{thermal}\right) dV. \tag{8}$$

The total strain energy stored within the beam given by:

$$W_{max} = \frac{1}{2}\iiint_V \sigma_{xx}\varepsilon_{xx}\, dV. \tag{9}$$

Finally, the thermoelastic damping is obtained from definition as [10]:

$$Q_{TED}^{-1} = \frac{1}{2\pi}\frac{\Delta W}{W_{max}}. \tag{10}$$

<u>**1D Analysis**</u>

Considering heat flow along the temperature gradient solely through the thickness of the beam, $\nabla \equiv \partial/\partial y$ in eq.(6). Assuming adiabatic boundary conditions at the opposite surfaces of the beam thickness, the steady state temperature field (which is harmonic and does not decay to zero as $t \to \infty$) is obtained as a complex quantity using the method of Green's functions [8] as:

$$\theta(x,y,t) = \theta^R(x,y,t) + i\theta^I(x,y,t). \tag{11}$$

Here the real and imaginary parts are:

$$\begin{aligned}
\theta^R &= \frac{2E\alpha T_0}{Ch}\frac{\partial^2 Y}{\partial x^2}\sum_{m=1}^{\infty}\frac{\omega_n^2}{\omega_n^2+\frac{k^2}{C^2}\beta_m^4}\int_{y'=0}^{h}\cos(\beta_m y')(y'-h/2)\,dy'\,\cos(\beta_m y)\\
\theta^I &= \frac{2E\alpha T_0}{Ch}\frac{\partial^2 Y}{\partial x^2}\sum_{m=1}^{\infty}\frac{\omega_n\frac{k}{C}\beta_m^2}{\omega_n^2+\frac{k^2}{C^2}\beta_m^4}\int_{y'=0}^{h}\cos(\beta_m y')(y'-h/2)\,dy'\cos(\beta_m y),
\end{aligned} \tag{12}$$

where the eigenvalues are $\beta_m = \dfrac{m\pi}{h}$. Using these results in eqs.(7 – 10), we obtain an expression for TED as:

$$Q_{TED}^{-1} = Q_{1D}^{-1} = 24\frac{E\alpha^2 T_0}{C}\frac{1}{h^4}\sum_{m=1}^{\infty}\frac{\omega_n\frac{k}{C}\beta_m^2}{\omega_n^2+\frac{k^2}{C^2}\beta_m^4}\eta_m^2, \tag{13}$$

where, $\eta_m = \int_{y'=0}^{h} \cos(\gamma_m y')(y' - h/2)\,dy'$.

Equation (13) does not contain the d^2Y/dx^2 curvature term. Hence, the TED predicted by the 1D model has no dependence upon the structural boundary conditions and mode shapes of vibration. Let us now consider only the first term of the series. Defining a thermal time constant, $\tau = h^2C/\pi^2k$, and recognizing that $(96/\pi^4) \approx 1$, eq.(13) reduces to:

$$Q_{1D}^{-1}\Big|_{m=1} = \frac{E\alpha^2T_0}{C}\frac{\omega_n\tau}{1+\omega_n^2\tau^2} = Q_{Zener}^{-1}. \tag{14}$$

Thus, Zener's classical formula for TED is obtained from our 1D analysis by retaining only the leading term of the infinite series. As additional terms are included in the series, the 1D model converges to the exact solution of Lifshitz and Roukes [3]. This feature is illustrated in Figure 1 for the case of single-crystal silicon which requires eight terms to converge to the exact 1D solution.

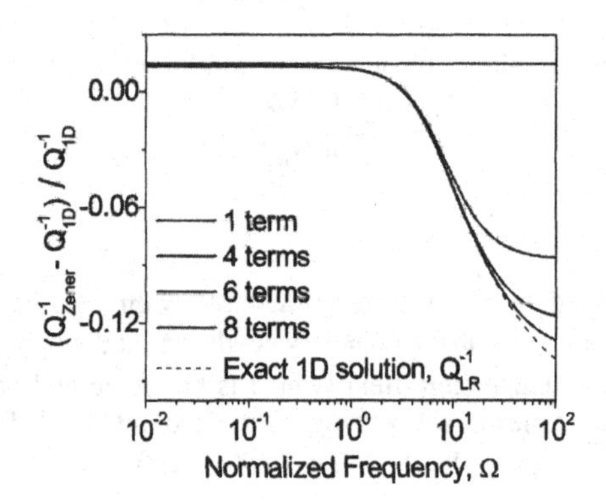

Figure 1: Comparison of the 1D model of eq.(13) with the formulas of Zener [2] and Lifshitz and Roukes [3].

2D Analysis

As the starting point for the 2D analysis, we set $\nabla \equiv \partial/\partial x + \partial/\partial y$ in eq.(6). Once again, using the Green's function approach [8] with isothermal boundary conditions at a fixed surface, and adiabatic boundary conditions at a free surface, the real and imaginary parts of the temperature field are obtained as:

$$\theta^R = \frac{4E\alpha T_0}{CLh} \sum_{m=1}^{\infty} \sum_{p=1}^{\infty} \frac{\omega_n^2}{\omega_n^2 + \frac{k^2}{C^2}\left(\beta_m^2 + \gamma_p^2\right)^2} \chi_m \, \eta_p \, \sin(\beta_m x) \, \cos(\gamma_p y)$$

$$\theta^I = \frac{4E\alpha T_0}{CLh} \sum_{m=1}^{\infty} \sum_{p=1}^{\infty} \frac{\omega_n \frac{k}{C}\left(\beta_m^2 + \gamma_p^2\right)}{\omega_n^2 + \frac{k^2}{C^2}\left(\beta_m^2 + \gamma_p^2\right)^2} \chi_m \, \eta_p \, \sin(\beta_m x) \, \cos(\gamma_p y), \tag{15}$$

where,

$$\chi_m = \int_{x'=0}^{L} \sin(\beta_m x') \frac{\partial^2 Y}{\partial x'^2} dx' \quad \eta_p = \int_{y'=0}^{h} \cos(\gamma_p y')(y' - h/2)\, dy'. \tag{16}$$

In the above, the eigenvalues are $\beta_m = (2m-1)\pi/2L$ and $\gamma_p = p\pi/h$ for a cantilever and, $\beta_m = m\pi/L$ and $\gamma_p = p\pi/h$ for a doubly-clamped beam. Finally, the expression for TED is obtained as:

$$Q_{TED}^{-1} = Q_{2D}^{-1} = 48 \frac{E\alpha^2 T_0}{C} \frac{1}{L\, h^4} \sum_{m=1}^{\infty} \sum_{p=1}^{\infty} \frac{\omega_n \frac{k}{C}\left(\beta_m^2 + \gamma_p^2\right)}{\omega_n^2 + \frac{k^2}{C^2}\left(\beta_m^2 + \gamma_p^2\right)^2} \frac{\chi_m^2}{\psi} \eta_p^2, \tag{17}$$

where

$$\psi = \int_{x=0}^{L} \left(d^2Y/dx^2\right)^2 dx. \tag{18}$$

The term (χ^2/ψ) survives in eq.(17), thereby retaining information about curvature in the 2D model. This enables the 2D theory to capture the effects of mode shape, and structural boundary conditions on TED.

RESULTS

This section presents results from the 1D and 2D theories for single-crystal silicon beams with various beam structural boundary conditions, mode shapes and *L/h* aspect ratios. In order to conform to Euler-Bernoulli theory, the aspect ratio is selected such that the distance between node points is greater than $10h$ in every case. Typically, 400 terms were required for the convergence of the 2D solution.

Figure 2 presents a series of graphs that quantify the relative errors between the 1D and 2D theories as functions of normalized frequency. Figure 2(a) compares 1D and 2D models for a cantilever with varying aspect ratio. For slender beams (*L/h*>40), the effect of longitudinal heat flux diminishes, and the results of the 1D and 2D models converge. Figure 2(b) examines the effect of structural boundary conditions for beams with an aspect ratio of 10, and shows that doubly-clamped beams are subjected to larger errors compared to cantilevered beams. Finally, the effects of mode shape are illustrated in Figure 2(c). Here, the *L/h* ratios were selected to represent the least slender beams that conform to Euler-Bernoulli requirements for a given mode number. This graph indicates that errors in the 1D theory diminish with increasing mode number.

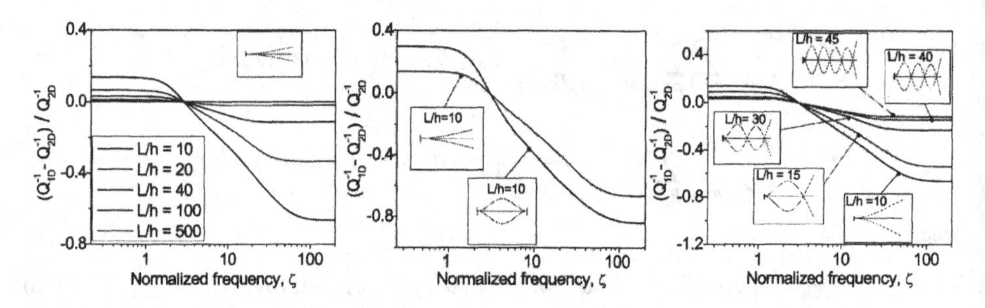

Figure 2: Effects of (a) beam aspect ratio (b) structural boundary conditions (c) flexural mode number on the errors inherent in the 1D model for TED.

SUMMARY

This paper compared the predictions of 1D and 2D theories of thermoelastic damping for flexural microresonators. The simple 1D models are accurate for long slender beams, but incur errors as large as 80% for other cases depending upon aspect ratio, mode shape, and boundary conditions. The detailed derivation of the 2D theory, accompanied by comparisons with experiments and finite-element analysis, will be described elsewhere.

ACKNOWLEDGEMENTS

We are indebted to the Canada Research Chairs program, *Le Fonds Quebecois de la Recherche sur la Nature et les Technologies* (FQRNT), and General Motors of Canada for financial support.

REFERENCES

[1] T. V. Roszhart, *Tech. Dig. Solid-State Sens. Actuator Workshop*, Hilton Head, 13-16 (1990).
[2] C. Zener, *Phys. Rev.*, **52**, 230-235 (1937).
[3] R. Lifshitz and M. Roukes, *Phys. Rev. B.*, **61**, 5600 -5609 (2000).
[4] V. T. Srikar and S. D. Senturia, *J. Microelectromech. Syst.*, **11**, 499-504 (2002).
[5] S.K. De and N.R. Aluru, *Phys. Rev.* B, **74**, 144305 (2006).
[6] P. Chadwick, *J. Mech. Phys. Solids.*, **10**, 99–109 (1962).
[7] J.B. Alblas, *J. Thermal. Stresses*, **4**, 333 – 355 (1981).
[8] M.N. Ozisik, *Heat Conduction*, 2nd ed. (John Wiley and Sons, New York, 1993) p. 214–257.
[9] B.A. Boley and J.H. Weiner, *Theory of Thermal Stresses*, (Wiley, New York, 1960) p. 1-40.
[10] A.S. Nowick and B.S. Berry, *Anelastic Relaxation in Crystalline Solids*, (Academic Press, New York, 1972)

Mater. Res. Soc. Symp. Proc. Vol. 1052 © 2008 Materials Research Society 1052-DD06-24

Compliant MEMS Motion Characterization by Nanoindentation

Joseph Goerges Choueifati[1], Craig Lusk[1], Xialou Pang[1,2], and Alex A. Volinsky[1]
[1]Mechanical Engineering, University of South Florida, Tampa, FL, 33620
[2]Department of Materials Physics and Chemistry, University of Science and Technology Beijing, Beijing, 100083, China, People's Republic of

ABSTRACT

Large out-of-plane displacements can be achieved when compliant mechanisms are utilized in MEMS. While mathematical and macroscopic modeling is helpful in building original designs, the actual MEMS device motion needs to be characterized in terms of the forces and displacements. A nanoindentation apparatus equipped with Berkovich diamond tip was used in an attempt to actuate and characterize the motion of the Bistable Spherical Compliant Micromechanism with a nonlinear (approximately cubic) mechanical response. Based on the obtained lateral force-displacement data it was concluded that the Berkovich diamond tip was too sharp, thus cutting through the polysilicon material of the MEMS device.

INTRODUCTION

The most common technique used in building MEMS is surface micromachining [1, 2] due its simplicity and low cost. A challenge in using surface micromachining is that it produces essentially two-dimensional products. The ratio of the length and width with respect to the thickness of the elements created is high, thus most MEMS have a planar working space, where the motion of their links traces a single plane [3]. In some applications such as active Braille [4] and micro-optical systems [5], it may be useful for MEMS to achieve accurate three-dimensional motion. This paper provides the results obtained from testing a bistable compliant MEMS device with out-of-plane motion using a nanoindenter. A description of the bistable spherical micromechanism is also presented.

Mechanisms that rely on elastic deformation of their flexural members to carry out mechanical tasks of transforming and transferring energy, force and motion are called *compliant mechanisms* [6]. Furthermore, compliant mechanisms combine energy storage and motion, thus eliminating the need for separate components of joints and springs [6]. Many products currently on the market such as nail clippers, shampoo cap hinges and mechanical pens utilize compliant segments in their designs. In addition, studies have shown that one of the main reasons behind MEMS failure is joints wear [7]; thus replacing rigid multi-pieces joints with compliant single member joints will likely increase the device's lifespan [8-10].

A bistable mechanism is a mechanism that has two stable equilibrium points within its range of motion. A mechanism is considered to be in stable equilibrium if it returns to its equilibrium position after being subjected to small forces or disturbances. A mechanism is in an unstable equilibrium when a small force causes the mechanism to change positions, usually to a position of stable equilibrium. According to Lagrange-Dirichlet theorem, an object is in a stable equilibrium when its potential energy is at its local minimum. We have designed a bistable device using compliant mechanism theory that has its first stable position in the plane of fabrication and its second stable position out of the fabrication plane. The first stable position of the mechanism is shown in Figure 1a and the second stable position of the mechanism is shown in Figure 1b.

EXPERIMENT

The design of the MEMS device followed the rules set by the Multi-User MEMS Processes (MUMPs) chosen for fabrication. The MUMPs is a three-layer polysilicon surface micromachining process, which has the following features:

1) polysilicon is used as the structural material;

2) deposited oxide (PSG) is used as the sacrificial layer and silicon nitride is used as electrical isolation between the polysilicon and the substrate [11].

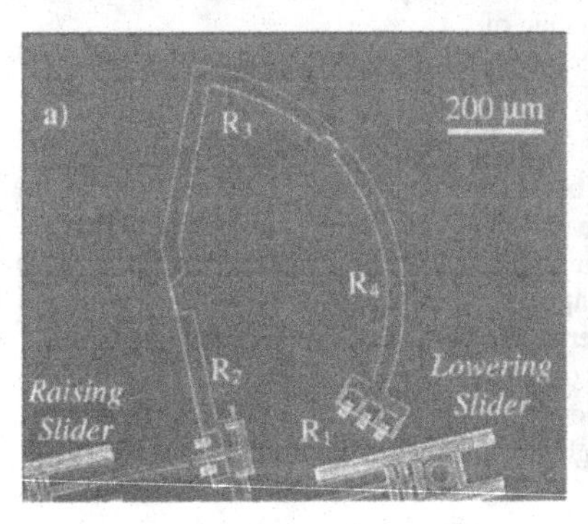

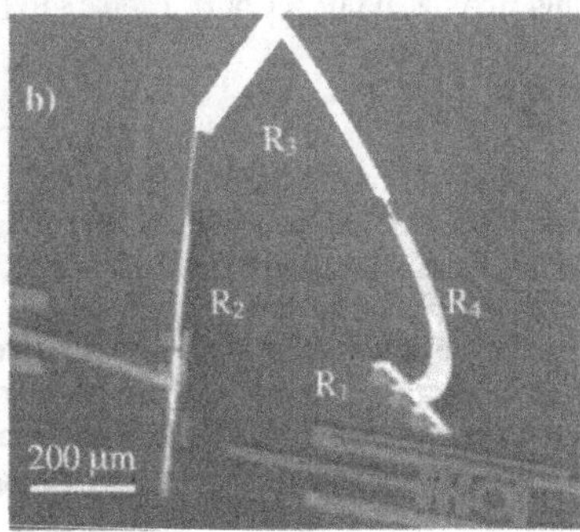

Figure 1. SEM micrograph of the Bistable Spherical Compliant Micromechanism (BSCM) in its stable positions: a) first stable position (as fabricated); b) second stable position (actuated).

Figure 1a is an SEM image of the MEMS device in its fabricated position. The Bistable Spherical Compliant Micromechanism (BSCM) has three basic components: two sliders and a compliant spherical four-bar mechanism with links R_1, R_2, R_3, and R_4. R_1 is the ground link, R_2 the input link, R_3 the coupler link and R_4 the follower link. Links R_2 and R_4 are joined to the substrate by a staple hinge [12] that allows 180° rotation. Link R_3 is connected to R_2 and R_4 by compliant joints as shown in Figure 1a. The axes of rotation of the four joints intersect at a single point. The sliders act as mechanical actuators and are all connected to the input link R_2 by staple compliant hinges. The mechanism in its as-fabricated position is shown in Figure 1a, which is its first stable equilibrium position. By moving the *Raising Slider* to the left, link R_2 will rotate and links R_3 and R_4 will move out-of-plane. In order to bring the mechanism back to its original position, the *Lowering Slider* would be moved to the right. Figure 1b shows the device in its second bistable position as actuated by the mechanical micromanipulator needle probe. The device was switched by the needle probe between the two bistable positions prior to scratch testing in the nanoindenter in order to avoid possible slider stiction. The lowering slider broke during the device actuation, and is no longer connected to the link R_2.

In order to acquire the nonlinear mechanical response of the MEMS device, it was actuated by the Hysitron Triboindenter operating in the scratch mode. The Berkovich diamond tip was used in attempt to actuate the mechanism by scratching and moving the *Raising Slider*. The MEMS device was subject to two consequent displacement and normal force-controlled scratch tests. The normal force and lateral displacement profiles for the first test are shown in Figure 1.

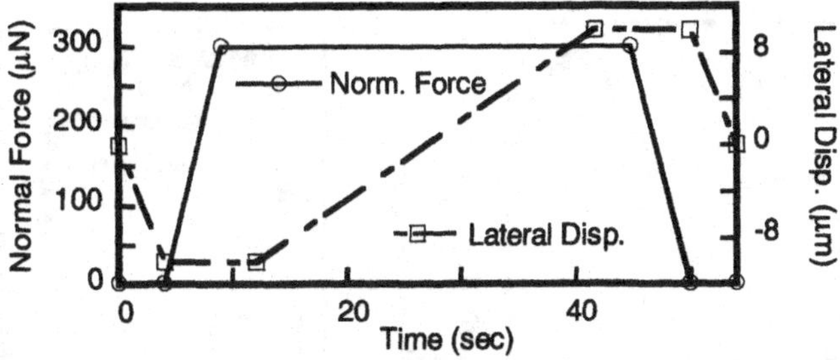

Figure 2. Normal force and lateral displacement profiles for the first scratch test.

In the first test, a maximum normal force of 300 μN and a lateral displacement of 20 μm were applied to the *Raising Slider* (Figures 2 and 3), and in the second test the *Raising Slider* was subject to a normal force of 1000 μN and a lateral displacement of 15 μm (Figure 4).

RESULTS

The two test results are plotted in Figures 3 and 4. The plots in Figure 3 and Figure 4 have similar trends, even though they represent data taken from two different consequent tests performed on the same MEMS device. Figure 3a shows the lateral force and displacement of the indenter tip moving through the hole in the *Raising Slider*, and Figure 3b shows the normal force and displacement.

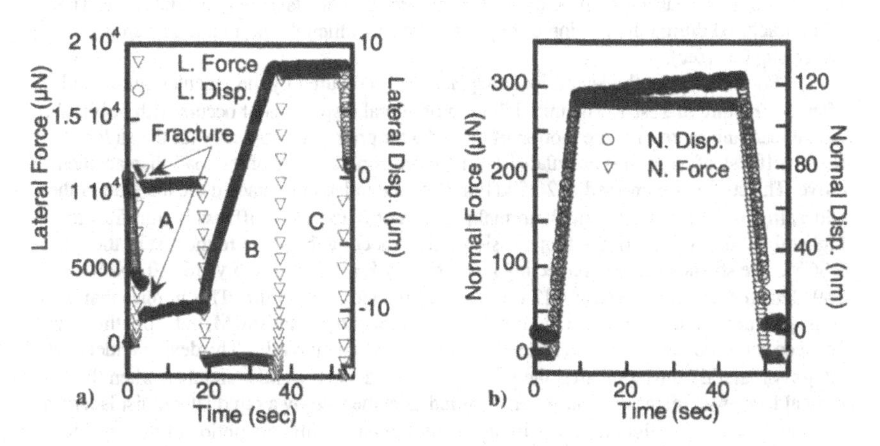

Figure 3. Test 1: a) Lateral force and displacement; b) normal force and displacement.

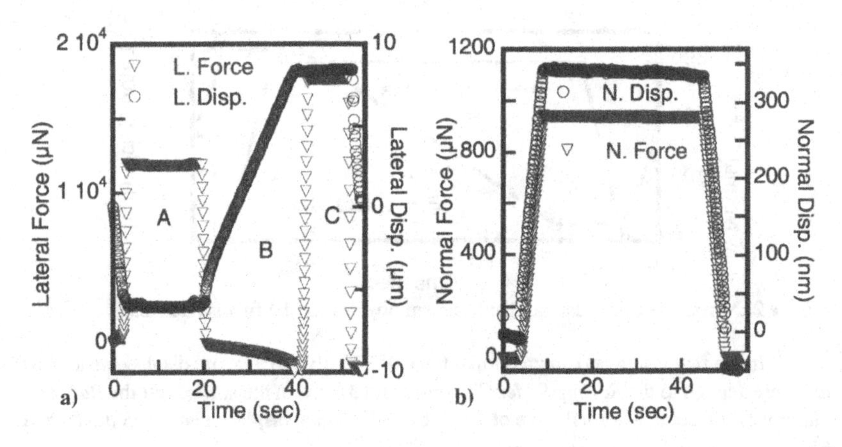

Figure 4. Test 2: a) Lateral force and displacement; b) normal force and displacement.

The plots obtained are significantly different form the expected cubic trend. The mechanical response of the MEMS mechanism could not be obtained. The lateral loads exerted on the *Rising Slider* by the indenter tip were not sufficient to cause the slider to move. On the other hand, the three-sided Berkovich tip edge was sharp enough that the force on the slider concentrated at a point caused its fracture due to high localized stresses. The slider fracture led to the discontinuity in displacement that occurred in part *A* of Figures 3. Here, the fracture point is clearly identified by the lateral force and displacement discontinuity. The slider fractured during the portion of the scratch test in which the normal load started increasing (at 4 sec).

Figure 5 shows the lateral load-displacement measured by the indenter before and after the fracture in Test 1. The initial 7.3 μm of lateral displacement occurs with minimal lateral load and represents the motion of the indenter prior to its contact with the slider. A small stiffness of 34 N/m is calculated from the pre-contact slope of the load-displacement curve. The stiffness increased to 26.5 kN/m when the indenter contacted the left side of the slider ring interior under the small normal pre-load of 2 μN. This stiffness is significantly higher than the stiffness of the transmission rod connecting the slider to the rest of the BSCM. The stiffness of the connecting rod is simply $k = EA/L$, which yields 10 kN/m for $E =$ 169 GPa, and $A = wt$, where $w = 12$ μm, $t = 1.5$ μm, and $L = 275$ μm. This implics that the resistance to the slider's motion did not come primarily from the BSCM, but from the forces between the slider and substrate and between the slider and the rails. The device underwent 0.5 μm of lateral translation after the indenter tip contacted the slider and then when the normal force was increased, the slider fractured as evidenced by a rapid lateral displacement of about 2 μm and a slight decrease in the lateral force resisting the motion of the indenter. The indenter then withdrew laterally from the slider, which gave an unloading stiffness of 51.6 kN/m over the 0.25 μm lateral displacement. After this the indenter tip moved to the right inside the slider ring under high normal load. Due to the pyramidal tip geometry high normal applied force caused a lateral force component, which is represented in the negative lateral force in Figure 5a. Upon complete unloading the MEMS device exerts a lateral force of 2 mN on the tip, which implies that the tip is in mechanical contact with the device. An optical micrograph of the non-damaged slider is shown in Figure 5b for comparison with the one fractured during the test in Figure 5c. Closer slider inspection reveals chipped regions on the slider ring interior. The white triangle drawn in the center of the slider ring suggests the shape and the orientation of the indenter tip.

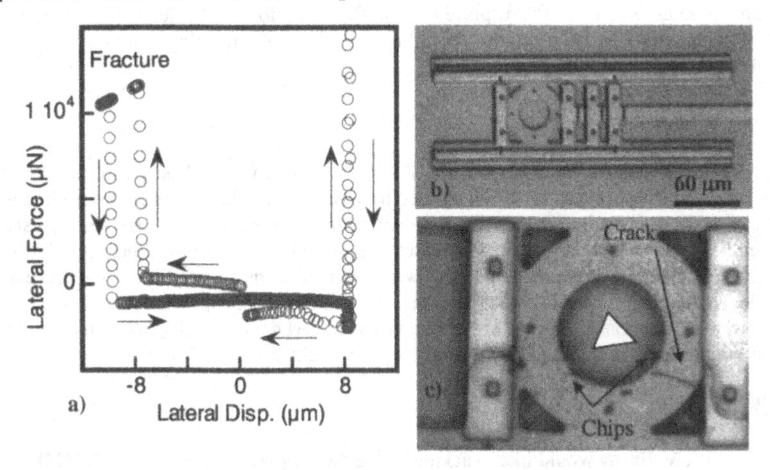

Figure 5. a) Lateral load-displacement profile showing slider fracture; b) non-actuated intact slider; c) Optical micrograph of the slider fracture. The white triangle represents the tip orientation. The inner slider ring diameter is 30 μm.

After leaving contact with the slider ring, the indenter tip moved across its interior and contacted the right side of the ring. On that side the stiffness measured by the indenter increased to 717 kN/m, while the unloading stiffness was 992 kN/m. The difference in stiffness of the slider ring opposite sides may be in part due to its geometry and in part due to the different orientation of the indenter tip when it came into contact with the other slider side.

Similar results where seen in the second test in which the normal force on the indenter tip was increased to a maximum of 1000 μN. A summary of the principal features of the load-displacement history seen in the tests is given in Table 1.

Table 1. Load-displacement history of the two consequent scratch tests.

	Test 1	Test 2
Max normal load	300 μN	1000 μN
Left side loading stiffness	26.5 kN/m	24.6 kN/m
Left side unloading stiffness	51.6 kN/m	56.6 kN/m
Right side loading stiffness	717 kN/m	407 kN/m
Right side unloading stiffness	992 kN/m	588 kN/m

The device loading stiffness is always less than its unloading stiffness, which implies that plastic deformation and/or fracture take place during loading, so the structure does not recover to its original shape. Thus, the structure reaches an unloaded condition with less displacement, which implies a higher stiffness.

The device was successfully switched between the two bistable positions with the needle probe, which has similar geometry to the conical indenter tip. In order to obtain an accurate representation of the mechanical response of the MEMS device, a conical indenter tip should be used, thus spreading the forces exerted on the slider and minimizing the risk of failure. In addition, the scratch displacement profile needs to be modified by removing the service segments A and C (designed for reducing drift while scratching a flat surface) to disable the tip reverse motion.

CONCLUSIONS

Force-displacement testing of a MEMS device was attempted using a nanoindenter. The localized stress caused by the Berkovich tip fractured the slider. The force required to actuate the device resulted in the stress concentration due to point contact between the sharp indenter tip and the slider, producing stresses that exceeded the fracture strength of the slider material. Load and displacement histories were recorded from the two tests, which reveal that the device requires a lateral force greater than 18 mN for its actuation and that the load-displacement relationship for the slider ring of the BSCM device is affected by the normal force magnitude and the indenter tip geometry and orientation.

ACKNOWLEDGEMENTS

Alex Volinsky would like to acknowledge the support from NSF under CMMI-0631526 and CMMI-0600231 grants. Xiaolu Pang would like to acknowledge the support from the State Scholarship Fund of China (No. 20063037).

REFERENCES

1. K. J. Gabriel, "Microelectromechanical Systems (MEMS) Tutorial", IEEE Test Conference (TC), 432-441 (1998).
2. M. Mehregany and M. Huff, "Microelectromechanical Systems", Proceedings of the IEEE Cornell Conference on Advanced Concepts in High Speed Semiconductor Devices and Circuits, 9-18 (1995).
3. C. Lusk, "Ortho-Planar Mechanisms for Microelctromechanical Systems", Ph.D. Dissertation, Brigham Young University, Provo, UT (2005).
4. A.H.F Lam, W.J. Li, Yunhui Liu; Ning Xi, 2002 IEEE/RSJ International Conference on Intelligent Robots and System, vol. 2, 1184 - 1189 (2002).
5. T. Fukushige, S. Hata, A. Shimokohbe, Journal of Microelectromechanical Systems **14**(2) 243-253 (2005).
6. G.K Ananthasuresh, L.L Howell, Larry, Journal of Mechanical Design **127**(4) 736-738 (2005).
7. B. Felton, "Better Robots Through Clean Living", Intec, May 2001.
8. R. Cragun and L. L. Howell, "A New Constrained Thermal Expansion Micro-Actuator", American Society of Mechanical Engineers, Dynamic Systems and Control Division (Publication) DSC, vol. 66, 365-371 (1998).
9. C. D. Lott, J. Harb, T. W. McClain, L.L. Howell, Technical Proceedings of the Fourth International Conference on Modeling and Simulation of Microsystems, MSM 2001, Hilton Head Island, South Carolina, 374-377 (2001).
10. L.L Howell, A. Midha A., ASME Journal of Mechanical Design **116**(1), 280-290 (1994).
11. D. Koester, R. Mahadevan, B. Hardy, and K. Markus, MUMPs Design Handbook. Research Triangle Park, NC: Cronos Integrated Microsystems (2001).
12. K.S.F Pister, M.W. Judy, S.R. Burgett, R.S. Fearing, Sensors and Actuators A, **33** 249-256 (1992).

Mater. Res. Soc. Symp. Proc. Vol. 1052 © 2008 Materials Research Society 1052-DD06-26

Effects of Supercritical Carbon Dioxide on Adhesive Strength between Micro-sized Photoresist Patterns and Silicon Substrates

Chiemi Ishiyama, Akinobu Shibata, Masato Sone, and Yakichi Higo
Precision and Intelligence Laboratory, Tokyo Institute of Technology, 4259 R2-18 Nagatsuta, Midori-ku, Yokohama, 226-8503, Japan

ABSTRACT

Adhesive bend testing of micro-sized photoresist components has been performed to clarify the effects of supercritical CO_2 ($ScCO_2$) treatment. Multiple microsized cylindrical specimens were fabricated on a silicon substrate using epoxy-type photoresist. The specimens were $ScCO_2$ treated at a temperature of 323K at a pressure of 15MPa for 30 min, and then decompressed to atmospheric pressure at two different rates. Double refraction appeared in the SU-8 specimens near the interface between the SU-8 specimens and the substrate after $ScCO_2$ treatment. The adhesive bend strength of $ScCO_2$ treated specimens based on the slow decompression process is approximately 60% higher than that of non- $ScCO_2$ treated specimens; however, the strength of the microsized photoresist was slightly degraded by $ScCO_2$ treatment with the quick decompression process. All the results suggest that $ScCO_2$ treatment can improve the adhesive strength between microsized photoresist components and a silicon substrate with control of the decompression process.

INTRODUCTION

In the last few decades, microfabrication for micro-electro mechanical system (MEMS) devices has made remarkable progress based on photolithography, which utilizes a fine photoresist pattern as a mask during etching process or as a mold for sputtering or plating. Recently, micro-sized elements for MEMS devices have become more complicated and now are often based on three dimensional shapes, therefore, it is desirable that the photoresist pattern is thicker with a high aspect ratio for MEMS devices. However it is difficult to completely clean the micro-sized photoresist patterns with complicated shapes and high aspect ratios using wet cleaning techniques, because the surface tension between the cleaning solvent and micro-sized photoresist patterns interferes with circulation of the fresh solvent throughout all the surfaces of the micro-sized patterns. Recently, supercritical carbon dioxide ($ScCO_2$) cleaning is being explored as an alternative to wet cleaning for sub-micron patterning [1]. $ScCO_2$ has near zero surface tension, high diffusivity and high solubility, thus, it can clean fine photoresist patterns completely. On the other hand, amorphous polymers absorb $ScCO_2$ and swell [2], which causes plasticization of these materials [3]. It has been suggested that the adhesive strength between micro-sized photoresist patterns and the Si substrate would be affected by the $ScCO_2$ treatment. However, there are few studies about the effects of $ScCO_2$ on the adhesive properties.

In this study, adhesive strength between microsized photoresist patterns and silicon substrates is quantitatively evaluated to clarify the effects of $ScCO_2$ treatment. Micro-sized cylindrical shapes were selected as adhesive test specimens, because the simple shape makes it easy to calculate bending stress and to avoid ambiguous loading. Epoxy type photoresist, SU-8, was selected for micro-sized cylinders with high aspect ratio; this photoresist can be manufactured by UV photolithography. Multiple microsized cylinders were fabricated on a silicon substrate and adhesive testing was first performed using a selection of the specimens before the $ScCO_2$ treatment for purposes of comparison with the results of the $ScCO_2$ treated specimens. The remaining specimens were testing after the $ScCO_2$ treatments. The results show the effects of $ScCO_2$ treatment on the adhesive strength of microsized SU-8 on a Si substrate.

EXPERIMENTAL DETAILS

The photoresist used in this study is an epoxy type, SU-8 3050, made by Kayaku Microchem. Multiple micro-sized cylinders of SU-8 were fabricated on a silicon substrate using UV photolithographic technique under the conditions shown in Table 1.

Table 1. Photolithographic conditions for SU-8 cylindrical specimens

Film Thickness		80-100 μm
Soft Baking	Step 1	at 338K for 6 min
	Step 2	at 368K for 60 min
Exposure Dose		300 mJ/cm^2
Post Exposure Baking (PEB)	Step 1	at 338K for 3 min
	Step 2	at 368K for 6 min
Development	Time	10 min
	Developer	SU-8 developer

Figure 1(a) shows a schematic diagram of multiple micro-sized SU-8 cylindrical specimens with 125μm diameter and a height of 80-100μm (film thickness) on a Si substrate. The aspect ratio of the cylindrical specimens is approximately 0.7. Specimen diameter and height were measured using a confocal scanning laser microscope before and after $ScCO_2$ treatment.

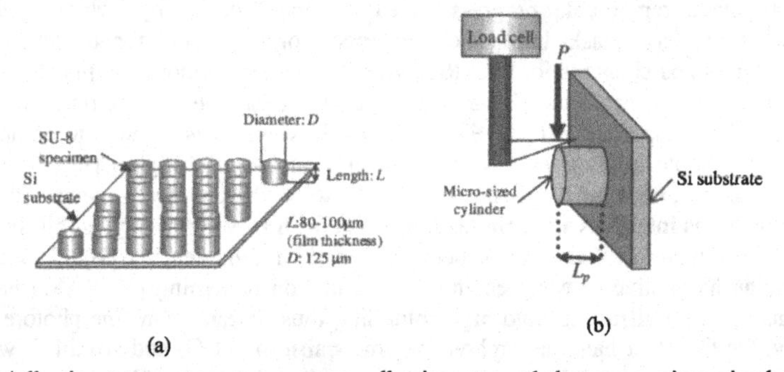

Figure 1. Adhesive test system to measure adhesive strength between micro-sized photoresist samples and substrate
(a) Micro-sized cylindrical specimens, which were fabricated using epoxy type photoresist, SU-8 on Si Substrate. (b) Micro sized adhesive testing system under bend loading condition.

$ScCO_2$ treatment were performed using a $ScCO_2$ treatment system, which composes of CO_2 cylinder, CO_2 delivery pump, a gastight enclosure in an oven and a back pressure regulator as shown in Figure 2. A silicon substrate with micro-sized cylindrical specimens was set on the specimen holder, which is shared with the mechanical testing machine for micro-sized materials, and it was put into a gastight enclosure.

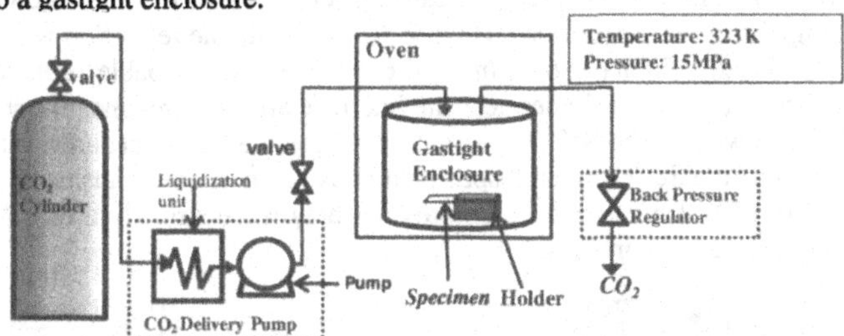

Figure 2. Supercritical CO_2 ($ScCO_2$) treatment system for micro-sized SU-8

A wafer with micro-sized cylindrical specimens was divided into two pieces to be subjected to two different $ScCO_2$ treatments. Table 2 shows the $ScCO_2$ treatment processes for SU-8 cylindrical specimens in this study. Temperature, pressure, and time for $ScCO_2$ treatment were the same for the both pieces, but decompression conditions after the treatment were different. The decompression rate after $ScCO_2$ treatment for specimen A was manual controlled at 0.02-0.03 MPa/s until reaching atmospheric pressure. Specimen B was decompressed at a burst by opening the valve at the back pressure regulator.

Table 2. $ScCO_2$ treatment conditions for SU-8 cylindrical specimens

Specimen Type		Non- $ScCO_2$ treated specimen (Reference)	$ScCO_2$ treated Specimen A	$ScCO_2$ treated Specimen B
Heating	Temperature	None	323K	
CO_2 Deliver	Step 1		at 5 ml/min until 14MPa	
	Step 2		at 2 ml/min	
$ScCO_2$ Treatment	Pressure		15 MPa	
	Time		30 min	
Decompression	Rate		0.02-0.03 MPa/ s	At a burst (Open valve)
	Pressure		Atmospheric pressure	

Micro-sized adhesive bend testing was performed under constant displacement mode using the mechanical testing machine for micro-sized materials, which was developed for a previous study [4]. All the tests were conducted at a crosshead speed of approximately 18μm/min at a temperature of 296 K in a clean room. First, adhesive bend testing was performed on some specimens before the $ScCO_2$ treatment as shown in Figure 1(b) to obtain reference data, and then, the rest of the specimens were tested after treatment.

RESULTS AND DISCUSSION

$ScCO_2$ treated specimen observation

Figure 3 shows a polarizing micrograph of the micro-sized cylindrical specimens before and after $ScCO_2$ treatment. The $ScCO_2$ treated specimens near the interface with Si substrate are clearly changed before the treatment. (Figure 3(d)-(f)). Double refraction was not observed in the non- $ScCO_2$ treated cylinder shown in Figure 3(d), however, double refraction appeared along the edge of the photoresist cylinder near the interface after the treatment as shown in Figure 3(e) and (f). This shows that the $ScCO_2$ treatment changes the SU-8 specimen interface. The double refraction area of the $ScCO_2$ treated specimen B expanded into the inside and it is brighter than that of Specimen A. The sudden decompression may more strongly affect the SU-8 specimens than slow decompression process.

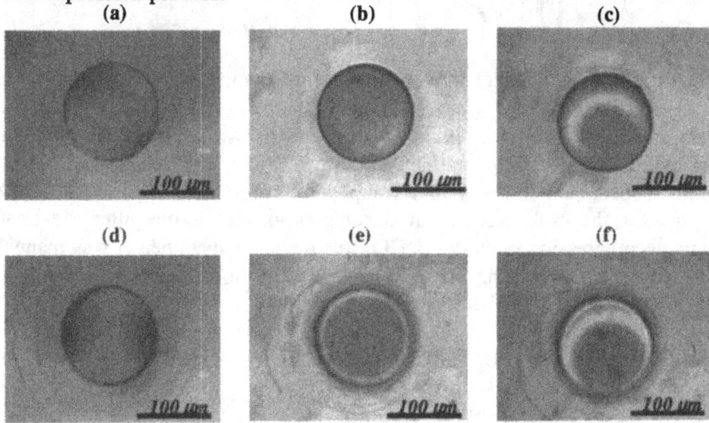

Figure 3. Polarizing micrographs of microsized cylindrical specimens
The above micrographs are focused on the top face of the cylindrical specimens. (a) Micro-sized cylindrical specimens without $ScCO_2$ treatment (non-$ScCO_2$ specimen), (b) $ScCO_2$ treated specimen A and (c) $ScCO_2$ treated specimen B. The below micrographs are focused on the Si substrate. (d) None $ScCO_2$ treated specimen, (e) $ScCO_2$ treated specimen A and (f) $ScCO_2$ treated specimen B.

Table 3 shows the dimensions of micro-sized SU-8 specimen before and after $ScCO_2$ treatment with different decompression rates. Specimen shape was little changed by the $ScCO_2$ treatments. This result shows that little swelling of the SU-8 occurred during the $ScCO_2$ treatment measured after decompression in this study.

Table 3. Effects of $ScCO_2$ treatment on the shape of SU-8 cylindrical specimens using a measuring system with scanning electron microscope

Specimen type	Cylindrical Diamater / μm		Cylindrical Height / μm	
	Before $ScCO_2$ treatment	Afrter $ScCO_2$ treatment	Before $ScCO_2$ treatment	Afrter $ScCO_2$ treatment
$ScCO_2$ treated specimen A	124.9	125.5	88.8	88.5
$ScCO_2$ treated specimen B	124.7	125.2	82.4	82.9

Effects of $ScCO_2$ treatment on adhesive bend strength between micro-sized SU-8 specimens and Si substrate

Figure 4 shows the adhesive bend strengths of the micro-sized SU-8 specimens with and without two different $ScCO_2$ treatments. The adhesive bend strength of $ScCO_2$ treated specimens from group A is significantly improved. The average strength after $ScCO_2$ treatment for this group is approximately 60% higher than that of the reference group. On the other hand, the strength of group B was slightly degraded.

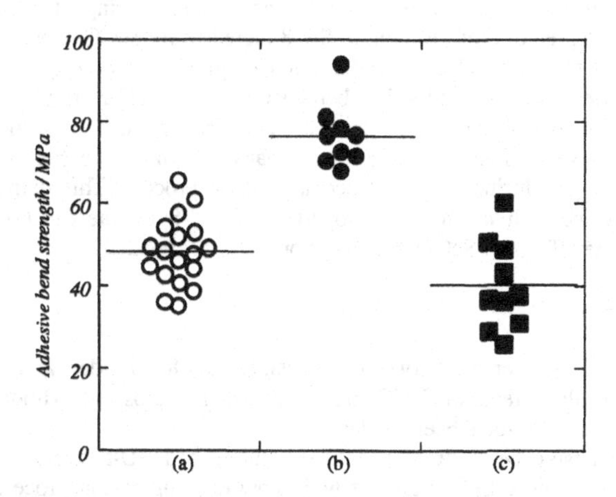

Figure 4. Effects of $ScCO_2$ treatment on adhesive bend strength between SU-8 cylindrical specimens and Si Substrate. Adhesive strength test results obtained from (a) non- $ScCO_2$ treated specimens, (b) $ScCO_2$ treated specimen group A and (c) $ScCO_2$ treated specimen group B. The lines in the plots show the average values for each group.

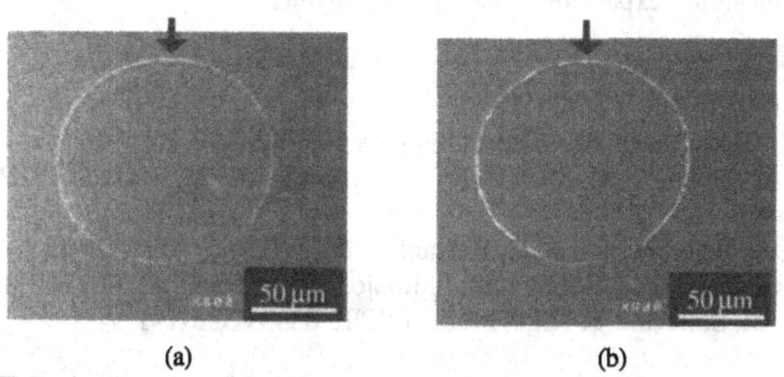

Figure 5. FE-SEM micrographs of the delamination surfaces of the SU-8 specimens after testing a) without $ScCO_2$ treatment and b) with $ScCO_2$ treatment at 323K at 15MPa for 30 min. ($ScCO_2$ treated specimen group A). Arrows in the pictures indicate crack propagation directions.

Figure 5 shows delamination surfaces of SU-8 cylindrical specimens. Fragments of the SU-8 remained on the wafer after testing of all the specimens. It appears that the delamination initiated from the SU-8 near the interface with the Si substrate.

The delamination of the micro-sized photoresist specimens suggests that all the adhesive strengths strongly reflect of the mechanical properties of SU-8. In other words, micro-sized SU-8 should become stronger by $ScCO_2$ treatment with a slow delamination process. This phenomenon is similar to that seen in heat-cured SU-8 specimens in our previous study [5]. The adhesive properties between micro-sized SU-8 specimens and a silicon substrate increased by 46% with increases in the degree of cross-linking by heat curing at 423K for 1hour It is suggested that the degree of cross-linking in SU-8 increases by $ScCO_2$ treatment or the treatment may give the photoresist some effects similar to annealing by heat curing.

On the other hand, the adhesive bend strength of $ScCO_2$ treated specimens slightly decreases with sudden decompression process. This may be due to some damage that was inflicted on the SU-8 specimens by the abrupt expansion, which is caused by the phase change from $ScCO_2$ to the gas during the rapid decompression process. This damage may cancel the effects of improvement in mechanical properties of SU-8 by the $ScCO_2$ treatment. Further researches about the effects of $ScCO_2$ treatment on SU-8 are needed.

CONCLUSIONS

Adhesive bend strength between microsized SU-8 cylinder and substrate has been evaluated to clarify the effects of $ScCO_2$ treatment using microsized cylinder shape specimens. The following conclusions have been made:

1. The adhesive bend strength between microsized SU-8 samples and a Si substrate remarkably improved by $ScCO_2$ treatment with slow decompression process. This may be the result of increases in the strength of SU-8 by the treatment, which may have an effect like heat curing.

2. The adhesive bend strength is slightly degraded by $ScCO_2$ treatment with an abrupt decompression process. It is believed that the sudden decompression may give some damage to the photoresist by abrupt expansion of CO_2 in the polymer.

REFERENCES

1. G. L.Weibel, C. K. Ober, Microelectro. Eng. 65,145 (2003)
2. B. Bonavoglia, G. Storti, M. Morbidelli, A. Rajendran and M. Mazzotti, J. Polym. Phys. Part B. 44, 1531 (2006)
3. S. Al-Enezi, K. Hellgardt and A. G. F. Stapley, Int. J. Polym. Anal. Charact. 12, 171(2007)
4. K. Takashima, Y. Higo, S. Sugiura, M. Shimojo, Mater. Trans. 42, 68 (2001)
5. C. Ishiyama, M. Sone and Y. Higo, Proc. of SPIE. 6533 65331F-1

Mater. Res. Soc. Symp. Proc. Vol. 1052 © 2008 Materials Research Society 1052-DD06-27

Protection Layer Influence on Capacitive Micromachined Ultrasonic Transducers Performance

Edgard Jeanne[1,2], Cyril Meynier[3,4], Franck Teston[3], Dominique Certon[3], Nicolas Felix[4], Mathieu Roy[2], and Daniel Alquier[1]
[1]Laboratoire de Microélectronique de Puissance, Université de Tours, Tours, 37071, France
[2]R&D, STMicroelectronics, Tours, 37071, France
[3]LUSSI CNRS FRE 2448, Université Francois Rabelais, Tours, 37000, France
[4]Vermon SA, Tours, 37000, France

ABSTRACT

For MEMS technology, reliability is of major concern. The implementation of a protection and passivation layer, that may easily enhance reliability of capacitive Micromachined Ultrasonic Transducers (cMUTs) must be done without degrading device performance. In this work, realization, simulation and characterization of passivated cMUT are presented. Two materials, SiN_x and Parylene C, were selected with regard to their mechanical and physical properties as well as their compatibility with device processing. Particular attention was paid on layer deposition temperature to avoid a structural modification of the top aluminium electrode and, hence, a membrane bulge. The characterization results are in good agreement with the simulations. The SiN passivation layer clearly impact device performance while Parylene C effectiveness is clearly pointed out even through ageing characterizations. If SiN_x layer can be used for passivation with particular precautions, Parylene is definitely an interesting material for cMUT passivation and protection.

INTRODUCTION

Capacitive Micromachined Ultrasonic Transducers (CMUTs) are a very attractive solution for high frequency ultrasound generation and, therefore, for medical ultrasound imaging. The first imaging assessment of the cMUT technology revealed a comparable image quality when faced to the standard piezoelectric ceramic technology [1]. A cMUT array element is composed of hundreds cells connected in parallel. A single cell can be assumed to be a parallel plate capacitor that consists of a highly doped polysilicon bottom electrode, a suspended membrane in silicon nitride over a vacuum gap, a top electrode in aluminum with eventually a final passivation layer [2]. To reach immersion requirements, a protection layer is mandatory. In order to efficiently protect the device, the chosen material must fill some essential mechanical and physical properties as well as compatibility with device manufacturing. In this work, after reviewing the protection layer requirements, two different types of materials will be investigated. First, silicon nitride (SiN_x), deposited by PECVD (Plasma Enhanced Chemical Vapor Deposition) can be a material of choice due to its large use in the semiconductor industry, low deposition temperature and its good passivation properties [3]. The parylene C, which appears as a new material for MEMS and packaging issues [4-7], may be an alternative material. Indeed, it exhibits interesting properties such as low Young's modulus, room temperature deposition and high hydrophobicity. Static and dynamic characterizations have been performed on arrays with and without protective layers. Finally, simulations of the key parameters will be compared with the experimental results.

EXPERIMENT

Passivation layer requirements

The passivation layer is used to achieve an efficient protection coating against any chemical and physical aggression in order to ensure the top aluminum electrode's integrity. The cMUT devices will be used in a medical field area and, hence, may be in contact with the human body, biological or living tissues and in liquid medium. In order to provide to the transducers an effective moisture protection, properties such as hydrophobicity and low water and gas permeation are widely recommended as well as a full biocompatibility. Moreover, the cMUT devices are processed in a microelectronic environment. This implies that the chosen material must fit the semiconductor industry requirements in terms of high volume and low cost manufacturing. To be completely implemented in the process flow, the passivation layer must be a conformal deposition technique and be easy to pattern with high selectivity as regard to the underneath material of the cMUT and particularly to aluminum. From a performance point of view, such a layer needs to have as little as possible influence on the device's key parameters. This largely depends on the mechanical material properties but also depends on a stress control of the deposited layer. Finally, the passivation layer must provide an efficient protection over time while being mechanically and thermally stable. As regard to these requirements, PECVD SiN_x and Parylene C seem to be suitable materials. Properties of these materials are summarized in the table I.

TABLE I. SiN_x and Parylene interesting properties for MEMS applications.

Properties	PECVD SiN_x	Parylene C
Mass density	2500 kg.m^{-3}	1289 kg.m^{-3}
Young's Modulus	160GPa	2.5 GPa
Poisson's ratio	0.253	0.4
Dielectric constant	7	2.95
Hydrophobicity	Hydrophilic	Highly hydrophobic

Considering Young's Modulus and the Poisson's ratio from Table 1, simulations have been performed, based on a model previously described in the case of square cell [8]. In this model, the plate equation with non-uniform flexion rigidity and electrostatics is solved using the finite difference method. The static regime is calculated in order to extract the electromechanical coefficient, the collapse voltage and the static capacitance. Then, the dynamic regime is modeled considering small signal amplitude as compared to bias voltage.

Protection layer deposition

cMUT Arrays have been processed on 6" wafers without a final protective coating. The cMUT process is described elsewhere [9]. Different designs have been implemented and tested on the wafers (square and rectangular membranes of various sizes). Two materials have been investigated as a passivation layer: SiN_x deposited by PECVD and Parylene C. The SiN_x deposition was realized at 400°C with a targeted thickness of 600nm. The stress of the film can be easily controlled thanks to a dual frequency deposition method [10]. The resulting stress was slightly tensile, measured after the deposition step to be below 5MPa. For Parylene C, a 1μm thick layer deposition was performed at room temperature after a surface treatment (silanization) in order to promote Parylene adhesion on the wafer. The starting raw material, in this case, is a powder that is vaporized and pyrolized at 680°C prior to cooling down to room temperature in a deposition chamber where Parylene condensates everywhere. It is important to mention that the coverage with Parylene is highly conformal while PECVD SiN_x requires a minimum thickness to ensure full coverage. After the deposition step, both SiN_x and Parylene have been patterned and reactive ion etching was applied in order to have again a contact on both top and bottom electrodes.

Characterization procedure

The membrane dynamic displacement was measured using a commercial laser Doppler vibrometer system (*POLYTEC OFV-3001*). The principle of measurement is based on phase comparison between a signal beam aiming the sample, and a reference beam. These measurements enable us to access the two cMUT key parameters: resonance frequency (f_r) and collapse voltage (V_c). The results, presented in this work, were extracted from measurements done on several cells of various arrays on 6" wafers.

To reach the cell natural frequency, the DC bias voltage dependence of the resonance frequency has to be taken into account. Indeed, the so-called "softening effect" induces a resonance frequency decrease when increasing the DC bias voltage. The linear dependence of the square of the resonance frequency versus the square of the DC bias voltage, that was demonstrated previously for several structures such as a circular shape membrane [11], was also found for square membranes. This allows determining the membrane natural frequency by extrapolation to 0V.

Finally, in order to check the stability of the protective layer, ageing characterizations have been performed. The structures were excited in air during 500 hours at 80% of the collapse voltage and at the corresponding resonant frequency representing around 30×10^{12} up and down cycles for each membrane. A complete mapping giving the homogeneity in both phase and amplitude within 12 cells was performed every 24 hours over 3 weeks.

RESULTS AND DISCUSSION

Simulation Results

The first simulations have been realized in order to understand the influence of the passivation layer. In Figure 1, we present the evolution of the natural frequency versus the thickness of the protection layer, for an initial 450nm thick membrane in LPCVD SiN_x.

The natural frequency evolution indicates a little bowing of the curve for thin layers both for SiN and Parylene with a different orientation. In first approximation, the cMUT can be simply described as a spring mass system. The natural resonance frequency f_0 of a circular membrane is then given by the following equation [12]:

$$f_0 \approx \frac{10.21}{16\pi}\sqrt{\frac{k}{m}} \quad (1)$$

Where k is the spring constant of the membrane, directly related to the stiffness of the material, and m is the mass of the membrane. In this case, for thin SiN_x layers, the natural frequency driving term seems to be the material stiffness and then the mass effect tends to balance out this first predominance. For Parylene, the opposite effect can be observed. Simulations have been performed for different configurations and designs in order to assess this model in comparison with experiments.

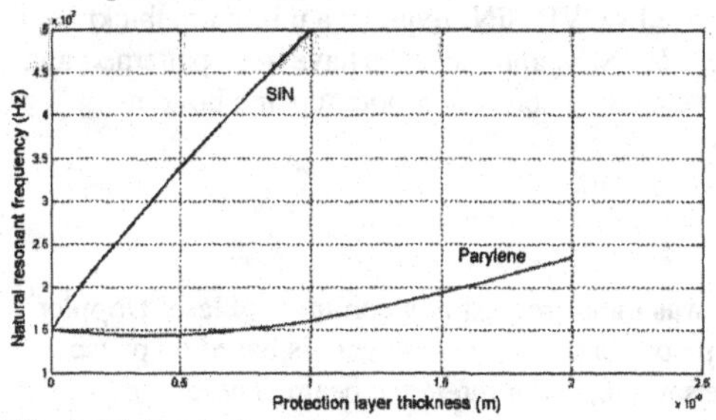

Figure 1. Evolution of the natural frequency versus the thickness of the protection layer

Structural characterization

First, characterizations with an optical microscope (Figure 2-a) show a bulge of the membrane with SiN_x passivation while no, or slight deflection is observed for both Parylene and unpassivated structures. Cell cross-sections have been then prepared by FIB (Focused Ion Beam) and observed on the SEM (Scanning Electronic Microscope). Figure 2-b points out a consistent increase of the gap height just above the top electrode and can be compared to the constant gap height of a cell without passivation in Figure 2-c.

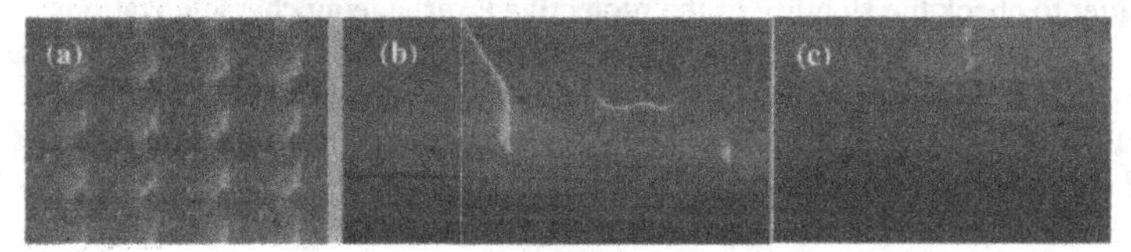

Figure 2. Top view optical image (a) and SEM cross-section of the processed membranes with SiN_x passivation (b) and of a standard membrane without protection layer (c).

The issue may be related to the cutting of the membrane which may have released some residual constraints and then generated an uplift of the second part of the membrane. In order to confirm the membrane shape after the SiN protection layer deposition, a non destructive characterization was performed using a commercial *Fogale nanotech* 3D optical profilometer. Figure 3 shows the obtained 3D profiles of a half elementary cMUT cell with (a) and without (e) SiN_x layer that confirm the shape modification membrane due to the passivation. In such a case, the final membrane is formed by a low-stress LPCVD SiN, a buried aluminum electrode and a low-stress PECVD SiN passivation layer deposited at a temperature higher than 250°C. This last step has probably structurally modified the aluminum layer. Indeed, when heating up, the aluminum stress changes with temperature and it may not return to its starting value when cooled down.

This phenomenon was reported by Modlinski et al. [13]. It indicates that the metal does not change elastically with temperature and even more that a plastic deformation takes place in the metal leading to a higher tensile stress that may reach up to 200MPa. The aluminum stress modification linked with a low membrane thickness results in the bulging of the structures. It is to be noted that since Parylene was deposited at room temperature, any added stress or structural modification of the membrane was produced by this supplementary layer.

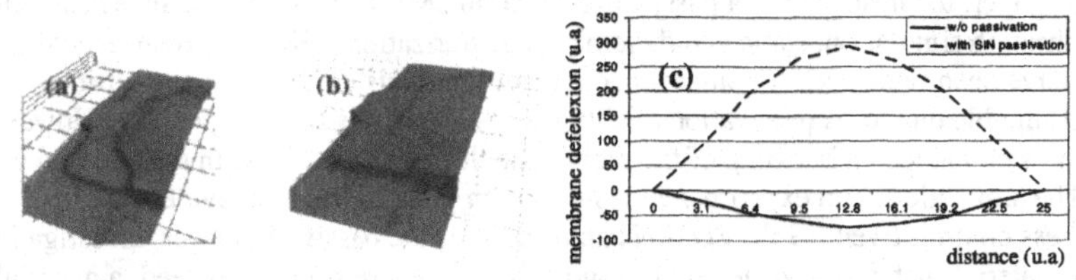

Figure 3. Cell Optical profile with (a) or without (b) SiN_x passivation (c) associated cut line.

Dynamic characterization

The resonance frequency was measured on two different square membranes (20 and 25 µm) and found respectively at 24.5MHz and 17.4MHz for SiN passivation and 16.1MHz and 11.9MHz for Parylene. Measured and simulated natural frequencies are in good agreement with around 10% error for a SiN passivation. As regards to the membrane deformation and aluminum stress modification, mismatch between theory and experiments is extremely low. Even though the stress in the aluminum has increased, in such a way that the shape of the membrane change from concave to convex, the stiffness of the membrane seems to be still low enough to allow membrane bending when applying a DC voltage, without affecting the resonance frequency. Concerning the protection realized by a Parylene layer, a very good agreement between theory and experiments has been achieved as less than 5% deviation has been noticed.

This work demonstrates that PECVD SiN_x can be used as passivation layer with few process adaptations. Nevertheless, several important drawbacks are noticeable. Among them, the evolution of the natural frequency versus layer thickness is extremely fast. PECVD temperature must be decreased as much as possible to overcome changes in the aluminum stress. This will evidently impact process and device homogeneity. In order to circumvent these issues, the investigations on softer materials such as Parylene have been undertaken and have exhibited very interesting results. Actually, in a first approximation, the resonance frequency is proportional to the square root ratio of the spring constant over the mass of the membrane, that is to say proportional to the square root ratio of the Young's modulus over the mass density. The lower this ratio, the lower the dependency of the top material on the membrane performances is. This assumption has been checked by simulation as well as experiments. As reported in Figure 1, the variation of the resonant frequency versus the Parylene thickness is low, which is the advantage of soft materials (low Young's modulus) which allows a fine control of the natural frequency according to the thickness and also offers a large process window.

Concerning the ageing of the structures, the objective is to assess the reliability of the devices with and without Parylene passivation. As regard the performance, a slight decrease of both the collapse voltage and the resonant frequency has been noticed for the structures without Parylene protection while none is observed with Parylene. Hence, Parylene coating stabilizes the cMUT's performance over time evidencing the mechanical stability of the added layer. Finally, no real influence on array homogeneity upon time can be observed by the addition of the protection layer promoting the idea of its integration in the process.

CONCLUSIONS

In this paper, the influence of a passivation layer on cMUT array performance has been enlightened through simulation, static and dynamic characterizations. First, the requirements of such a layer have been reviewed. Among semiconductor materials, PECVD silicon nitride seemed to be suitable due to its passivation quality and its low deposition temperature. However, this deposition step induced a buckling of the membrane through an increase in the stress in the aluminum. The simulations and experimental results are in good agreement, evidencing the possible process control. Even if PECVD SiN can be used as the passivation layer, investigations led on softer material such as Parylene seem to show that it is more suitable. Indeed, a material with a low Young's modulus helps to better control the natural frequency according to its thickness. Moreover, materials, deposited at room temperature have no thermal influence on the stress of the aluminum top electrode. This study points out that Parylene highly improves cMUT's reliability. Finally, it can be concluded that Parylene is definitely a better candidate than the PECVD SiN, exhibiting very high barrier functions with minimized and highly controllable impact on CMUT array behavior without any modification of design rules.

ACKNOWLEDGMENTS

The authors would like to thank Dr. J. TERRY and Dr. L. HAWORTH from the Scottish Microelectronics Centre for their knowledge in Parylene. This work was done within MEMSORS EM89 project labeled by the EUREKA's cluster program EURIMUS and funded by the French Ministry of Finances / Direction Générale des Entreprises (DGE).

REFERENCES

1. N. Felix, in *Smart System Integration 2007*, edited by T. Gessner, 99-105 (2007).
2. B. Belgacem, N. Yaakoubi, E. Jeanne, M. Roy, R. Jerisian, D. Alquier, *Proc. Eurosensors XIX*, WPb 28 (2005).
3. P. L. Ong, J. Wei, F. Tay, C. Iliescu, *J. Phys.: Conference Series*, **34**, 764-769 (2006).
4. T. J. Yao, X. Yang and Y.-C. Tai, *Sens. Actuators A*, **98**, 771-5 (2002).
5. T. J. Yao, *Parylene for MEMS applications*, Ph.D. thesis, California Institute of Technology, CA, (2002).
6. C. Karnfelt, C. Tegnander, *IEEE Trans. Microwave Theory and Techniques*, **54**, 8 (2006).
7. J. M Chen and J. J. Zhao, *Solid-State Integ. Circuit Tech.*, 569-571 (2006).
8. D. Certon, F. Teston, F. Patat, *cMUT Special Issue in IEEE-UFFC*, **52**, 12, 2199-2210 (2005).
9. B. Belgacem, D. Alquier, P. Muralt, J. Baborowski, *J. Micromech. Microeng.*, **14**, 299-304 (2004).
10. E. Cianci, A. Schina, A. Minotti, V. Foglietti, *Sens. Actuators A*, **127**, 80-87 (2006).
11. P. Attia, P. Hesto, *Mat. Res. Soc. Symp. Proc.*, **518**, 3-8, (1998).
12. B. Belgacem, PhD. Thesis, Tours, France (2004).
13. R. Modlinski, *Microelectron. Reliab.*, **44**, 1733-1738 (2004).

Mater. Res. Soc. Symp. Proc. Vol. 1052 © 2008 Materials Research Society 1052-DD06-29

Fabrication of C54-$TiSi_2$ Thin Films Using Cathodic Arc Deposition and Rapid Thermal Annealing

Hui Xia[1], William R. Knudsen[2], and Paul L. Bergstrom[1,3]

[1]Department of Materials Science and Engineering, Michigan Technological University, Houghton, MI, 49931

[2]College of Engineering, Michigan Technological University, Houghton, MI, 49931

[3]Department of Electrical and Computer Engineering, Michigan Technological University, Houghton, MI, 49931

ABSTRACT

An enabling material for high density microelectronics technologies, C54-$TiSi_2$ thin films can be used in many related integrated microsystem technologies to reduce the RC delay and improve the dynamic performance due to its low electrical resistivity and high thermal stability. In this paper, C54-$TiSi_2$ thin films were prepared for the first time using cathodic arc deposition with rapid thermal annealing. The impact of energetic ion bombardment on the film microstructure and subsequent C49-C54 phase transformation during annealing were studied. The $TiSi_2$ compound was used as the cathode material and substrate bias was varied to control the ion energy during the film growth. Rutherford backscattering spectrometry and transmission electron microscopy were utilized to characterize the film composition and microstructure. The composition of the resultant $TiSi_x$ thin films varied from x=2.4 to x=1.4 when the substrate bias was varied from a floating self-bias to –200V. The films deposited at room temperature were amorphous with a phase separation at the nano scale. The Si atoms were seen to segregate on the boundary of Ti-rich domains and the domain size increased with the magnitude of the substrate bias. For a 90nm-thick $TiSi_2$ film deposited on a SiO_2/Si substrate, the kinetics of the C49-C54 phase transformation was studied by measuring the change of film resistivity upon rapid thermal annealing. It was found that the C49-C54 phase transition temperature was higher (>900°C) for the arc-deposited $TiSi_2$ thin films compared to evaporated or sputtered films. The activation energy of the C49-C54 transformation was calculated to be 6.1±0.2eV.

INTRODUCTION

Due to its low electrical resistivity and high thermal stability, C54-$TiSi_2$ is a promising material in some MEMS applications, such as RF MEMS, to reduce the RC delay and improve the dynamic performance. Compared to heavily doped polysilicon, C54-$TiSi_2$ has a much lower resistivity (15–20μΩ·cm) and similar mechanical properties, thus it can serve as either a highly conductive coating or structural material in surface micromachining. C54-$TiSi_2$ is usually prepared by annealing a thin layer of Ti in contact with Si substrates or through the crystallization of co-deposited amorphous $TiSi_2$ thin films [1]. In both cases, the stable orthorhombic face-centered C54-$TiSi_2$ is formed through the phase transformation from a metastable orthorhombic base-centered C49-$TiSi_2$ during high temperature annealing. The C49-C54 phase transformation is a critical step in the fabrication of C54-$TiSi_2$ thin films and it has been reported to be influenced by many factors such as the film thickness, film substrates, C49-$TiSi_2$ grain size, and film deposition methods [2-5].

To date, most $TiSi_2$ thin films have been prepared by sputtering or evaporation. Little study has been carried out to synthesize C54-$TiSi_2$ using cathodic arc film deposition prior to this reporting. Cathodic arc deposition is a low-voltage, high-current plasma discharge in which the electric current is conducted through plasma consisting of ionized material emitted from the cathodic electrode. As an ion-beam based technique, cathodic arc film deposition can be used to prepare materials with modified microstructure and properties [6]. In this study, $TiSi_2$ thin films were fabricated for the first time using cathodic arc deposition to study the impact of energetic ion bombardment on the film microstructure and subsequent C49-C54 phase transformation during thermal annealing.

EXPERIMENTAL DETAILS

$TiSi_x$ thin films were deposited using an arc deposition apparatus as shown schematically in Figure 1. A $TiSi_2$ (compound) sputtering target with a purity of 99.5% (Kurt Lesker, Co.) was used as the cathode and the anode was the grounded chamber. A stainless steel substrate holder was positioned opposite to the cathode at a distance of ~22cm from the target. A DC power supply (DCG-10DA, Shindengen Electric Mfg. Co. Ltd., Tokyo, Japan) was connected to the substrate holder to adjust the substrate bias from a floating self-bias (–20~ –50V) to –200V. More details on the arc apparatus were described in [7]. The arc was ignited by a momentary mechanical trigger contact of a tungsten (W) rod (held at the anode potential) with the cathode surface. During the film deposition, the arc was operated in a pulsed mode with the conditions: 100A peak current, 40A background current, 300 pulses per second (PPS), and 2ms pulse width. The deposition chamber was evacuated to a base pressure below 2.0×10^{-6}Torr. A low flow of high purity (99.998%) Ar gas was introduced in the chamber during the deposition process to enhance the arc stability. The $TiSi_x$ thin films were deposited at room temperature on a thermal oxide (450nm) coated Si substrate with an Ar gas pressure of 3mTorr. Rutherford backscattering spectrometry (RBS) was used to measure the film composition and thickness. Transmission electron microscopy (TEM) was utilized to characterize the film microstructure. Rapid thermal annealing (RTA) of the $TiSi_2$

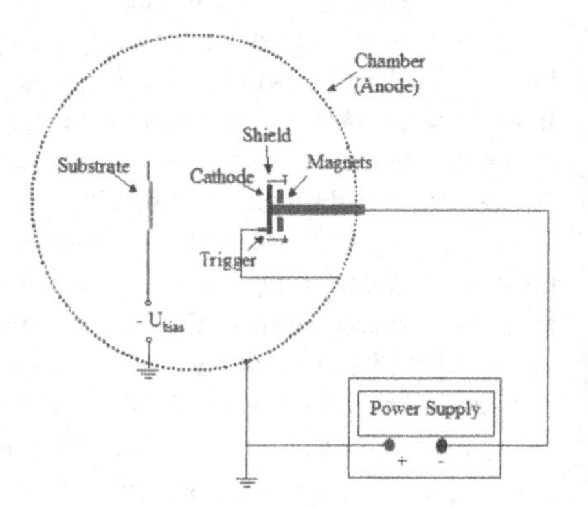

Figure 1. Schematic diagram of the pulsed cathodic arc film deposition apparatus.

film deposited at –100V was carried out in two steps in forming gas (95%Ar + 5%H_2) with an AG Associates Heatpulse 610 system. The heating rate was controlled at 5°C /s. First, the film was annealed at 650°C for 120s in order to obtain the C49-$TiSi_2$ phase. The phase conversion to C54-$TiSi_2$ was investigated by performing isothermal annealing at different temperatures between 915~950°C. The annealing was interrupted for sheet resistance measurements of the annealed films with a four-point probe at different time.

RESULTS AND DISCUSSION

Film deposition

The dependence of the $TiSi_x$ film composition on the substrate bias is shown in Figure 2. It was found that the Si content in the films decreased significantly with increasing the magnitude of the substrate bias. A film deposited at a floating substrate bias contained a higher Si content (x=2.40±0.05) than the $TiSi_2$ cathode. At a substrate bias of –200V, the ratio of Si/Ti in the films dropped to x=1.40±0.05. The stoichiometric $TiSi_2$ film was obtained at a substrate bias of approximately –100V. Figure 3 shows the film deposition rate versus substrate bias. It was found that the film deposition rate decreased with increasing the substrate bias, indicating that a fraction of the adatoms on the film surface were ablated instead of being deposited under the ion bombardment of the incident material flux.

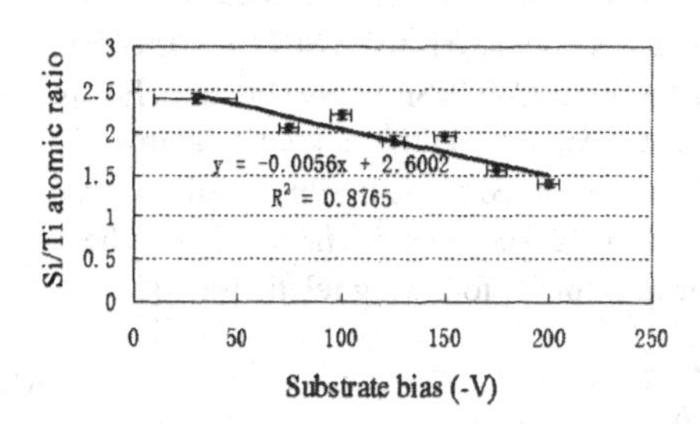

Figure 2. $TiSi_x$ film composition vs. substrate bias.

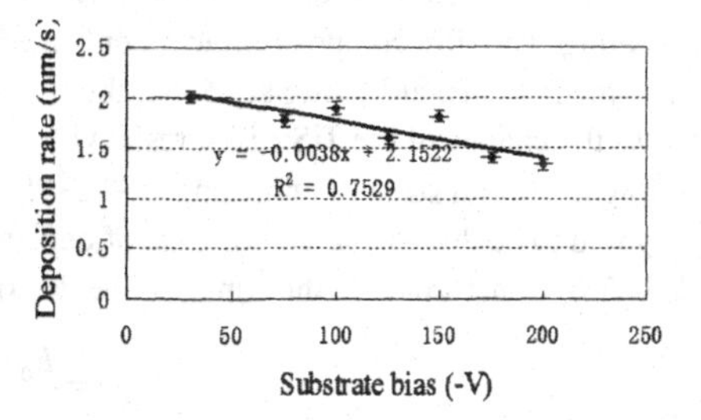

Figure 3. Film deposition rate vs. substrate bias.

Figure 4 shows high-resolution top view TEM images of the $TiSi_x$ thin films deposited at different substrate biases at room temperature. The as-deposited $TiSi_x$ films are amorphous. There is an obvious phase separation within the amorphous phase with increasing the substrate bias, as seen by the existence of nano domains with alternating dark and bright contrast. From the shape and distributions of the domains, it is surmised that the dark regions are rich in Ti and the light regions are rich in Si due to the atomic mass contrast. Compared to sputter deposited or evaporated films, the observed phase separation in the amorphous $TiSi_x$ films is interesting and unique. It is believed that the appearance of the phase separation is closely related to the increased atom surface mobility and the local atomic heating during the arc deposition.

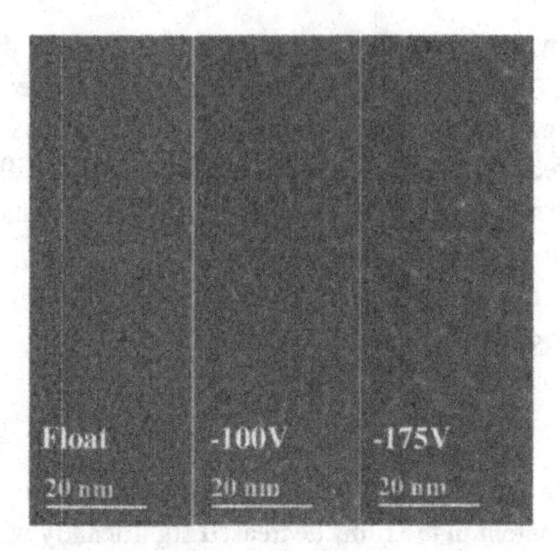

Figure 4. High-resolution TEM images of the $TiSi_x$ thin films deposited at room temperature with different substrate bias, showing phase separation within the amorphous phase. The dark and bright contrast in the image is mainly due to the atomic mass contrast, corresponding to the Ti- and Si-rich regions, respectively.

Film annealing

A 90nm-thick $TiSi_x$ film was deposited at –100V substrate bias for the study on the film annealing. The film composition was confirmed by the RBS analysis to be Si:Ti = 2.02±0.05. After the first step annealing at 650°C, the film had a single C49-$TiSi_2$ phase structure. Figure 5 shows the change of the $TiSi_2$ film resistivity versus anneal time at different temperatures. During the initial stage of annealing at 915~950°C, the rapid reduction in film resistivity was proposed to be due to a large decrease of defects in the C49-$TiSi_2$ phase. The fraction of the C54-$TiSi_2$ transformed in the films can be determined from the following relationship [2]

$$X(t) = \frac{R_0 - R(t)}{R_0 - R_f} \qquad (1)$$

where $X(t)$ is the fraction of the film transformed into C54-$TiSi_2$, R_0 is the film resistivity of the C49-$TiSi_2$ before the transition, R_f is the value of the C54 phase, and $R(t)$ is the measured time-dependent film resistivity. Using R_0=72μΩ·cm and R_f=20μΩ·cm, the calculated C54-$TiSi_2$ phase fraction versus anneal time at three different anneal temperatures is plotted in Figure 6. The Johnson-Mehl-Avrami analysis [8] was used to study the kinetics of the C49-C54 phase transformation. Figure 7 is the Arrhenius plot of the time to transform half of the $TiSi_2$ film into the C54 phase in the temperature range of 915~950°C. The effective activation energy of the C49-C54 phase transformation was determined from the slope of the plot to be 6.1eV±0.2eV. The nucleation mode can be extracted from the slopes of the log-log plots, as shown in Figure 8, resulting in n =1.35±0.11. This value is close to n =1, indicating that the nucleation of C54-$TiSi_2$ occurred mostly at the C49 grain boundaries [2].

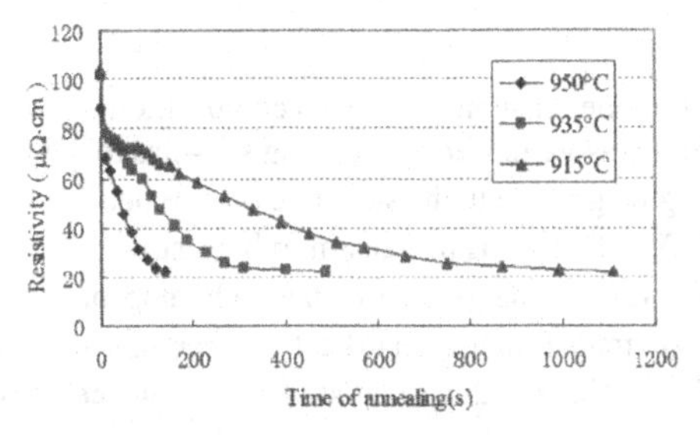

Figure 5. Film resistivity vs. time during the C49- to C54-$TiSi_2$ phase transformation.

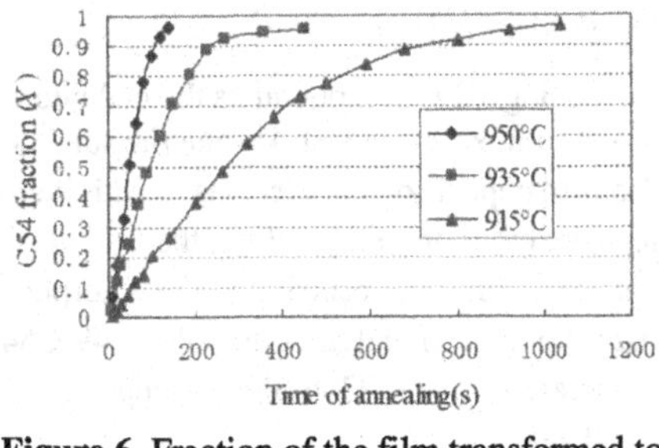

Figure 6. Fraction of the film transformed to C54-$TiSi_2$ vs. time during annealing.

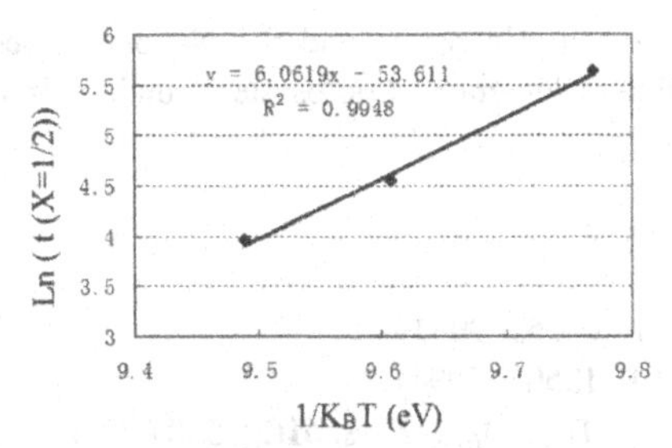

Figure 7. Arrhenius plot of the time to transform half of the film into the C54 phase.

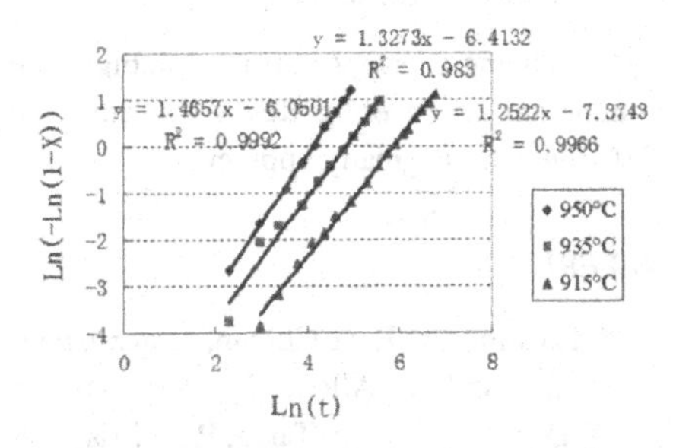

Figure 8. Log-log plot to determine the nucleation mode of the transformation.

The C49-C54 phase transformation in $TiSi_2$ thin films is reported to occur at 750~850°C with an activation energy spread in the range of 3.5~6eV, depending on the details of fabrication process [2-4]. For a $TiSi_2$ film of similar thickness and deposited on a SiO_2 substrate, Thompson et al. [3] reported that the C49-C54 transformation occurred at 800°C for a co-evaporated $TiSi_2$ film with an activation energy of 4.55±0.05eV. In this study, we found that the C49-C54 phase transformation in the arc-deposited $TiSi_2$ films was impeded to higher annealing temperatures (>900°C) with a larger activation energy. This could be related to film characteristics associated with the cathodic arc deposition. Under the ion bombardment during the arc deposition, surface atoms of the films are knocked into the interstitial positions beneath and within the film surface. These closely packed atoms in the films could be inherited in the C49-$TiSi_2$ that transformed from the as-deposited films after annealing. The interstitials could slow down the diffusion of atoms during the C49-C54 structure transformation.

CONCLUSION

Using $TiSi_2$ compound as the cathode material, the resultant $TiSi_x$ film composition varied from x=2.4 to x=1.4 when the substrate bias was varied from a floating self-bias to –200V. The film deposition rate decreased with increasing magnitude in the substrate bias. When deposited at room temperature, the as-deposited $TiSi_x$ thin films had an amorphous structure with a phase separation occurring in the amorphous phase at the nano scale with an inhomogeneous distribution of Ti and Si atoms. The C49-C54 phase transition was retarded to higher annealing temperatures (>900°C) in the arc-deposited $TiSi_2$ thin films with a calculated activation energy of 6.1±0.2eV.

ACKNOWLEDGEMENTS

Financial support from Michigan Technological University and the National Science Foundation Engineering Research Center for Wireless Integrated Microsystems under contract EEC-9986866 is greatly appreciated.

REFERENCES

1. J. P. Gambino, E. G. Colgan, *Mater. Chem. and Phys.*, **52**, 99 (1998).
2. Z. Ma and L. H. Allen, *Physical Review* B **49**(19), 13501 (1994).
3. R. D. Thompson, H. Takai, P. A. Psaras, and K. N. Tu, *J. Appl. Phys.* **61**(2), 540 (1986).
4. H. J. W. V. Houtum, I. J. M. M. Raaijmakers, and T. J. M. Menting, *J. Appl. Phys.* **61**(8), 3116 (1987).
5. J. Chang, G. B. Kim, D. S. Yoon, H. K. Baik, D. J. Yoo and S. M. Lee, *Applied Physics Letters*. **75**, 2900 (1999).
6. I. G. Brown, *Annu. Rev. Mater. Sci.* **28**, 243 (1998).
7. H. Xia, Y. Yang, and P. L. Bergstrom in *Low-Temperature Silicon Films Deposition by Pulsed Cathodic Arc Process for Microsystems Technologies*, edited by G. Ganguly, M. Kondo, E. A. Schiff, R. Carius, R. Biswas, (Mater. Res. Soc. Symp. Proc. **808**, Pittsburgh, PA, 2004) pp. A9.38.1.
8. J. W. Christian, *The Theory of Transformation in Metals and Alloys,* Part I, Pergamon, Oxford. (1975).

Mater. Res. Soc. Symp. Proc. Vol. 1052 © 2008 Materials Research Society 1052-DD06-30

New Experimental Approach for Measuring Electrical Contact Resistance With an Accurate Mechanical Actuation, Evaluation of the Performances of Gold Micro-Switches

Cedric Seguineau[1,2], Adrien Broue[1], Fabienne Pennec[3], Jérémie Dhennin[1], Jean-Michel Desmarres[4], Arnaud Pothier[5], Xavier Lafontan[1], and Michel Ignat[2]
[1]NOVA MEMS, 10 Avenue de l'Europe, Ramonville, 31520, France
[2]SIMaP, INPG, UMR 5266, Université Joseph Fourier, St Martin d'Heres, 38402, France
[3]LAAS - CNRS, 7 Avenue du Colonel Roche, Toulouse, 31401, France
[4]DCT/AQ/LE, CNES, 18 Avenue Edouard Belin, Toulouse, 31401, France
[5]Minacom dpt, XLIM Laboratory, 123 Avenue Albert Thomas, Limoges, 87060, France

ABSTRACT

A specific experimental setup combining nanoindentation and electrical inputs has been developed in order to determine the reliability and the performances of Micro-ElectroMechanical Systems (MEMS) like micro-switches. The evolution of the electrical resistance with respect to a mechanical solicitation applied on the contact, is henceforth available. The description of the set-up goes with a brief overview of the tests performed on a gold ohmic switch. A discussion is developed considering the mechanisms involved in the contact response. A confrontation among the experimental results, the analytical modeling and also finite-element analysis is presented.

INTRODUCTION

Micro-switches always elicit a great interest among micro- and nano- technologies. Yet, despite the considerable efforts put on this field, their Technology Readiness Level [1] is still globally lower than other MEMS devices like accelerometers. The best compromise between reliability and electrical performances strongly depends on the mechanical properties of the materials and surfaces forming the electrical contact. The establishment of a database based on the same models as those made in classical microelectronics is still problematic to implement. Currently, reliability testing can then only be addressed with extensive and precise experimental characterizations. In this context, rightly customized nanoindentation experiments can provide convenient results, by combining the use of a diamond tip as a mechanical micro-actuator and electrical connections for a 4-wire resistance measurement. Such a technique presents several advantages in comparison with traditional electrostatic or magnetic actuations. First, the actuation load is more reproducible, and better controlled. Secondly, the mechanical and electrical effects can also be handled separately in order to better understand the physics involved in the contact evolution. The mechanical behavior of the material at the contact surfaces must be accurately determined, and properly related to the electrical current transmitted. The analysis described here is accordingly based on three complementary approaches. The experimental characterization provides electrical and mechanical measurements, while an analytical modeling of the experiments would allow to determine which physics governs the behavior of the contact. Finally a new finite-element procedure taking into account the true distribution of contact asperities has been tested, validating the analytical assumptions. The experiments shown here have been realized on gold ohmic switches made by XLIM laboratory in the frame of SMARTIS project, EURIMUS labeled.

EXPERIMENT

Samples description

The switches are obtained through 6 main steps shown on figure 1. They are made on a polished alumina substrate using standard micromachining and UV photolithography techniques. The devices are cantilevers with two 10x10μm "contact fingers" ensuring the electrical contact between the free-standing part and the other electrode.

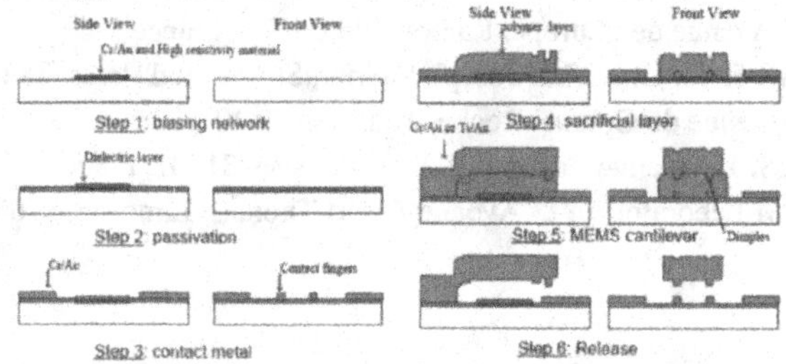

Figure 1: 1) Cr/Au evaporated and layer patterned to form biasing lines and pull-down electrodes. 2) insulating alumina layer (PECVD technique). 3) first metallization (Cr/Au bilayer evaporated) patterned by UV lithography to define contact finger and access RF line, followed by selective metallization electroplating. 4) sacrificial layer deposited and patterned (2 steps). 5) Ti/Au bilayer evaporated (150nm) before thick gold electroplating layer. 6) sacrificial layer removed and supercritical CO_2 drying of the switch

Experimental set-up

Nanoindentation has been developed to characterize the intrinsic mechanical properties of thin structural materials [2]. The experiment is based on a diamond tip driven into the surface of the sample. Properties are calculated from two measurements: the load applied on the tip and the resulting displacement. Yet, the accuracy of the system, respectively 10^{-9}N and 10^{-10}m, make this set-up valuable for mechanical actuations of micro-switches. Indeed, the load applied on the electrical contact by the nanoindenter tip can then be accurately determined if the location of the latter above the sample is closely controlled. Thus it can be used to substitute the actual electrostatic or magnetic actuation load by a more reproducible and better controlled mechanical load. The latter is correlated with the contact pressure as soon as information on the actual area of contact is available. The hard point is rightly the location of the tip. A solution can be found in the literature to get round that issue: one can use the nanoindenter tip column as one of the two electrodes [3]. A specific tip is processed and coated with the first material, and the sample is constituted by a wafer coated with the other one. Nonetheless, that kind of characterization makes any test rather far from the actual conditions of use of the MEMS. On the contrary, our experiment which uses the standard tip in order to bend the free-standing electrode is more suitable for the characterization of actual switches. The issue of the tip location has been sorted out by using a piezoelectric actuated sample-holder [4]. Once the switch is correctly positioned, the tip is brought in contact with the surface of the mobile part, and then lowered until the mechanical contact occurs. The stiffness of the membrane can be measured in this step [4]. The load applied in addition, F_c, is then linked with the pressure on the contact.

In parallel, the electrical measurements are performed by the use of a 4-point probe: a current flow I is applied and the potential drop V is independently measured. Cold and Hot switching are achievable by controlling the voltage compliance U_{comp}. A schematic view of the set-up is shown on figure 2, as well as the main inputs and outputs used for this experiment. The results will be commented in the discussion.

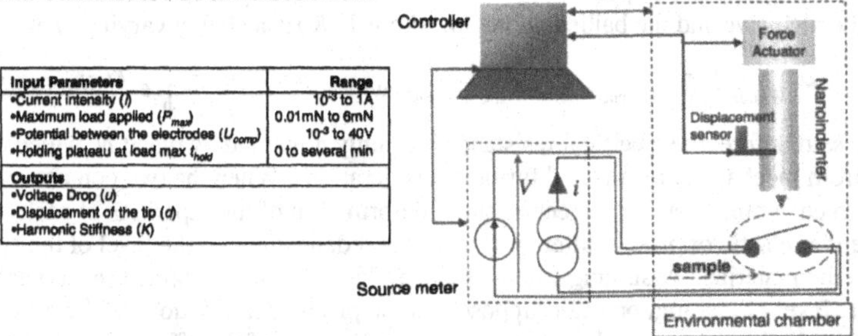

Input Parameters	Range
•Current intensity (I)	10^{-3} to 1A
•Maximum load applied (P_{max})	0.01mN to 6mN
•Potential between the electrodes (U_{comp})	10^{-3} to 40V
•Holding plateau at load max t_{hold}	0 to several min
Outputs	
•Voltage Drop (u)	
•Displacement of the tip (d)	
•Harmonic Stiffness (K)	

Figure 2: schematic view of the experiments and main I/O parameters. The nanoindenter is a XP Nanoindenter (MTS Nano Inst.) equipped with DCM, CSM and nanopositionning options. The source meter is a Keithley 2420. For the experiments on the XLIM switch, I = 1mA (Cold Switching), Fc_{max}=30µN, and t_{hold}=40s (several actuations, Fc_{max}=60µN, have been done before)

DISCUSSION

Analytical approach for the electrical contact resistance

An analytical modeling of the electrical contact has been realized, based on different existing theories. A more detailed description of this approach can be found in [5]. The aim is to estimate the electrical resistance of contact R_c, which depends, on one hand, on the mechanical behavior (which is directly related to the area of contact), and on the other hand on electrical assumptions. Indeed, R_c depends on the area of contact but also on the way the electrons are flowing through. The latter is closely dependent to the ratio between the size of the contact spot and the electron mean free path of the material. That is why assumptions must be done in order to provide an estimation of the contact area from the measurements. Three kinds of transportation mechanisms of the electron have been identified (ballistic, quasi-ballistic or ohmic transport)

If considering a single circular spot of contact, "Ohmic contact" means that the contact size a, that is to say its radius, is at least one order of magnitude higher than the mean free path l_e of the electrons in the material ($l_e << a$). The contact resistance is then dominated essentially by a diffuse scattering mechanism, and is given by the Maxwell spreading resistance formula:

$$R_{C\,Maxwell} = \frac{\rho}{2a} \qquad (1)$$

where ρ is the resistivity of the metallic material. As the current actually flows through multiple asperities, three different models can be used to find approximate solutions: Holm's, Greenwood's and Boyer's by taking into account the interactions between the asperities.

On the contrary, the "ballistic contact" model may be applied when l_e is much larger than a ($l_e >> a$). The conduction is then dominated by the semi-classical Sharvin's resistance:

$$R_{C\,Sharvin} = \frac{4\rho K}{3\pi a} \quad \text{where } K = \frac{l_e}{a} \qquad (2)$$

There is no interaction between the contact spots, and then a is then the effective radius of contact (for N asperities, $a_{effective} = N.a$)

Neither of these two models is applicable directly on MEMS because generally l_e and a are of the same order of magnitude ($l_e \sim a$). Wexler proposes a solution of the Boltzmann's equation by using variational principle to maintain the continuity of the conduction behavior between the diffusive and the ballistic domain, where $\Gamma(K)$ is a slowly varying Gamma function:

$$R_{C\,Wexler} = R_{C\,Sharvin} + \Gamma(K) R_{C\,Maxwell} \quad \text{where} \quad \Gamma(K) \approx \frac{2}{\pi} \int_0^{\infty} e^{-Kx} Sinc(x)dx \tag{3}$$

As said before, the electrical resistance is mainly governed by a and N, which can be obtained from mechanical results and further considerations. When the two contact surfaces collide with each other, the main mechanism of deformation of the asperities of each contact could be either elastic or plastic (after an elastic stage) depending on the level of the stresses applied on the materials. Assuming that the surfaces have already sustained several actuations with higher level of loading, one can suppose that the plastic deformations of the asperities undergone under harsh cyclic solicitations lead to an increase of the effective area of contact and then to a reduction of the actual sustained stress, until the behavior become mainly elastic (stress of the same order of magnitude as the increased Yield point).

The contact area and the contact load F_c can be linked to the radius of the contact spot a and to the vertical displacement d by Hertz's theory:

$$a = \sqrt[3]{\frac{3f_c R}{4E^*}} = \frac{3f_c}{4E^* d} = \frac{3F_c}{4NE^* d} \quad where \quad d \approx \frac{a^2}{R} \tag{4}$$

with f_c is the load applied on one asperity, R its radius and E^* is the effective Young's modulus. A rough estimation is made on the distribution of the asperities (same size and pressure).

We are then able to compare direct measurements of R_c with analytical modeling from materials properties E^*, and experimental data d, F_c and N.

Comparison with the theoretical modeling of the contact resistance

In order to validate the deformation mode sustained by the asperities on one hand, and the regime of electronic conduction on the other hand, the experimental data has been compared with the several modeling of the contact resistance. Figure 3 presents the comparison of experimental Rc and ohmic, ballistic and quasi-ballistic models, by using Hertz' theory:

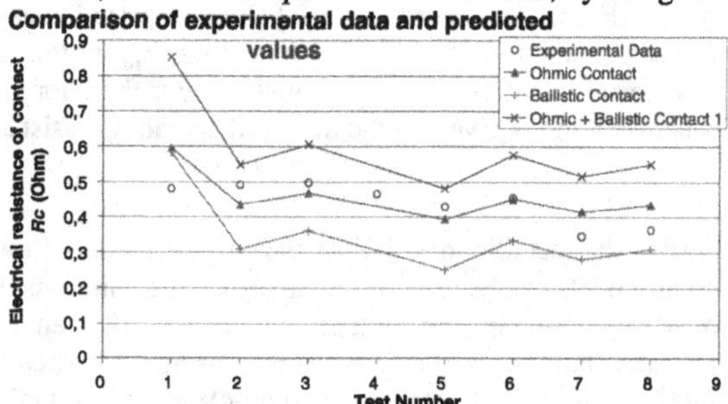

Figure 3: a) Electrical resistance of contact R_c, estimated by several models and compared with experimental data for 7 actuations.

Considering figure 3, we can observe that the ballistic model underestimates the actual experimental results, as expected if considering that the area of contact is in this case of the same order of magnitude than the mean free path l_e of the electrons. More surprising, while one would expect to get the best correspondence between Wexler's theory and experimental data, it turns out it is the ohmic approach which seems to provide the best correlation. The average gap between Wexler's theory and experimental data is about 35%. In fact, one can identify several factors which could explain such results. First, the elastic assumption does not seem sufficient to reflect the actual behavior: a small decrease of the contact resistance can be observed along the actuations, that is to say that a small part of the deformations must be irreversible. The use of an elastic-plastic modeling like the CEB model [6] could be here relevant. The second point concerns the assumption made on the distribution of the contact asperities. An approach taking into account a more accurate distribution should be use. That is why a numerical approach has been made too, based on AFM measurements of the relief of the fixed electrode.

Finite Element simulation of RF MEMS contacts

With the increase of computation capabilities, the surface topography can be indeed included in finite element simulations. Here is described a new method that allows the simulation of the DC contact of RF MEMS devices through finite element simulation and surface characterization. This numerical method is used to predict the real contact area as a function of the applied load starting from real shape of the surfaces that come in contact.

The novel approach relies on a new method used to generate the real shape of the surface. Figure 4 describes the full method developed on ANSYS platform [7]. MATLAB functions is used to convert the AFM topography of the contact surface to an ASCII file compatible with ANSYS Parametric Design Language (APDL). Next, the rough surface is obtained by creating key points from the imported file. Since the key points are not co-planar, ANSYS uses Coons patches to generate the surface. The block volume is finally created with the rough surface on the top. The finite element contact analysis is performed using the augmented Lagrangian method. The post–processing generates the distribution of the contact pressure on the contact surface.

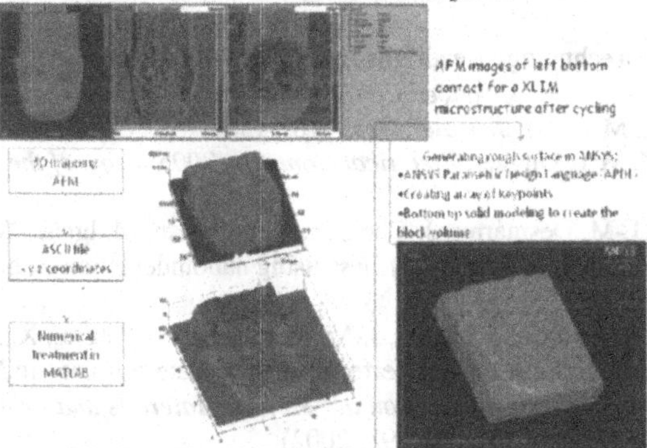

Figure 4 AFM micrographs and schematic views of the process of generation of the contact area

To correlate the experimental measurements with finite element simulations, the surface topography of XLIM gold contact is captured (figure 1). Then the model consists in a gold layer defined as a flexible material, with the generated real surface on the top. The target surface (indentor) is assumed smooth and flexible. From the pressure distribution and size of contact spots, an analytical expression is used to extract the electrical contact resistance using Maxwell spreading resistance. The first simulations give a contact resistance of 45,8mΩ for an applied load of 26μN, whereas experimental measurements show a contact resistance close to 350mΩ. These discrepancies are probably related to a limitation of the modelling in the mesh size and to the fact that the roughness of the beam is not taken into account. The contact area is thus over-evaluated and contact resistance underestimated. Additional simulations are on going to overcome these issues and should show results that tend more toward the experimental values.

CONCLUSIONS

The rise of the TRL level of micro-switches will go along with extensive experimental characterization. The purpose of this work was to develop a new method combining experiments and several modeling in order to address reliability issues. The experimental set-up based on a nanoindentation apparatus with its closely controlled mechanical actuation, provides an accurate estimation of the performances of tested switches. Moreover, such an experiment is convenient in order to improve the understanding of the physics involved in the electrical contact, and then to optimize parameters in the purpose to improve the performances but also the reliability of the devices. This method is based on an analytical modeling completing the experimental data and leading to assume several hypotheses on the physics (electron transport, mechanical deformation mechanism ...). In order to improve the correlation between experimental and analytical results, a numerical approach has been made based this time on a finite element analysis. Particularly, the actual roughness of the surface can be accurately taken into account in the modeling.

Whatever, first results obtained on the XLIM switch are very promising. One can be confident on the results delivered by the developed approach, in order to study the performances and reliability of MEMS technology.

REFERENCES

1. J.C. Mankins, "Technology Readiness Levels: A White Paper", *NASA, Office of Space Access and Technology, Advanced Concepts Office*
2. W.C. Oliver, G.M. Pharr, *J. Mater. Res.*, 7(6), 1564-1583 (1992)
3. D.J. Dikrell III, M.T. Dugger, *Electrical Contacts, 2005. Proc. of the 51st IEEE Conf.*, 255-258 (2005)
4. C. Seguineau, J.-M. Desmarres, J. Dhennin, X. Lafontan, M. Ignat, "MEMS reliability : Accurate measurements of beam stiffness using nanoindentation techniques", *Proc. of CANEUS 2006*
5. C. Seguineau, A. Broue, J. Dhennin, J.-M. Desmarres, A. Pothier, X. Lafontan, M. Ignat, "Mechanical cycling for electrical performances of materials used in MEMS: application to Gold micro-switches", *for publication in "Metals, Materials and Processes"*
6. I.A Green, *Trib Trans*; 45(3), 284-293 (2002)
7. D. Peyrou et Al., "Multiphysics Softwares Benchmark on Ansys / Comsol Applied For RF MEMS Switches Packaging Simulations", *Eurosime 2006, Côme (Italia), 24-26 Ap. 2006*, 8p
8. Mikrajuddin A., Shi F.G., Kim H.K., Okuyama K., *Material Science in Semiconductor Processing*, Vol. 2, 321-327 (1999)

Mater. Res. Soc. Symp. Proc. Vol. 1052 © 2008 Materials Research Society 1052-DD06-31

A Comparative Study of the Strength of Si, SiN and SiC used at Nanoscales

Tuncay Alan, and Pasqualina M. Sarro
Electronic Components, Technology and Materials, Delft University of Technology, Delft, 2600GB, Netherlands

ABSTRACT

Microelectromechanical systems (MEMS) are being used in many critical applications that require very high stress levels. To properly design MEMS components, mechanical properties should be characterized testing relevant sized samples that are fabricated with the same procedures as the final structure. In this paper we use atomic force microscopy (AFM) experiments to study the fracture strength statistics of polycrystalline SiC and SiN nanobeams, and compare their mechanical performance with the performance of previously tested Si nanostructures. Using the same AFM method and similar sample shape and sizes, allows a direct comparison to be made, which will be useful in determining the best material for different mechanical applications and also to validate the theoretical limits.

INTRODUCTION

Over the past decade, microsystem technology has attracted a significant commercial interest, and Microelectromechanical systems (MEMS) have been used in a wide variety of novel science and engineering products. At MEMS size scales the fracture properties of materials differ significantly from the well known bulk properties: They are influenced significantly by the sample size [1] and often by small changes in the fabrication processes [2] and by environmental effects [3].

So far, numerous studies (with novel in-situ experimental procedures) have been performed to characterize the mechanical properties and fracture reliability of materials used in MEMS size scales [4]. Yet, due to differences in experimental methods, geometric properties of the samples that were used and fabrication procedures, there is a significant scatter in the strength values reported by different groups. It is generally quite challenging to build the complex experimental set-ups that are required, which makes it difficult to repeat the tests performed by other groups. To generate a database of mechanical properties at small scales, and to effectively compare the properties of different materials, a test procedure that is easily repeatable (and not requiring a specially built equipment) is necessary. Here, we use atomic force microscopy (AFM) based bending tests [2,5] to compare fracture strength distribution of poly crystalline SiN and SiC samples with the strength of previously tested single crystal Si nanobeams.

EXPERIMENT

Sample Preparation

The test samples were 200 nm thick doubly clamped beams that have an effective length of 5 μm and widths changing between 0.5 μm and 1 μm. The samples were fabricated using standard lithography and wet etching procedures. A 200 nm thick layer of SiN or SiC was deposited on a (100) oriented Si wafer with low pressure chemical vapor deposition (LPCVD). The film was patterned through photolithography and the beam shape was defined in the deposited layer through dry etching. The dry etch was further continued to create 5 micron deep trenches in the Si substrate. The wafers were then diced and individual chips were immersed in a 25% (w/w) potassium hydroxide (KOH) solution at 82°C for 5 minutes to completely release the test structures. Typical test beams are shown in Figure 1 a), b) and c). The rounded beam corners and the small undercut that originates from the wet etch were designed to minimize geometric stress singularities and to ensure that maximum stresses always occur at the center of the beam.

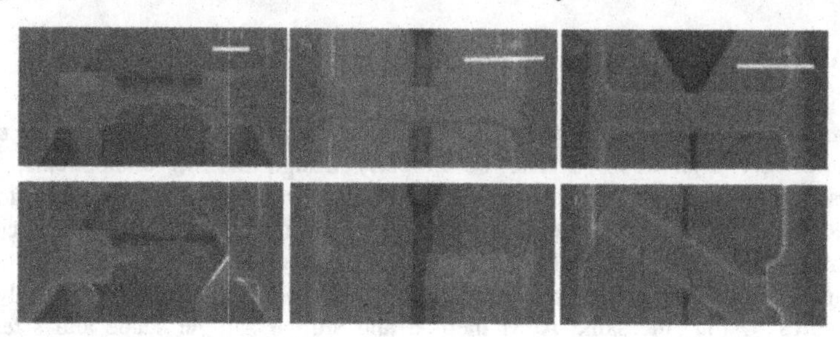

Figure 1. a, b, c) SEM images of typical Si, SiN and SiC test beams respectively, d,e,f) fractured test beams

Experimental Procedures

The released beams were loaded by an uncoated, stiff, single-crystal Si AFM cantilever (Veeco TAP525) until they fractured (Figure 2). The details of the experimental procedures are discussed in detail in [2] and [6]. Briefly, the applied force, F, and the deflection of the beam, δ_{beam}, during loading were calculated as

$$\delta_{beam} = \delta_{piezo} - \delta_{cantilever} , \quad (1)$$

$$F = k\delta_{beam} . \quad (2)$$

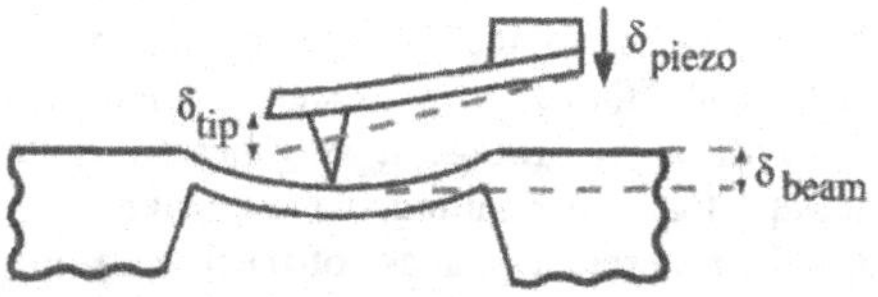

Figure 2. Schematic drawing showing an AFM cantilever loading a test beam.

The cantilever stiffness, k was calculated to be 282 N/m with a previously described calibration procedure [2]. Figures 2-d), e) and f) show SEM micrographs of typical fractured test beams.

During loading, the beams undergo large deflections due to small film thicknesses, hence resulting in nonlinear load-deflection curves. The elastic moduli of the beams were determined from the linear parts of the individual load deflection curves. The intrinsic stresses were separately measured by wafer curvature method [7], using similar wafers processed at the same time as the tested samples [3]. To account for the intrinsic and geometric nonlinearities, the beam strengths were determined from a geometrically nonlinear finite element analysis (FEA), which takes into account the undercuts at the both ends of the beam (measured from SEM images in Figure 1) and uses the calculated elastic modulus and intrinsic stress values (Table 1). Both the SiN and SiC samples were assumed to be isotropic and a Poisson's ratio of 0.25 was used. For Si samples, full anisotropic material properties (which are still applicable at the considered length scales) were used in the simulations. This allowed us to accurately compare simulated and experimental load-deflection curves (Figure 3). The tensile stresses corresponding to the peak loads were inferred as the fracture strength of the tested beam.

Preliminary Results

The fracture strength data from repeat tests were characterized with Weibull probability distribution, which assumes that the whole sample fails when a critically sized defect is encountered anywhere in the sample [8]. Figure 4 displays the Weibull strength distributions for SiN and SiC beams as well as the previously reported [6] strength distribution of single crystal Si beams.

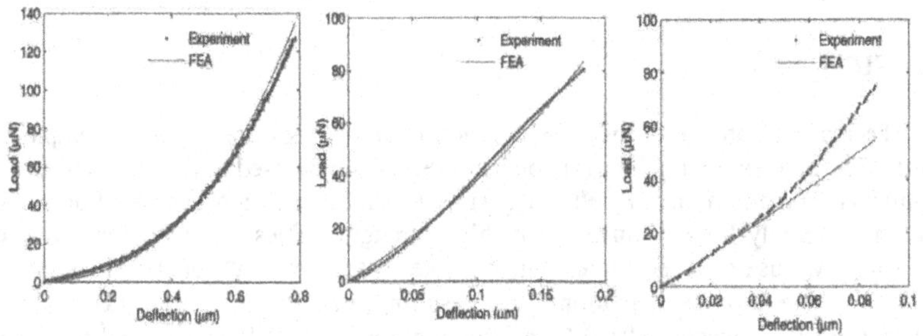

Figure 3: Typical load-deflection curves for a)Si, b)SiN, and c)SiC beams

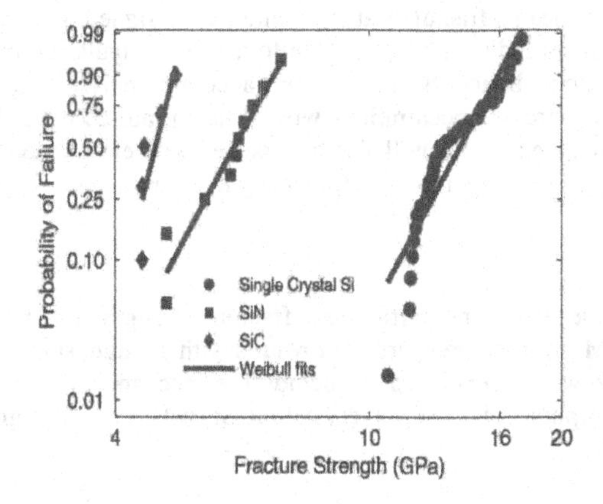

Figure 4. Weibull failure probability plots for SiC, SiN and single crystal Si. The estimated Weibull parameters for different materials are displayed in Table 1.

Previously reported Young's moduli of SiN samples vary between 86 and 370 GPa and the strengths change from 0.9 to 12.2 GPa [4]. For poly SiC, Nemeth, et al measured a Young's modulus value of 308 GPa and fracture strength of 4.5 GPa [4], Similarly, Dao et. al [9] reported a strength value as high as 23 GPa considering an elastic modulus of 710 GPa. Clearly, there is a wide scatter in the values reported by different groups and further experimental investigations that use similar test methods and fabrication procedures are necessary for a more accurate comparison.

Table 1. Calculated material properties and estimated Weibull parameters for different materials. For Si, the known anisotropic constants were considered.

Material	Intrinsic Stress (MPa)	Elastic Modulus (GPa)	Weibull strength (GPa)	Weibull modulus
SiC	550	400 +/- 13.6	4.7	18.5
SiN	160	329 +/- 35	6.4	8.5
Si	-	-	14.4	8.6

DISCUSSION

The main advantage of the AFM method is that it does not require a complex set-up. A standard AFM and commercially available cantilevers can be used to repeat the tests in different environments. The preliminary results suggest that both SiC and SiN samples fracture at stresses that are considerably lower than the very high strength values observed for single crystal Si beams, which we use as a reference. Yet, we note that the strength of the Si samples is highly dependent on the process conditions and resulting surface properties. (The samples were fabricated from 0.1° miscut, >1000 Ωcm, floating zone, Si (111) wafers and the resulting rms roughness of the beam surfaces was 1.5 nm [6].)

The tested SiN beams fractured at the center, as designed, but the SiC beams fractured from the rounded corners indicating failure due to a stress singularity, possibly at a fabrication induced defect or a grain boundary. For a more accurate analysis, the grain structure of the materials, and possible stress concentrations will be taken into account. To further improve the accuracy of our results, the analysis will also be repeated with elastic modulus and intrinsic stress values obtained independently by resonant frequency measurements.

CONCLUSIONS

We studied the elastic properties and fracture strength distributions of SiC and SiN nanobeams with AFM tests and compared our results with strength statistics of Si nanobeams. In the future, this work will be extended to include the mechanical characterization of different materials such as amorphous SiC and polycrystalline Si (with controlled grain structures).

ACKNOWLEDGMENTS

This work was sponsored in part by STW project DTC.6663. The SiC and SiN test samples were fabricated at the Delft Institute for Microsystems and Nanoelectronics and the experiments were performed at the facilities of the Surfaces and Interfaces Department, at Delft University of Technology. The authors are grateful to Dr. Rafael Pujada for his help with the AFM facilities.

REFERENCES

1. T. Tsuchiya, O. Tabata, J. Sakata, Y. Taga, J. of Microelectromech. Sys., 7, 106 (1998).
2. T. Alan, M. A. Hines, A. T. Zehnder,Appl. Phys. Lett. 89, 091901 (2006).
3. T. Alan, A. T. Zehnder, D. Sengupta, M. A. Hines, Appl. Phys. Lett. **89**, 231905 (2006)
4. O.M Jadaan, N.N. Nemeth, J. Bagdahn, W. N. Sharpe, J. Mat. Sci., 38, 4087 (2003).
5. C. Serre, A. Perez-Rodriguez, J.R. Morante, P. Gorostiza, J. Esteve, Sens. Act. A, 74, 134 (1999).
6. T. Alan, Ph.D. Dissertation, Cornell University (2007).
7. L.B. Freund and S. Suresh, Thin Film Materials, Cambridge University Press (2003).
8. A. McCarty, I. Chasiotis, Thin Sol. Films, 515, 3267 (2007).
9. D. Gao, C. Carraro, V Radmilovic, R.T. Howe, R. Maboudian, J. Microelectromech. Syst., 13, 972 (2004).

Mater. Res. Soc. Symp. Proc. Vol. 1052 © 2008 Materials Research Society 1052-DD06-32

Compressive magnetostriction of FeSm alloy film

Ryo Nakano[1], Yoshihito Matsumura[2], and Yoshitake Nishi[2]
[1]Metallurgical Engineering, Graduate School of Engineering, Tokai University, 1117 Kitakaname, Hiratsuka, 259-1292, Japan
[2]Science and Technology, Graduate School of Science and Technology, Tokai University, 1117 Kitakaname, Hiratsuka, 259-1292, Japan

ABSTRACT

The compressive magnetostriction values of $Fe_{2.4}Sm$ alloy thin film prepared on silicon, copper and titanium substrates (300 μm thickness) by direct current magnetron sputtering process were investigated. When the residual gas pressure before argon sputtering and the sputtering pressure of argon gas (5 N) were below 3.2×10^{-4} Pa and 2.0×10^{-1} Pa, respectively, the thickness of the $Fe_{2.4}Sm$ films deposited was about 3 μm. Compressive magnetostriction of $Fe_{2.4}Sm$ alloy film deposited on titanium sheet generates large bending motion, compared to those deposited on silicon wafer and copper substrates. High magnetostrictive susceptibility of the films was observed at low magnetic field.

INTRODUCTION

The RFe_2 type cubic Laves phases of rare earth metals (R) and iron (Fe) often show giant magnetostriction [1], which may have applications for sensors and actuators [2,3]. The giant mover strain of Terfenol-D ($Tb_{0.3}Dy_{0.7}Fe_2$), which is larger than that of commercial piezo-ceramics such as PZT, has been found with low driving potential, high power generation, high responsiveness and wireless operation by a magnetic field [4-8]. In order to create tiny acoustic devices and sensors and actuators driven by low intensity magnetic fields, giant magnetostrictive films have been studied [9,10]. We have reported that a giant magnetostrictive alloy shows a superior characteristic as a thin film actuator. Bending strain in mover devices often depends on substrate thickness. Large motions have been found in mover devices based on of hydrogen storage alloy film on copper thin sheet [11]. In this study, $Fe_{2.4}Sm$ alloy films on silicon, copper and titanium substrates (300 μ m thickness) have been prepared by a D.C. magnetron sputtering process. The compressive magnetostriction and its susceptibility are evaluated.

EXPERIMENT

The $Fe_{2.4}Sm$ alloy thin film was prepared on a substrate (300 μm thickness) by using D.C. magnetron sputtering. The minimum gas pressure before argon sputtering and the sputtering pressure of argon gas (5 N) were 3.2×10^{-4} Pa and 0.2 Pa, respectively. The substrate temperature was 423 K (see Table 1). The mean leak rate of the chamber was about 1.96×10^{-5} $Pa\cdot m^3/s$. The sputtering power, deposition time, sputtering distance between target and sheet were 200 W, 3.6 ks and 90 mm, respectively. Mean thickness of the $Fe_{2.4}Sm$ films deposited was about 3 μm. The film composition was analyzed by Energy dispersive X-ray spectroscopy (EDX). Crystalline structures of the prepared film were determined by α-2θ method of thin film X-ray diffraction (Cu-Kα; X'Part-MRD, PHILIPS), where irradiation angle to sample plane was 0.5 deg. The magnetostriction ($\Delta\lambda_{//}$) of the film was measured using the bending cantilever method using a He-Ne LASER under magnetic fields from -1200 to 1200 kA/m and was estimated by a following equation [8-10],

$$\lambda = d \cdot t_s^2 \cdot E_s(1+\nu_f)/\{3t_f \cdot l^2 \cdot E_f(1-\nu_s)\} \qquad (1)$$

where, t_s and t_f were the thickness values (300 μm thickness) of the substrate and $Fe_{2.4}Sm$ films, respectively. d and l were the bending displacement of the sample and the distance from the clamp to the laser spot on the film, respectively. The E_s and E_f were values of the Young's modulus of substrate and $Fe_{2.4}Sm$ film [12].

Table 1. Conditions of the film formation by D.C.magnetron sputtering

Target	FeSm
Substrate	Si, Cu, Ti
Base pressure	3.2×10^{-4} Pa
Gas pressure	2.0×10^{-1} Pa
Substrate temperature	473 K

RESULTS and DISCUSSION

·Magnetostriction (λ)

Figure 1 shows the applied magnetic field dependence on the magnetostriction (λ) of $Fe_{2.4}Sm$ alloy films prepared on different substrates of silicon, copper and titanium. The largest negative value of compressive magnetostriction (λ) at 400 kA/m of magnetic field was found on the titanium substrate.

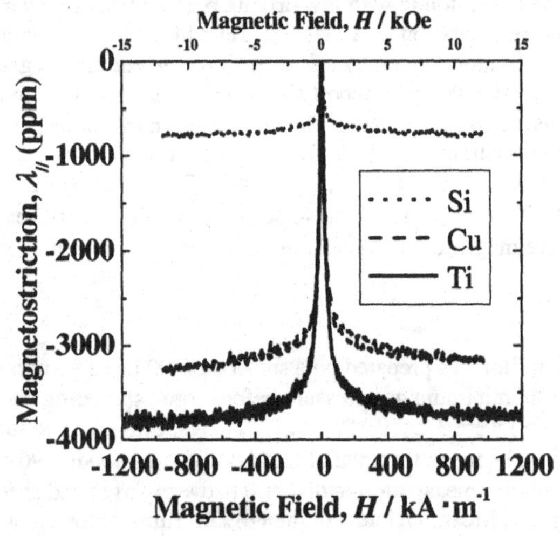

Fig 1. The applied magnetic field dependence on the magnetostriction (λ) of $Fe_{2.4}Sm$ alloy films prepared on different substrates.

The deformation resistance, which depends on elasticity and thickness, contributes to the bending motion. Figure 2 shows the applied deformation resistance of the substrate dependent magnetostriction (λ) of $Fe_{2.4}Sm$ alloy film prepared on different substrates. The decrease in deformation resistance of the substrate enhances the compressive magnetostriction (λ). The largest value of compressive magnetostriction at 400 kA/m of $Fe_{2.4}Sm$ film was 3650 ppm on the titanium substrate.

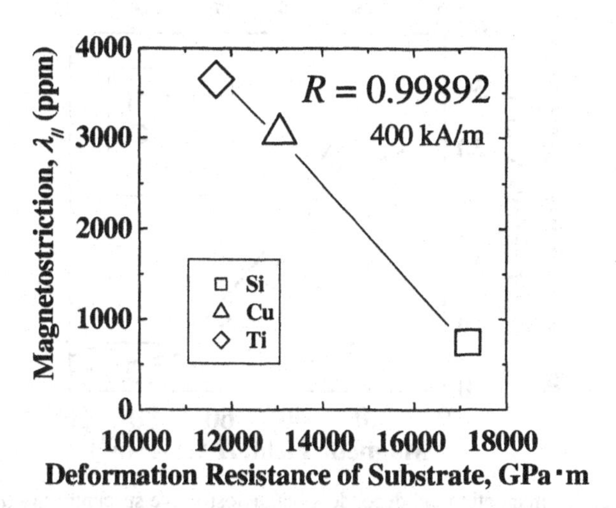

Fig 2. Dependence of the applied deformation resistance of the substrate on magnetostriction (λ) of $Fe_{2.4}Sm$ alloy film prepared on different substrates.

·Magnetostrictive susceptibility ($d\lambda/dH$)

Figure 3 shows applied magnetic field dependent magnetostrictive susceptibility ($d\lambda/dH$) of $Fe_{2.4}Sm$ alloy film prepared on different substrates. The magnetostrictive susceptibility ($d\lambda/dH$) depends on the deformation resistance of the substrates. The decrease in deformation resistance of the substrate enhances the magnetostrictive susceptibility ($d\lambda/dH$). The largest value of magnetostrictive susceptibility ($d\lambda/dH$) at 10 kA/m of $Fe_{2.4}Sm$ film was 49 ppm/kAm^{-1} on the titanium substrate.

CONCLUSIONS

The influences of deformation resistance of substrates (300 μ m thickness) on negative (compressive) magnetostrictive susceptibility of $Fe_{2.4}Sm$ alloy thin films prepared by direct current magnetron sputtering process were investigated.

1: Magnetostriction (λ) and magnetostrictive susceptibility ($d\lambda/dH$) strongly depends on the deformation resistance of the substrate.
2: The decrease in deformation resistance of the substrate enhances the magnetostriction (λ) and magnetostrictive susceptibility ($d\lambda/dH$).

3: The largest value of magnetostriction at 400 kA/m of $Fe_{2.4}Sm$ film was 3650 ppm on the titanium substrate.

4: The largest value of magnetostrictive susceptibility *(dλ/dH)* at 10 kA/m of $Fe_{2.4}Sm$ film was 49 ppm/kAm^{-1} on the titanium substrate.

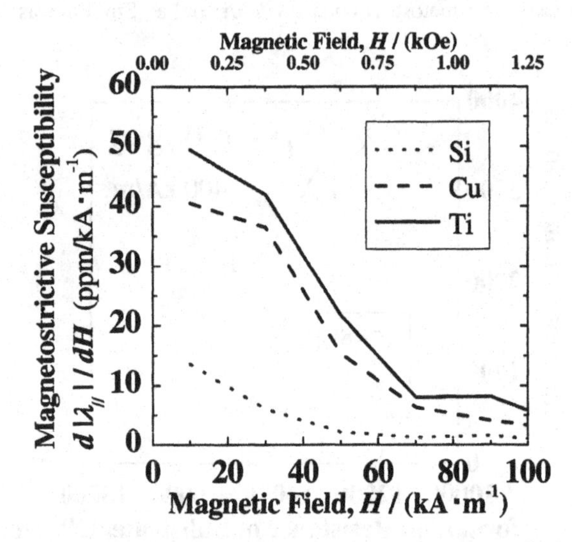

Fig 3. Applied magnetic field dependent magnetostrictive susceptibility *(d λ /dH)* of $Fe_{2.4}Sm$ alloy film prepared on different substrates.

REFERENCES

1. A. E. Clark and H. S. Belson: Phys. Rev. B 5 (1972) 3642-3644.
2. M. Sahashi and T. Kovayashi: J Acoustical Soc. Japan 466 (1990) 591-599.
3. V. Koeninger, Y. Matsumura, H. H. Uchida and H. Uchida: J. Alloy. Compd. 211/212 (1994) 581-584.
4. Y. Nishi and H. Yabe: J. Appl. Electromagnetics Mech. 10 (2002) 394-399.
5. M. Wada, H-H. Uchida, Y. Matsumura, H. Uchida and H. Kaneko: Thin Solid Films 281-282 (1996) 503-506.
6. M. Wada, H. Uchida and H. Kaneko, J. Alloy and Compound, 258 (1997) 143-148.
7. H. Yabe and Y. Nishi: TETSU-TO-HAGAME 89 (2003) 93-98.
8. H. Yabe and Y. Nishi: Jpn. J. Appl. Phys. 42 (2003) 96-99.
9. H. Uchida, M. Wada, A. Ichikawa, Y. Matsumura, H.H. Uchida, Proc. ACTUATOR 96, 26-28 June 1996, Bremen, Germany (1996) 275-278.
10. H. Uchida and H. Kaneko, Proc. ACTUATOR 96, 26-28 June 1996, Bremen, Germany (1996) 262-267.
11. K.Numazaki, H.H.Uchida, Y.Nishi: J. Japan. Inst. Metals 69 (2006) 751-754.
12. A. E. Clark in E. P. Wohlfarth (ed): Ferromagnetic materials, 1, Chap. 7. (North-Holland, Amsterdam, 1980).

MEMS Materials and Processes I

Mater. Res. Soc. Symp. Proc. Vol. 1052 © 2008 Materials Research Society 1052-DD07-03

Influence of Materials on the Performance Limits of Microactuators

Prasanna Srinivasan, and S. Mark Spearing
School of Engineering Sciences, University of Southampton, Southampton, SO171BJ, United Kingdom

ABSTRACT

The selection of actuators at the micro-scale requires an understanding of the performance limits of different actuation mechanisms governed by the optimal selection of materials. This paper presents the results of analyses for elastic bi-material actuators based on simple beam theory and lumped parameter thermal models. Comparisons are made among commonly employed actuation schemes (electro-thermal, piezoelectric and shape memory) at micro scales and promising candidate materials are identified. Polymeric films on Si subjected to electro-thermal heating are optimal candidates for high displacement, low frequency devices while ferroelectric thin films of Pb-based ceramics on Si/ DLC are optimal for high force, high frequency devices. The ability to achieve ~10 kHz at scales < 100μm make electro-thermal actuators competitive with piezoelectric actuators considering the low work/volume obtained in piezoelectric actuation (~ 10^{-8}J.m^{-3}.mV^{-2}). Although shape memory alloy (SMA) actuators such as Ni-Ti on Si deliver larger work (~ 1 J.m^{-3}K^{-2}) than electro-thermal actuators at relatively low frequencies (~ 1 kHz), the critical scale associated with the cessation of the shape memory effect forms the bounding limit for the actuator design. The built-in compressive stress levels (~ 1GPa) in thin films of Si and DLC could be exploited for realizing a high performance actuator by electro-thermal buckling.

INTRODUCTION

The choice of materials has a strong influence on the performance of Micro-Electro-Mechanical-Systems (MEMS). The use of traditional MEMS materials is predominantly driven by their compatibility with existing micro-fabrication processes [1]. As a result, these materials are not often optimal from the performance point of view. This paper presents a comparative analysis of the performance limits of commonly employed MEMS actuation schemes (electro-thermal/ piezoelectric/ shape memory). The results presented herein serve as a guide for the selection of actuation principle and the relevant candidate materials for various applications. The structural response of the actuators is assumed to follow Euler-Bernoulli beam theory and the thermal response is described by a lumped heat capacity formulation. The critical performance metrics which impact the actuator design are tip slope (deflection), blocked moment (force), work per volume and actuation frequency.

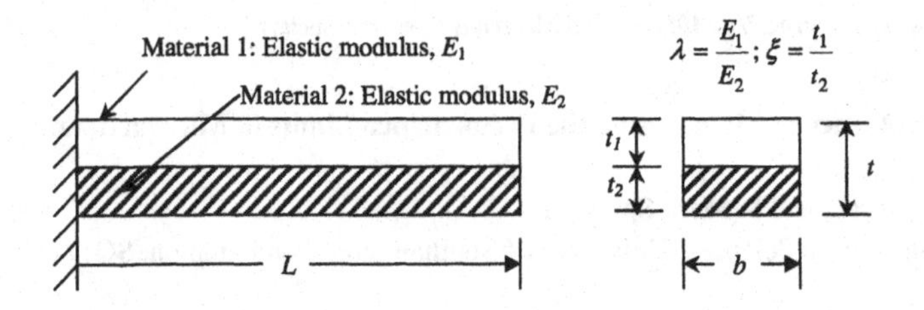

Figure 1 Schematic of a cantilever bending actuator in bi-material architecture

MECHANICS OF A BI-MATERIAL ACTUATOR STRUCTURE

Bi-material actuators are very common in MEMS due to the simplicity of their design and their ability to achieve out of plane actuation. The capability of present micromachining processes to grow thin films of arbitrary materials on various substrates presents an opportunity to optimize the candidates in order to achieve maximum performance. The geometry of the basic actuator structure is defined by a beam of length, L with a uniform width, b and constant thickness, t. The thicknesses of the bi-layers are t_1 and t_2 respectively. Applying the mechanics of a bimetallic strip developed by Timoshenko [2], the free end slope (Θ_f), blocked moment (M_{blk}) and work/volume (W) of the actuator structure normalized by the width, b are obtained which can be expressed as a function of λ and ξ for a constant actuation strain, ε_a with respect to a given substrate [3]. The actuation strain depends on the difference in thermal expansion coefficients ($\Delta\alpha$), d-coefficients (d_{31}) and recovery strain (ε_{rec}) for electro-thermal, piezo-electric and shape memory actuators respectively in bi-material architecture. It can be shown that the optimal condition for maximum performance is given by [3]:

$$\lambda \xi_o^2 = 1 \qquad (1)$$

where ξ_o is the optimal thickness ratio for a given pair of materials for a fixed geometry. Using equation (1), maximum performance for any given material pair can be obtained.

OPTIMAL CANDIDATE MATERIALS FOR VARIOUS ACTUATION SCHEMES

In order to make an effective selection of actuation scheme, it is necessary to compare the performance achieved by the three actuation methods. This is feasible if contours of equal performance are plotted on an "Ashby selection map" [4] in the domain of the relevant material properties. It should be noted that linear elastic behavior is assumed for all classes of materials which may not be an accurate prediction for certain classes of materials especially, for polymers. Nevertheless, this approach yields a sufficiently accurate estimate by which to compare the actuation principles and in some cases to perform preliminary design. Silicon (E= 165 GPa, α = 2.24 μm/mK) is one of the commonly preferred materials for MEMS applications. Hence this work focuses on the performance of different materials with respect to a Si substrate. The approach could similarly be applied to any combination of materials.

Figure 2a shows the contours of thermo-elastic performance for different materials on Si. For the Ni-Ti (Nitinol) - Si combination the actuator performance achievable due to shape memory effects is higher than that obtained by thermal expansion within the transformation temperature range (40-80°C). Since the material properties governing the performance of electro-thermal (α) and shape memory (ε_{rec}) actuators are not identical, a uniform basis need to be established to compare performance. This is feasible if the actuation strain due to the phase change in Ni-Ti ($\varepsilon_{rec} = 0.4-0.8\%$) is represented as an equivalent strain due to thermal expansion corresponding to the temperature limits associated with the austenite-martensite phase transformation as shown in Figure 2a. This is a reasonable representation considering the operating temperature difference of most electro-thermal actuators (~ 100°C).

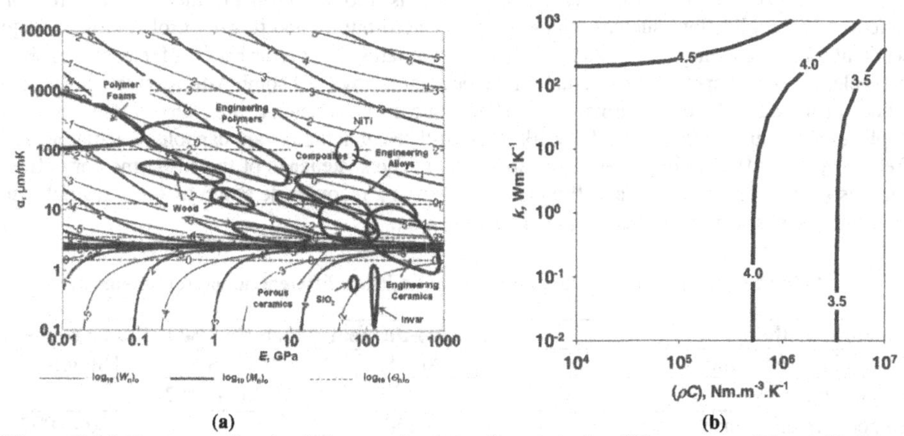

(a) (b)

Figure 2 (**a**) Contours of optimal thermo-elastic performance for different materials on Si plotted on Ashby's [4] selection chart (**b**) Contours of equal actuation frequency ($\log_{10}(f)$) for different materials on Si for an achievable temperature difference (~ 150°C) for L= 60μm

Figure 2b shows contours of thermal actuation frequency, $\log_{10}(f)$ for different materials on Si in the domain of thermal conductivity (k) and specific heat (ρC). High thermal actuation frequencies of the order of ~ 10 kHz can be achieved at small scales (L< 60μm) for an achievable temperature difference of ~ 150°C using engineering alloys/ceramics substrates [5]. Polymers such as PMMA, PS, PP and PDMS on Si are optimal candidates for high displacement, low frequency (Θ_o~ 10^{-4} K^{-1} and f ~ 100 Hz) actuator structures while thin films of Zn, Al, Ni, Zr_2O_3, Mg, Cu on Si/SiO_2 and Invar substrates are promising for high moment/work (~ 10^5Nm.m^{-3}.K^{-1} / 0.1 J.m^{-3}K^{-2}) at relatively high frequencies (~ 1-10 kHz). Similar contours plotted in the domain of d-coefficients and elastic moduli for commonly employed piezo-active materials on Si [6] reveal Pb-based ceramics such as PZT, PZNPT and PMNPT on Si are potential candidates for large force, although the work/volume delivered is small compared to that of electro-thermal/shape memory actuators. Furthermore, the ability of piezoelectric actuators to operate at mechanical resonance results in much higher actuation frequencies (~ 100 kHz) than electro-thermal and shape memory actuators. Candidates such as PZT, PMN-PT, PZN-PT and $BaTiO_3$ on Si/DLC substrates are optimal combinations for high force, high frequency (M_{blk}~ 10 Nm.m^{-3}.mV^{-1}and f ~ 100 kHz) actuator structures.

SELECTION OF ACTUATORS FOR VARIOUS FUNCTIONAL REQUIREMENTS

The intersection of the performance domain and the functional requirements domain decides the suitability of a particular actuator for an application. For instance, an application such as a micro-mirror device requires a large displacement at frequencies of a few hundred Hz, which automatically makes electro-thermal actuation (polymers on Si) more attractive than other schemes for meeting the functional requirements. For applications such as micro-fluidic valves, a large force at high frequency is the primary functional requirement and piezoelectric actuation (Pb-based piezo-ceramics on Si/DLC substrates) is more promising than other actuation schemes. For applications such as micro-grippers, micro-pumps and flow control devices, a large work at high frequency is desirable. Table 1 compares the performance of electro-thermal, piezoelectric and shape memory actuation for optimal bi-material combinations with respect to a silicon substrate. Electro-thermal and shape memory actuators are very promising for applications requiring large specific work. Electro-thermal actuators are capable of actuating at a frequency of ~ 10 kHz for scales less than ~ 60μm. The sustenance of the shape memory effect for reasonable life times (e.g. a million cycles) even at a recoverable strain of ~ 1% restricts the frequency of shape memory actuators to about ~ 1 kHz.

Table I A comparison on the performance limits of different actuation schemes

Actuation method	**Electro-thermal**	**Piezoelectric**	**Shape memory**
Bi-material combinations	Al – Si	PZT – Si	Nitinol-Si
Actuator dimension	60 μm x 30 μm x 2 μm		
Temperature difference, ΔT in C	~ 150	-	~ 30 - 40°C
Electric field, E_p in MV/m	-	5	-
Recoverable strain	-	-	0.8 %
Power dissipated, P in W	~ 0.1	-	~ 0.01
Dielectric energy/volume stored, E_s in J/m^3	-	~ 143 x 10^{-3}	-
Tip deflection in μm	~ 5	~ 0.5	~ 11
Blocked moment, M_{blk} in Nm	~ 5 x 10^{-9}	~ 5 x 10^{-10}	~ 12 x 10^{-9}
Maximum work/volume, W in J/m^3	~ 27410	~ 293	~ 150021
Actuation frequency, f in kHz	~ 10	~ 250	~ 1
Loss coefficient, χ	-	~ 10^{-3}	-
Actuation efficiency, η	~ 10^{-4}	~ 0.1	~ 10^{-3}

Generally piezoelectric actuation has been preferred for applications such as boundary layer flow control [7, 8] due to its ability to operate at mechanical resonance (~ 100 kHz). The results presented herein suggest that electro-thermal actuators at small scales (~ 60 μm) are competitive with piezoelectric actuators at ~ 10 kHz. Unlike electro-thermal/piezoelectric actuators, there is a critical length scale for shape memory actuators which is dependent on the grain size of the alloy system. There has been a wide speculation on this limiting scale associated with the shape memory effect [9, 10, 11] which depends on the processing routes. Hence electro-thermal actuation could be a better alternative at such scales.

The performance limits of electro-thermal actuators can be further improved using a clamped beam configuration actuated by electro-thermal buckling. For candidates possessing large values of $E\alpha$ (~1.0 MPa/K), the temperature difference, $(\Delta T)_c$ required to buckle the structure is small which has a significant bearing on the actuation efficiency and thermal actuation frequency. Materials in particular, polycrystalline Si [12] and DLC [13] are capable of being grown as thin films with high intrinsic compressive stress, σ_c which can be exploited in reducing the temperature difference required to buckle the structure. Since the energy required for buckling is high compared to bending, the work/volume delivered by electro-thermal buckling actuators (~ 543 kJ/m^3) is on a par with that for shape memory actuators.

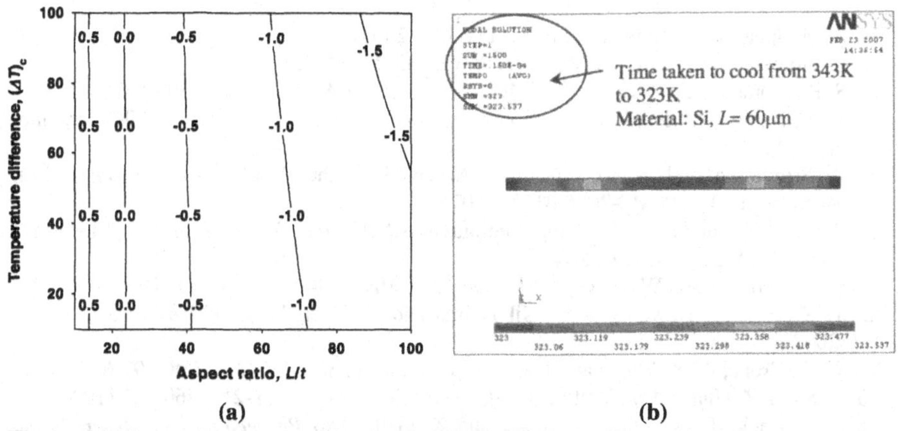

(a) (b)

Figure 4 **(a)** Iso-stress contours, $\log_{10}(\sigma_c)$ in GPa for Si electro-thermal buckling actuator for a range of aspect ratios and critical temperature differences **(b)** Transient thermal response of Si electro-thermal buckling actuator.

Figure 4a shows contours of $\log_{10}(\sigma_c)$ as a function of L/t and $(\Delta T)_c$ for silicon. It is evident from the plot that for a structure with L/t varying from 20 - 80 the temperature difference can be reduced to a few tens of degrees if the intrinsic compressive stress is suitably varied between 1GPa -100MPa. Figure 4b shows the transient thermal response of an electro-thermal buckling actuator structure made of silicon by numerical simulation using ANSYS. Actuation frequencies up to ~ 50 kHz could be achieved for temperature amplitude of 20K for Si which can be further increased to ~ 400 kHz if DLC is employed.

CONCLUSIONS

The performance limits of different actuators depend on the governing physical mechanisms which are dictated by the relevant properties of materials. Analytical solutions based on simple beam theory assumptions were used to evaluate the structural performance and a lumped heat capacity formulation was applied to estimate the thermal performance. Our analysis suggests that electro-thermal actuation is promising for large displacement, low frequency applications and that engineering polymers on Si are optimal candidates for such actuators. For applications requiring a large force at high frequency, piezoelectric actuation is promising and material combinations such as PZT, PZN-PT, PZM-PT and $BaTiO_3$ on Si/DLC substrates are optimal candidates. Shape memory actuators (Nitinol on Si) are capable of delivering high work/volume at relatively low frequencies (<1 kHz). Although the work/volume delivered by bi-material electro-thermal actuators is lower than that of the shape memory actuators by an order of magnitude, electro-thermal actuators are capable of actuating at ~ 10 kHz range at scales less than 100μm. Electro-thermal buckling actuators are on par with shape memory actuators in delivering large work/volume at high frequencies (~ 10-100 kHz).

REFERENCES

1. S. M. Spearing, *Acta Materialia*, **48**, 179-196 (2000)
2. S. Timoshenko, *J. Opt. Soc. of America*, **11**, 233-255 (1925)
3. S. Prasanna and S. M. Spearing, *J. Microelectromech. Syst.*, **16**, 248-259 (2007)
4. M. F. Ashby, "Materials selection in mechanical design". 1st ed. Oxford, UK, Pergamon press, 45 (1993)
5. P. Srinivasan and S. M. Spearing, Accepted in the *J. Microelectromech. Syst.*, Manuscript ID: JMEMS-2006-0177 (2007)
6. P. Srinivasan and S. M. Spearing, Submitted to *J. of Microelectromech. Syst.*, Manuscript ID: JMEMS-2007-0047 (2007)
7. S. A. Jacobson and W. C. Reynolds, *J. of Fluid Mech.,* **360**, pp.179-211 (1998)
8. T. Segawa, Y. Kawaguchi, Y. Kikushima and H. Yoshida, *J. of Turbulence*, **3**, 1-14 (2002)
9. W. C. Crone, A. N. Yahya and J. H. Perepezko, *Matl. Sci. For.*, **386-388**, 597-602 (2002)
10. Q. Su, S. Z. Hua and M. Wuttig, *J. Alloys and Compounds*, **211-212**, 460-463 (1994)
11. C. P. Frick, T. W. Lang, K. Spark and K. Gall, *MRS Proceedings on Materials and Devices for Smart System II*, **888**, 2005
12. R. T. Howe and R. S. Muller, *J. of App. Phys.*, **54**, 4674-4675 (1983)
13. T. A. Friedman, J. P. Sullivan, J. A. Knapp, D. R. Tallant, D. M. Follstaedt, D. L. Medlin, and P. B. Mirkarimi, *App. Phys. Letters*,**71**, 3820-3822 (1997)

Mater. Res. Soc. Symp. Proc. Vol. 1052 © 2008 Materials Research Society 1052-DD07-05

Science and Technology of Piezoelectric/Diamond Hybrid Heterostructures for High Performance MEMS/NEMS Devices

Orlando Auciello,[1,2] Anirudha Sumant,[2] Jon Hiller,[3] Bernd Kabius,[3] and Sudarsan Srinivasan,[2] *
[1]Materials Science Division, [2]Center for Nanoscale Materials,
[3]Center for Electron Microscopy, Argonne National Laboratory, Argonne, IL 60439
*Now at INTEL

ABSTRACT

Most current micro/nanoelectromechanical systems (MEMS/NEMS) are based on silicon. However, silicon exhibits relatively poor mechanical/tribological properties, compromising applications to some devices. Diamond films with superior mechanical/tribological properties provide an excellent alternative platform material. Ultrananocrystalline diamond (UNCD®) in film form with 2-5 nm grains exhibits excellent mechanical and tribological properties for high-performance MEMS/NEMS devices. Concurrently, piezoelectric $Pb(Zr_xTi_{1-x})O_3$ (PZT) films provide high sensitivity/low electrical noise for sensing/high-force actuation at relatively low voltages. Therefore, integration of PZT and UNCD films provides a high-performance platform for advanced MEMS/NEMS devices. This paper describes the bases of such integration and demonstration of low voltage piezoactuated hybrid PZT/UNCD cantilevers.

INTRODUCTION

Most current micro/nanoelectromechanical systems (MEMS/NEMS) are based on silicon. However, the relatively poor mechanical/tribological properties of Si compromise applications to particular devices, specifically those trequiring high Young modulus, such as resonators, and low or no stiction, such as switches. Alternatively, diamond, which exhibits high Young's modulus and negligible surface adhesion, on contact, can yield high-performance MEMS/NEMS devices, such as resonators and switches. Fabrication of diamond-based devices requires growth of diamond films on appropriate substrates followed by photolithography and etching to release moving structures (e.g., cantilevers, beams). Ultrananocrystalline diamond (UNCD®) developed at Argonne National Laboratory (ANL), exhibits hardness (98 GPa) and Young modulus (980 GPa) close to values for single crystal diamond (100 GPa and 1100 GPa, respectively), extremely low friction coefficient (0.02-0.04) and force of adhesion for surfaces in contact (~ 30mJ/m), smoothest surface morphology (~ 4-7 nm rms roughness), compared to other films with the sp^3 bonding structure of single crystal diamond (notice that the comparison is not with diamond-like (DLC) or ta-C films, which exhibit lower roughness than UNCD but also lower hardness and Young modulus than UNCD and other "diamond" films[1]). In addition, UNCD is currently the only "diamond" film that can be grown at the lowest deposition temperature demonstrated today (~ 400 °C), compared with other process producing microcrystalline diamond (MCD) and nanocrystalline diamond (NCD) films. DLC films can be deposited at room temperature, but they need to be annealed to >> 400 °C to release stresses, which impede, currently, their integration with CMOS devices for monolithically integrated diamond-MEMS/CMOS devices.

UNCD films are grown on substrates exposed to Ar-rich CH_4 microwave plasmas[2], which yields 2-5 nm diamond grains and 0.4 nm wide grain boundaries with sp^3 and sp^2 bonding[2]. This nanostructure provides the name UNCD for distinction from NCD films grown with H-rich/CH_4 plasmas that yield films with 30-100 nm grain sizes. UNCD exhibits a unique combination of high fracture strength (~ 5.4 GPa) and Young modulus (~ 990 GPa), low stress (~ 50-80 MPa)[3] negligible stiction,[4] exceptional chemical inertness, high electric field-induced electron emission, and surface functionalization that makes it bio-inert/biocompatible for application to biosensors[5], and biomedical devices[6]. However, MEMS/NEMS devices based on UNCD and other materials, such as Si, are generally actuated electrostatically, which requires high voltages (≥ 10 volts).

On parallel scientific/technological paths, piezoelectric $Pb(Zr_xTi_{1-x})O_3$ (PZT) films, with excellent piezoelectric and electromechanical coupling coefficients and high remanent polarization, are being investigated for application to MEMS devices[7,8]. Relevant to this paper, the high-force generated from a piezo-layer upon application of voltage between the top and bottom electrode sandwiching that layer yields efficient micro/nanoactuators[8,9]. The piezoelectric coupling coefficients of PZT films are much higher (d_{31}= -59 pC/N, $e_{31,f}$= -8 to-12 C/m^2 for epitaxial films) than those of other piezoelectric materials, such as ZnO ($e_{31,f}$= -1C/m^2 (SAW experiments), $e_{31,f}$= -0.6 C/m^2 MEMS deflective measurements) and AlN ($d_{31,f}$= -1.98 pC/N, $e_{31,f}$= -1.05 C/m^2) (ref. 8 and references there in). Piezoelectric actuation, using PZT films, can be achieved with low voltages (≤ 10 volts) depending on film thickness. However, PZT has much lower Young modulus (80 GPa)[10] than UNCD (980 GPa)[3], which is critical for applications such as MEMS resonators addressed in this paper.

The work reported here focused on developing the integration of an oxide piezoelectric material such as PZT with a carbon-based material such as diamond. Few attempts have been made in the past to integrate piezoelectric and diamond layers for surface acoustic wave (SAW) devices[11,12], and limited attempts at investigating piezoactuated hybrid PZT/diamond MEMS structures, using conventional microcrystalline diamond (MCD) cantilevers. However, MCD layers with relatively high surface roughness contribute to diminishe the piezoelectric constant of PZT films.[13]

Theoretical calculations performed by two of the present authors,[14] indicated that diamond is the best platform material for integration with PZT as the piezoelectric actuation layer and UNCD as the structural support layer. Application of an external electric field $\overrightarrow{E_3}$ across the PZT layer results in bending of the heterostructure (deflection at the free end). The calculations, accounting for different elastic moduli for non-piezoelectric materials, showed that the product resonance frequency/dynamic displacement at resonance is the largest for the PZT/UNCD hybrid compared to the other PZT-based hybrids, including Si, nitrides, metals and insulators. In addition, UNCD films provide the best support material for resonators, due to the highest Young modulus.

Reliable integration of PZT films with diamond layers, for MEMS, is difficult, due to high thermal expansion mismatch between PZT and diamond ($\alpha_{PZT} = 3.5\times10^{-6}$/K ; $\alpha_{diamond} = 1.1\times10^{-6}$/K) and stress in the PZT layer. Also, deposition of PZT films in oxygen at ≥ 500 °C on diamond results in chemical etching due to oxygen/carbon atoms reaction, forming volatile CO/CO_2 species. Room temperature sputter-deposition of PZT films followed by rapid thermal annealing in N_2 may be provide partial solution to obtaining PZT films with reasonable piezoelectric properties, although not as good as for PZT films grown by sputter-deposition or MOCVD in appropriate oxygen environment.

The research discussed here focused on developing a robust PZT/UNCD integration and processes for the fabrication of hybrid oxide piezoelectric/diamond MEMS/NEMS devices. The work reported here demonstrated that the PZT/UNCD integration can be achieved by using robust TiAl or TaAl (named TA) layers with dual functionality as oxygen diffusion barrier, to avoid interaction of oxygen with the carbon-based UNCD, and adhesion layers interposed between the PZT and UNCD layers. The TA barriers were chosen based on thermodynamic arguments, which indicate that oxygen atoms react preferentially with Ti, Ta and Al to form stable oxides due to the lowest energy of oxide formation for these elements with respect to all other elements in the periodic table.[15]

EXPERIMENT

The experiments described here involved the following steps: (1) growth of ~ 1 μm thick UNCD layer on a Si (100) substrate; (2) fabrication of UNCD (10 to 140 μm long x 1- 10 μm wide) cantilevers, using a Focused Ion Beam (FIB) system with a 30 keV Ga^+ ion beam to etch the UNCD layer to define the cantilevers; (3) growth of a ~ 10 nm thick TaAl barrier layer on the UNCD film; (3) growth of ~ 180 nm thick Pt layer on top of the TaAl barrier; (4) growth of a 60 nm thick $PbZr_{0.47}Ti_{0.53}O_3$ piezoelectric layer, via sputter-deposition at ~ 600 °C in 100 mTorr of oxygen; (5) growth of the top 50 nm thick Pt layer to complete the capacitor-like structure needed for piezo-actuation via voltage application between the top and bottom Pt electrode layers. Pt/PZT/Pt capacitors were produced on the same UNCD film used for MEMS structures, to measure the polarization properties of the PZT layers integrated with the UNCD films.

X-ray diffraction (XRD) analysis of the PZT layer, grown on the Pt/TaAl/UNCD heterostructure, showed a polycrystalline structure with a slight (001) enhanced orientation (Fig. 1). This PZT layer yielded capacitors with well-saturated polarization (~ 29 $\mu C/cm^2$) in the 5-9 volt range. Figure 2 shows a cross-section SEM image of the $Pt/PZT/Pt/TaAl/UNCD/SiO_2/Si$ heterostructure. Cross section TEM studies published elsewhere[16] showed sharp interfaces between all layers, indicating no interlayer diffusion processes.

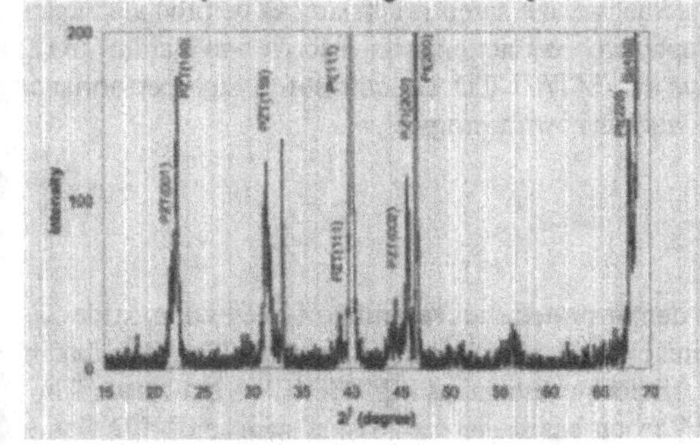

Figure 1. XRD spectrum of PZT layer grown on Pt/TiAl/UNCD using the conditions described in the text above.

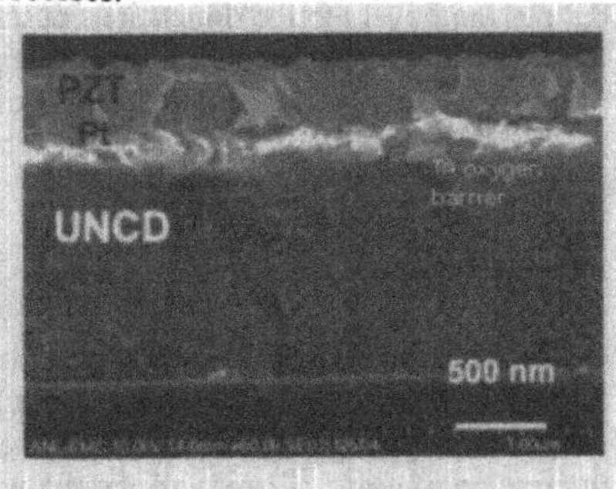

Figure 2. Cross-section SEM of a PZT/Pt/TA/UNCD heterostructure grown using the conditions described in the text above

DISCUSSION

A key part of the work discussed in this paper is the demonstration of piezoelectric actuation of diamond-based MEMS structure. Figure 3 show SEM pictures from frozen frames of a movie recorded when actuating an array of hybrid PZT/UNCD cantilevers (length= 20 μm, width=500 nm) with 3 Volts (AC) excitation in the range 1 Hz to 1 MHz applied between the top and bottom Pt layers sandwiching the PZT piezoelectric film. Fig. 3 (a) shows the array of hybrid cantilevers at rest, while Fig. 3 (b) shows the cantilevers vibrating at 357 kHz (six resonances were demonstrated with oscillation amplitudes in the range 1-4 μm).

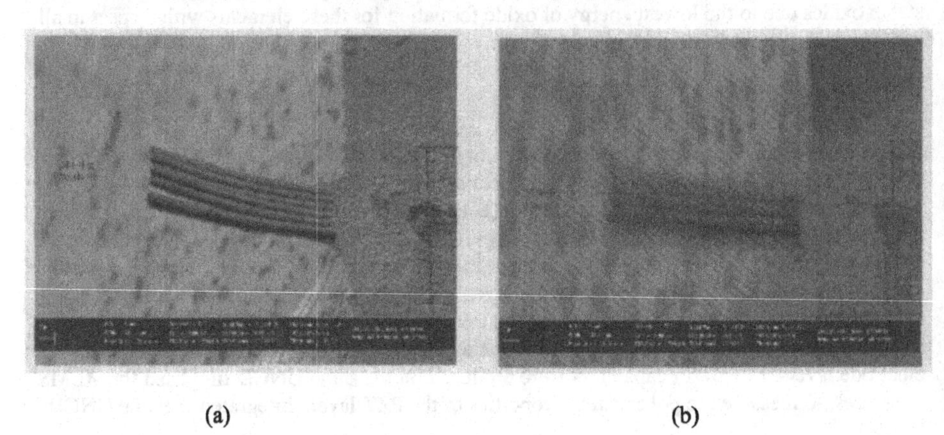

(a) (b)

Figure 3. (a) Array of Pt/PZT/Pt/TaAl/UNCD piezoelectrically actuated heterostructures fabricated by FIB sputter-cutting from a wide cantilever structure; (b) Same cantilever array shown in (a) at vibrating resonance (357 kHz, 3 V between top and bottom Pt electrodes).

The cantilevers were driven up to 1 billion cycles, demonstrating the robustness of the PZT/UNCD hybrid structure. Although the PZT/UNCD cantilevers were fabricated using FIB to cut the cantilevers to the dimensions described above, the same structures can be produced using industrial processes, involving photolithography in conjunction with reactive ion etching (RIE) in oxygen plasmas to produce large arrays of PZT/UNCD structures for high-performance MEMS/NEMS piezo-actuated devices. This work is now in progress.

CONCLUSIONS

The work described in this paper has demonstrated the feasibility of integrating oxide piezoelectric layers with high actuation efficiency with mechanically robust UNCD layers that exhibit high Young modulus to enable high efficiency resonating MEMS/NEMS structures. The integration of oxide piezoelectric layers with carbon-based diamond layers such as UNCD was achieved via robust oxygen diffusion barriers such as those made with TiAl or TaAl alloys, which components (Ti, Ta, and Al) exhibit the lowest energy of oxide formation among all elements of the periodic table. Hybrid PZT/UNCD based cantilevers were fabricated using FIB process, and piezoelectric actuation was demonstrated, using low voltage (3 V) applied between the tip and bottom Pt electrodes sandwiching the PZT layer. This work opens the way for the development of low voltage piezoelectrically actuated diamond-based MEMS/NEMS devices such as resonators and RF switches, which could be monolithically integrated with low voltage CMOS devices as drivers.

ACKNOWLEDGMENTS

This work was supported by the US Department of Energy, BES-Materials Sciences; and the use of the Center for Nanoscale Materials was supported by the US Department of Energy, office of Science, Office of Basic Energy Sciences, both under Contract DE-AC02-06CH11357 and by DARPA/MTO.

REFERENCES

1. H. Espinosa, B. Peng, M. Moldovan, F. Friedmann, X. Xiao, D.C. Mancini, O. Auciello, J.A. Carlisle, A. Zorman, M. Merhegany, App. Phys. Lett., **89** (2006) 073111.
2. J. Birrell, J.E. Gerbi, O. Auciello, J.M. Gibson, J. Johnson and J.A. Carlisle, *Diam Relat. Mater.* **14**, 86 (2005).
3. O. Auciello J. Birrell, J.A. Carlisle, J.E. Gerbi, X. Xiao, B. Peng and H.D. Espinosa, *J. of Physics: Condensed Matter.* **16**(16), R539 (2004).
4. A. V. Sumant, D. S. Grierson, J. E. Gerbi, J. Birrell, U. D. Lanke, O. Auciello, J. A. Carlisle and R. W. Carpick, *Advanced Materials* **1**, 1039 (2005).
5. W. Yang O. Auciello, J. E. Butler, W. Cai, J.A. Carlisle, J.E. Gerbi, D.M. Gruen, T. Knickerbocker, T.L. Lasseter, J.N. Russell, L.M. Smith and R.J. Hamers, *Nature Mater.* **1**, 253 (2002).
5. X. Xiao J. Wang, C. Liu, J.A. Carlisle, B. Mech, R. Greenberg, D. Guven, R. Freda, M.S. Humayun, J. Weiland and O. Auciello, *J. Biomedical Materials* 77B(2), 273 (2006).
7. D. Polla, in MRS Bulletin," Electroceramic Thin Films, Part 2:Device Application," O. Auciello, R. Ramesh (Guest Editors), **21**(6), (June 1996).
8. P. Muralt, Piezoelectric Micromachined Ultrasonic Transducers For Rf Filtering and Ultrasonic Imaging. International Journal of Computational Engineering Science **4(2)**, 163 (2003).
9. S. Trolier McKinstry and P. Muralt, J. of Electroceramics **12**, 7 (2004).
10. J. Akedo and M. Lebedev, Appl. Phys. Lett. 77 (11), 1710 (2000).
11. H. Du, D. W. Johnson, Jr., W. Zhu, J. E. Graebner, G. W. Kammlott, S. Jin, J. Rogers, R. Willett, and R. M. Fleming, *J. Appl. Phys.* **86**(4), 2220 (1999).
12. Q. Wan N. Zhang, L. Wang, Q. Shen and C. Lin, *Thin Solid Films* **415**, 64 (2002).
13. T. Shibata, K. Unno, E. Makino, and S. Shimada, Sensors and Actuators A: Physical **114**, 398 (2004).
14. S. Srinivasan and O. Auciello (unpublished).
15. *Thermochemical Properties of Inorganic Substances, 2nd Edition*, edited by O. Knacke, O. Kubaschewski, and K. Hesselmann, (Springer, Berlin, 1991).
16. S. Sudarsan, J. Hiller, B. Kabius, and O. Auciello, Appl. Phys. Lett. **90**, 134101 (2007).

Mater. Res. Soc. Symp. Proc. Vol. 1052 © 2008 Materials Research Society 1052-DD07-06

Relationship between film stress and dislocation microstructure evolution in thin films

Ray S. Fertig, and Shefford P. Baker
Materials Science and Engineering, Cornell University, Bard Hall 214, Ithaca, NY, 14853

ABSTRACT

Metals with one or more dimensions in the submicron regime are widely used in MEMS devices. Device stresses often exceed the strength of the corresponding bulk material by an order of magnitude and can lead to a variety of mechanical failures. At moderate temperatures, high stresses occur partly because complex dislocation behavior, such as junction formation, annihilation, and nucleation, prevents dislocation motion. In this report, we present results from analytical models, cellular automata simulations, and large-scale dislocation dynamics simulations of thin films to examine the relationship between dislocation interactions and material strength. Our results reveal a complex relationship between dislocation interactions and stress inhomogeneity that arises from the stress fields of the dislocations. We show that the stress inhomogeneity increases both the likelihood of interactions and acts to increase the strain hardening rate.

INTRODUCTION

Thin metal films are used in applications ranging from chemical barriers to MEMS devices. However, the stresses supported by thin films often exceed bulk yield stresses by an order of magnitude [1, 2], which increases the likelihood of device failure and makes metal films not viable for some MEMS applications [3]. To increase reliability and expand the usefulness of metal films in MEMS applications, residual stresses must be reduced. For this to be realized, the origin of high film stresses must be better understood.

At moderate temperatures, stress relaxation occurs via dislocation motion; in thin films, relaxation occurs via motion of threading dislocations (threads), which extend through the thickness of the film. As a thread moves it creates a misfit dislocation (misfit) or surface step, which reduces the film stress. Thus, high film stress depends on preventing thread motion.

Three mechanisms that prevent thread motion have been identified. First, the onset of thread motion is controlled by a dimensional constraint, whereby a thread may not move until some channeling stress τ_{ch} (or strain ε_{ch}), which is inversely proportional to the film thickness, is exceeded [4, 5]. Second, the thread may be impeded by interaction with grain or twin boundaries. Finally, a thread may be stopped by interaction with other dislocations [6]. For example, a misfit can stop a moving thread up to a film stress of about $1.3\tau_{ch}$ [6, 7]. Most models of film stress only consider interactions of threads with misfits [7]. At higher stresses, the thread breaks free and continues moving through the film.

Three other important features of film relaxation have been observed but their effect on film strength has not been studied. First, in metal films on amorphous substrates misfits are observed to disappear over time [8], indicating dislocation core spreading into the interface [9]. Second, irregularity of the dislocation structure gives rise to inhomogeneity in the stress field [10].

Finally, threads are known to interact with other threads to form junctions and dipoles and to annihilate [11].

In this paper, we examine two features of film relaxation: stress inhomogeneity and dislocation interactions. We study these features in two ways. First, three dimensional dislocation dynamics (DD) simulations are used to determine the interactions that stop dislocations during the course of relaxation. The film stresses, both average and local, are also monitored. Any correlations between the types of interactions and the local stresses can also be observed. The second method to study the effects of stress inhomogeneity on dislocation interactions is a cellular automata simulation. This is explicitly designed to look for correlations of stresses with interactions between threads. Finally, we show that a more complex picture of film relaxation emerges from the results from these two simulation techniques.

SIMULATION

Dislocation dynamics simulations

DD simulations [12] were used to model a 200 nm thick passivated single crystal fcc film with (001) film normal. Because the passivation and substrate were assumed to be infinitely thick and have the same elastic properties as the film, no image forces were considered. Infinite planar dimensions were enforced by use of periodic boundaries, with a periodic unit cell of 4 μm × 4 μm. Initially 70 glide loops, providing 140 threads, were seeded at random locations in the unit cell and were distributed randomly over the eight slip systems that can relax an applied biaxial strain in a (001) film.

Strain was applied incrementally to the film, beginning with a strain of $1.3\varepsilon_{ch}$ and increasing to $3.3\varepsilon_{ch}$ in increments of $0.5\varepsilon_{ch}$, where $\varepsilon_{ch} = 0.00155$ for this film. The strain was held constant at each strain level until all threads had stopped moving. After the threads stopped moving, the interaction stopping each thread was determined. Film stresses were calculated at the midplane of the film on a 125 × 125 grid, the spatial correlation between this field and the location of the stopped threads was also calculated. Further simulation details are given elsewhere [13].

Cellular automata simulations

Results from an analytical statistical model [14] and from the DD simulations (described below) suggest that stress inhomogeneity plays an important role in determining where threads stop. To study this effect, a two dimensional cellular automata simulation was constructed. A square simulation cell was divided into a 100 × 100 grid; 500 objects, representing threads, were placed randomly on grid points. The spacing between grid points can be thought of as a thread interaction radius r_{int} (i.e. if threads are within this radius, they will form a junction or annihilate). Each grid point was assigned a stress τ_i. This stress is actually the excess stress [15], which is the stress above the τ_{ch}. A thread j located on grid point i was assumed to move with a velocity $v^i_j = M\tau_i$, where M is the thread mobility. Therefore the thread spends time $t^i_j = \frac{r_{int}}{M\tau_i}$ at grid point i. For all the simulations reported here $M = \frac{r_{int}}{\tau_{ch-s}}$ so that $t^i_j = 1$ for $\tau_i = \tau_{ch}$. The stresses at the grid points were assigned randomly so as to follow a normal distribution except for a cutoff of $\tau_c = 0.03\,\tau_{ch}$, which ensured that all threads would be moving with some velocity.

To mimic travel on a specified slip plane, each thread was assigned one of two perpendicular directions before the first time step; the thread was allowed to travel only in this direction. Two threads are considered to be interacting when they reside on the same grid point. The mean normalized stress $\left\langle \frac{\tau}{\tau_{ch}} \right\rangle$ was fixed at unity. The normalized standard deviation $\frac{\sigma}{\tau_{ch}}$ of the stress distribution was varied from 0.0 to 1.0. For each standard deviation, 100 different simulations were performed, each one running to a time of 50 s. The number of interacting threads was recorded at each time step for each run and the results of the runs were averaged.

RESULTS

Dislocation dynamics

Both thread-misfit (TM) and thread-thread (TT) interactions were observed to stop threads, but the quality of the interactions was different. Threads stopped by TT interactions often remained in those interactions upon successive strain increments. Threads stopped by TM interactions usually broke free upon increasing the strain, so that the threads continued moving through the film. These observations are consistent with the high strength of TT interactions and low strength of TM interactions [6]. However, even when the average film stress exceeded the strength of TM interactions, they were still observed to play a significant role in stopping threads. This was possible because the stress field arising from the misfit structure was strongly inhomogeneous, with local regions of stress lower than the TM interaction strength.

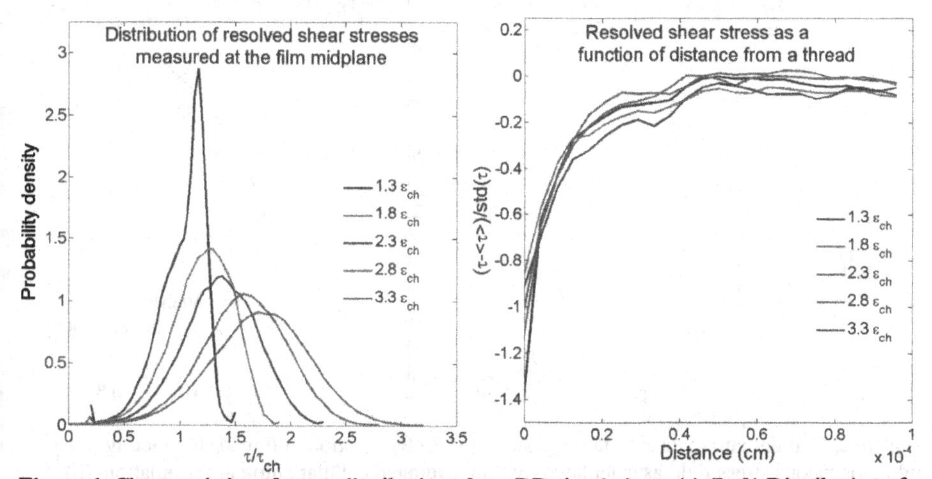

Figure 1. Characteristics of stress distributions from DD simulations: (a) (Left) Distribution of film stresses measured at the mid-plane of the film with increasing applied strain and subsequent relaxation. (b) (Right) Distribution of resolved shear stresses τ around stopped threads. Threads are always stopped in regions of locally low stress.

Figure 1a shows the distribution of resolved shear stress τ averaged over the eight active slip systems as obtain from DD simulations. The distribution becomes normal at higher applied strains. The stress inhomogeneity, as measured by the standard deviation of stress, increased with increasing strain. Figure 1b shows the distribution of resolved shear stresses averaged around each thread. This is the most important result from the DD simulations: *all* threads stopped in regions of low stress. That threads stopped by TM interactions are found in regions of low stress is not surprising; when the average film stress exceeds their interaction strength the only location they can occur is in regions of stress lower than the mean. However, even threads stopped by TT interactions were stopped in regions of low stress. This suggests that inhomogeneous stresses concentrate threads in particular regions of the film, which increases the interaction likelihood. Thus, stress inhomogeneity appears to facilitate *both* TM and TT interactions.

Cellular automata

The purpose of these simulations was to quantify the enhancement, if any, of stress inhomogeneity on the likelihood of TT interactions. Figure 2a shows the fraction of interacting threads f with time, normalized by the interacting fraction f_h in a homogeneous field. The most important feature of these data is that stress inhomogeneity clearly increases the likelihood of TT interaction. The interacting fraction initially increases with time because the initial distribution of threads is random and the influence of the inhomogeneous field takes some time to develop. As the threads move through the film their distribution approaches a steady-state distribution dictated by the stress field.

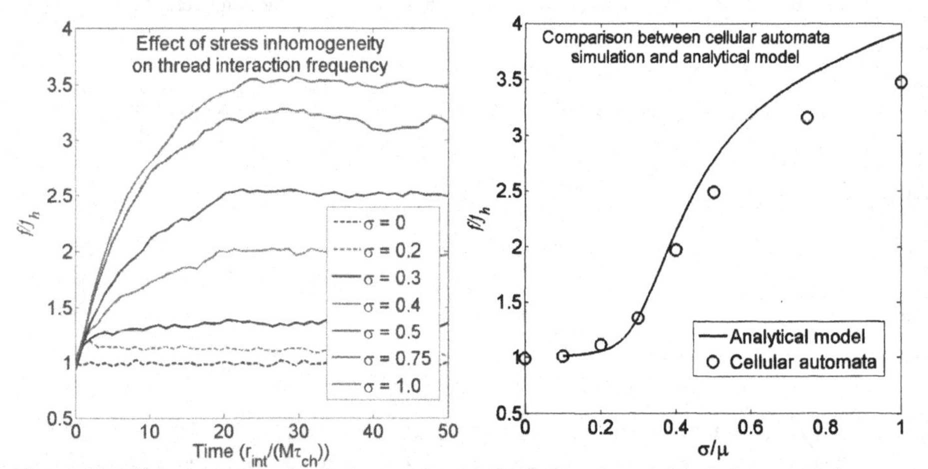

Figure 2. Cellular automata simulation results: (a) (Left) Fraction of threads interacting in an inhomogeneous stress field as calculated by time-averaged cellular automata simulation. (b) (Right) Comparison of analytic model [14] with plateau values of cellular automata simulation as a function of standard deviation σ divided by mean μ.

An analytical model based on a statistical analysis [14] has predicted that for threads moving in a *random* direction in a random field, the fraction of interacting threads can be described by the solid line in Figure 2b. This model predicts that the fraction of threads f interacting in an inhomogeneous stress field τ, varies as

$$f = \rho_{TD} r_i^2 \left(\frac{\langle \tau^{-2} \rangle}{\langle \tau^{-1} \rangle^2} - \rho_{TD} r^2 \frac{\langle \tau^{-3} \rangle}{\langle \tau^{-1} \rangle^3} \right), \tag{1}$$

where ρ_{TD} is the density of threads. The plateau levels of each curve from Figure 2a are plotted in Figure 4 along with the analytical model prediction and the two predictions are in good agreement. Both the analytical model and cellular automata simulation predicts an increase in fraction of interacting threads that becomes significant when $\sigma/\mu \approx 0.25$.

DISCUSSION

Our results showed that stress inhomogeneity affects both TT and TM interactions and that both TT and TM interactions play significant roles in stopping threads during relaxation. In order to discuss these effects on film stress τ, we recall a simple relationship between the motion of threads and stress relaxation for a constant applied strain,

$$d\tau = -\gamma \rho_{TD} dx, \tag{2}$$

where γ is a constant that includes geometric factors, biaxial modulus, and Burgers vector; dx is the average incremental distance that a moving threads travels, and ρ_{TD} is the thread density. As the thread density decreases the stress relaxation with each successive increment of motion diminishes—so the effect of reducing the thread density is a decrease in stress relaxation resulting in an increase in film strength.

One way to reduce the mobile thread density is through TM interactions. Threads can stop in TM interactions only when these interactions occur where the local stress is less than the TM interaction strength. Therefore, an upper bound for the average distance λ that a thread travels before being stopped is the distance between these local regions of low stress, since threads can only travel a shorter distance if they also interact with other threads. The number of moving threads stopped by TM interactions during some average increment of dx is then proportional to dx/λ. For average film stresses above the strength of a TM interaction, increasing stress inhomogeneity causes a decrease in λ, increasing the likelihood of a TM interaction during the increment dx. From this we conclude that the effect of stress inhomogeneity on TM interactions serves to increase the film strength.

The other way to reduce the mobile thread density is through TT interactions. The results of the cellular automata simulations show that increasing stress inhomogeneity causes an increase in the likelihood of TT interactions by concentrating them in regions of low stress. This was qualitatively observed in the DD simulations. Here too, the effect of stress inhomogeneity on TT interactions serves to increase the film strength.

Since the effect of stress inhomogeneity on both TM and TT interactions leads to an increase in film strength over the case of a homogeneous stress field, we posit that the overall effect of stress inhomogeneity in films is to increase the strain hardening rate. In order for a constitutive model for film stress to be developed from fundamental dislocation behaviors, two phenomena need to be better understood. First, in real films the mean stress will decrease and the standard deviation will increase as threads move. In the cellular automata simulation the motion of the threads did not affect the stress field. To capture this feature, the relationship between misfit spacing and stress inhomogeneity needs to be quantified. Second, this relationship needs to be augmented with a quantitative description of the *spatial* fluctuation in the stress field so that the average distance that a dislocation may travel before arriving at a low-stress region can be determined.

CONCLUSION

We have used DD simulations and cellular automata simulations to show that stress inhomogeneity and interactions between threading dislocations are intertwined. The DD simulations showed that during film relaxation *all* threads stop in regions of low stress. The cellular automata simulation demonstrated that inhomogeneous stresses concentrate threads, increasing the likelihood of interactions between two threads. Finally, we showed that the effects of stress inhomogeneity on both TT and TM interactions lead to an increase in film stress. As such, stress inhomogeneity must be considered in any model of film stress relaxation by dislocations.

ACKNOWLEDGEMENTS

This research was supported by the National Science Foundation, DMR-0311848. The dislocation dynamics simulations were conducted using the resources of the Cornell University Center for Advanced Computing, which receives funding from Cornell University, New York State, the National Science Foundation, and other public agencies, foundations, and corporations.

REFERENCES

[1] S. P. Baker, Mater. Sci. Eng. A **319-321**, 16 (2001).
[2] S. P. Baker, R. M. Keller-Flaig, and J. B. Shu, Acta Mater. **51**, 3019 (2003).
[3] J. R. Stanec, C. H. Smith, I. Chasiotis, and N. S. Barker, J. Micromech. Microeng. **17**, N7 (2007).
[4] L. B. Freund, J. Appl. Mech. **54**, 553 (1987).
[5] W. D. Nix, Metall. Trans. A **20**, 2217 (1989).
[6] P. Pant, K. W. Schwarz, and S. P. Baker, Acta Mater. **51**, 3243 (2003).
[7] L. B. Freund, J. Appl. Phys. **68**, 2073 (1990).
[8] T. S. Kuan and M. Murakami, Metall. Trans. A **13**, 383 (1982).
[9] S. P. Baker, L. Zhang, and H. J. Gao, J. Mater. Res. **17**, 1808 (2002).
[10] M. A. Phillips, R. Spolenak, N. Tamura, W. L. Brown, A. A. MacDowell, R. S. Celestre, H. A. Padmore, B. W. Batterman, E. Arzt, and J. R. Patel, Microelec. Engr. **75**, 117 (2004).
[11] K. W. Schwarz, Phys. Rev. Lett. **91**, 145503 (2003).
[12] K. W. Schwarz, J. Appl. Phys. **85**, 108 (1999).
[13] R. S. Fertig, P. Pant, K. W. Schwarz, and S. P. Baker, In preparation (2007).
[14] R. S. Fertig and S. P. Baker, In preparation (2007).
[15] L. B. Freund and R. Hull, J. Appl. Phys. **71**, 2054 (1992).

Mater. Res. Soc. Symp. Proc. Vol. 1052 © 2008 Materials Research Society 1052-DD07-07

Systematic Characterization of DRIE-Based Fabrication Process of Silicon Microneedles

Jochen Held[1], Joao Gaspar[1], Patrick Ruther[1], Matthias Hagner[2], Andreas Cismak[3], Andreas Heilmann[3], and Oliver Paul[1]

[1]Department of Microsystems Engineering (IMTEK), University of Freiburg, Georges-Koehler-Allee 103, Freiburg, D-79110, Germany

[2]Department of Physics, University of Konstanz, Universitätsstr. 10, Konstanz, D-78457, Germany

[3]Department of Biological materials and interfaces, Fraunhofer Institute for Mechanics of Materials Hal, Walter-Hülse-Straße 1, Halle (Saale), D-06120, Germany

ABSTRACT

This paper reports on the systematic characterization of a deep reactive ion etching based process for the fabrication of silicon microneedles. The possibility of using such microneedles as protruding microelectrodes enabling to electroporate adherently growing cells and to record intracellular potentials motivated the systematic analysis of the influence of etching parameters on the needle shape. The microneedles are fabricated using dry etching of silicon performed in three steps. A first isotropic step defines the tip of the needle. Next, an anisotropic etch increases the height of the needle. Finally, an isotropic etch step thins the microneedles and sharpens their tip. In total, 13 process parameters characterizing this etching sequence are varied systematically. Microneedles with diameters in the sub-micron range and heights below 10 µm are obtained. The resulting geometry of the fabricated microneedles is extracted from scanning electron micrographs of focused ion beam cross sections. The process analysis is based on design-of-experiment methods to find the dominant etch parameters. The dependence of the needle profiles on process settings are presented and interpolation procedures of the geometry with processing conditions are proposed and discussed.

INTRODUCTION

For drug development and disease studies, electrical recording of cells is a fundamental method. However, the classical patch-clamp methods available for intracellular measurements are time-consuming and require experienced staff [1]. On the other hand, methods like patch-on-chip systems enable the parallel examination of a larger number of cells [1]. However, the patch-on-chip method is restricted to cells in suspension, as the cells have to be positioned and fixed through small holes in the measurement chip. So far, cell monitoring chips with adherent cell cultures have allowed the detection of extra-cellular potentials only [2].

To provide a method that enables the measurement of intracellular cell potentials of adherently growing cells, a novel cell chip design comprising microneedle-based electrodes has been developed [3,4]. The cells are cultured on these electrodes which are introduced into the cytoplasm using electroporation [4]. Due to the low radius of curvature of the electrodes below 1 µm, a voltage pulse on the order of a few hundred millivolts up to a few volts is sufficient for this purpose [4]. The profile of the needles is important for a successful electroporation and subsequent intra-cellular recording. To optimize and predict the microneedles shape, the influence of 13 relevant process parameters on the profile of the needles is systematically evaluated in this study. In view of the relatively large number of parameters, the analysis of their respective influence on the geometry of the needles, based on design-of-experiment (DoE) is adopted [5].

FABRICATION

Figure 1 schematically shows the process sequence used to produce out-of-plane microneedles required for the realization of cell penetrating electrodes. The process starts with single-side polished (100) silicon wafers oxidized to a resulting thermal oxide thickness of 300 nm patterned using reactive ion etching (RIE) [Figure 1 (a)]. This oxide layer is used as an etch mask for the subsequent dry etching steps performed in an inductively coupled plasma (ICP) etcher (STS, Surface Technology Systems). The three-dimensional profile of the microneedles is defined in three individual etching steps: (i) isotropic, (ii) anisotropic and (iii) isotropic dry etching of silicon [Figure 1 (b-d)]. The first etch step realizes the tip of the needle; the silicon substrate is patterned with an isotropic plasma etch using a SF_6 plasma [Figure 1 (b)]. The anisotropic etching step uses DRIE to adjust the height of the needles [Figure 1 (c)]. The last isotropic step sharpens the needle tip and thins the needle [Figure 1 (d)]. In a final step, the partially suspended oxide etch mask is removed in buffered hydrofluoric acid (HF) solution [Figure 1 (e)].

RESULTS AND DoE ANALYSIS

Figure 2 shows scanning electron micrographs (SEM) of cross sections obtain from focused ion beam (FIB) cuts of typical microstructures realized using the as described process sequence. In case of the needles shown in Figure 2 (a-c) the mask size is varied from 5 µm to 10 µm while using standard process parameters for the etching sequences. Figure 2 (d-f) illustrates as a further example the effect of the duration of the second isotropic etch step on the diameter of the microneedles; it is varied from 1 min to 2 min.

In addition to variations of the mask size as given above, process parameters such as etch duration, platen power, SF_6 gas flow and etching time of both isotropic etching steps, and etching and passivation times, platen power and SF_6 and C_4F_8 gas flows of the anisotropic step, are varied in a systematic way. In total a number of 13 process parameters i_n ($n = 1 \ldots 13$) as listed in Table 1 with their center values and respective variation ranges are studied with respect to their influence on the microneedle geometry. Each input is varied around its center value while keeping the remaining process parameters constant.

The microneedle geometry used for the DoE analysis is defined by points along the FIB cross sections extracted using SEM micrographs. The general profile of a microneedle thus may be characterized by the four points P_1 through P_4 defined by coordinates (o_i, o_{i+1}) and two

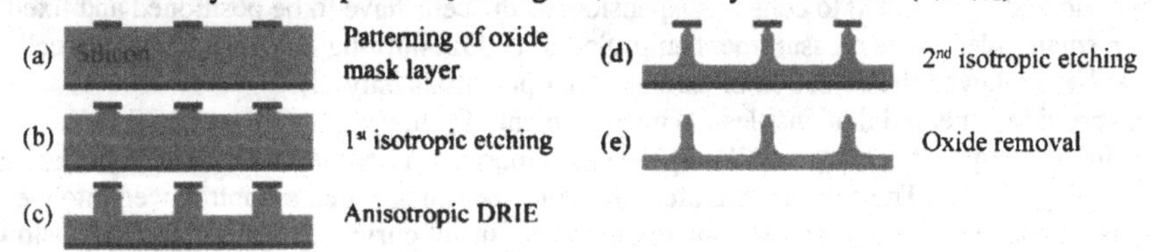

Figure 1. Fabrication schematics of stepwise isotropic anisotropic isotropic process of microneedle arrays using dry etching of silicon.

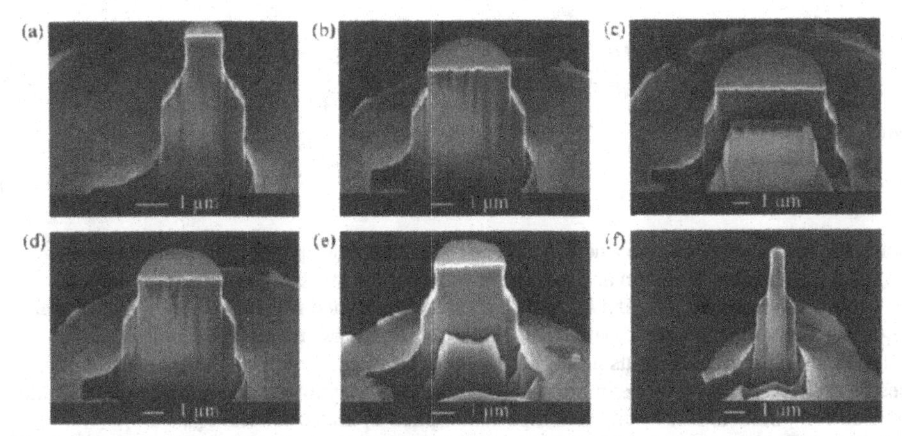

Figure 2. SEM micrograph obtained from FIB cross sections of fabricated microneedles; (a-c) different mask sizes i_1 of 5 µm, 7.5 µm and 10 µm and (d-f) different etch times of the 2nd isotropic etching step 1 min, 1:30 min and 2 min, respectively.

functions f_1 and f_2 between these points, as illustrated in Figure 3. As a consequence of the applied etch sequence, points P_2 and P_3 are assumed to be vertically aligned. The function f_1 between points P_1 and P_2 is a cubic function (Equation 1) while f_2 defined between points P_3 and P_4 is a polynom of seventh order as given in Equation 2. In order to extract the functions f_1 and f_2 describing the needle profile for each set of process parameters i_n additional points are introduced in the respective FIB cross sections as exemplarily shown in Figure 3. The coordinates of the points P_1 to P_4 as well as the coefficients of the profile functions are assigned with the output variables o_1 to o_{20}, respectively.

$$f_1(x) = o_9 + o_{10}x + o_{11}x^2 + o_{12}x^3 \tag{1}$$

$$f_2(x) = o_{13} + o_{14}x + o_{15}x^2 + o_{16}x^3 + o_{17}x^4 + o_{18}x^5 + o_{19}x^6 + o_{20}x^7 \tag{2}$$

To first order approximation, the correlation between output parameters o_n ($n = 1...20$) and input parameters i_i ($i = 1...13$) is described by

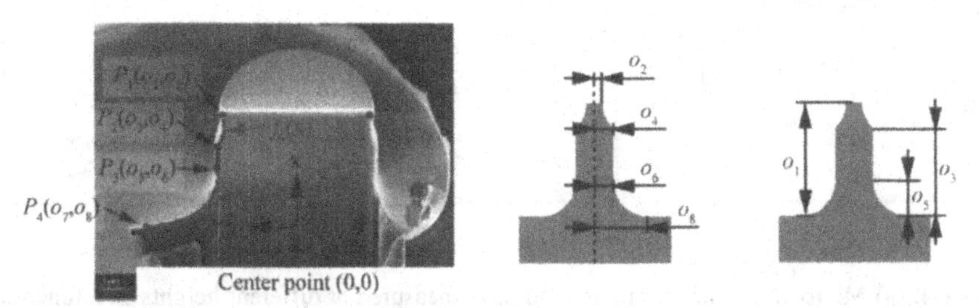

Figure 3. Schematics of microneedle shape with relevant geometrical dimensions extracted as a function of the process parameters.

$$o_{L,n} = \underbrace{b_{0,n}}_{\text{offset}} + \underbrace{\sum_{i=1}^{13} b_{i,n}\bar{i}_i}_{\text{main effect}} + \underbrace{\sum_{i=1,\, j\neq i}^{13,13} b_{ij,n}\bar{i}_i\bar{i}_j}_{\text{2-way interacton terms}} \qquad n = 1\ldots 20 \tag{3}$$

with

$$\bar{i}_i = 2\frac{i_i - (i_{i,\max} + i_{i,\min})/2}{i_{i,\max} - i_{i,\min}} \tag{4}$$

being the normalized input parameter describing the relative weight of input parameter, where $i_{i,min}$ and $i_{i,max}$ are the minimum and maximum values of the input parameter i_i, respectively.

The coefficients $b_{0,n}$, $b_{i,n}$ and $b_{ij,n}$ in Equation 3 are obtained from multidimensional linear regression fits of the measured outputs o_n. While coefficient $b_{0,n}$ represents the offset of the solution, the sum over all inputs multiplied by the coefficient $b_{i,n}$ reflects the main effect on needle profile . The third terms in Equation 3 describes 2–way interaction terms through the sum over all combinations of two different inputs multiplied by the coefficient $b_{ij,n}$.

The DoE analysis is extended through higher order solutions as given in Equations 5 and 6. The quadratic solution given Equation 5 contains the sum of pure quadratic terms multiplied by coefficients $b_{ii,n}$ added to the linear solution described in Equation 3.

$$o_{Q,n} = o_{L,n} + \underbrace{\sum_{i=1}^{13} b_{ii,n}\bar{i}_i^2}_{\text{quadratic terms}} \quad n = 1\ldots 20 \tag{5}$$

The cubic solution is represented in Equation 6. It contains 3–way interactions as well as the cubic terms.

$$o_{C,n} = o_{Q,n} + \underbrace{\sum_{i=1,\, j\neq i,\, k\neq j\neq i}^{13,13,13} b_{ij,n}\bar{i}_i\bar{i}_j\bar{i}_k}_{\text{3-way interaction terms}} + \underbrace{\sum_{i=1}^{13} b_{iii,n}\bar{i}_i^3}_{\text{cubic terms}} \qquad n = 1\ldots 20 \tag{6}$$

As an example, Figure 4 (a-b) illustrates the influence of mask size, i.e. input parameter i_1, on the needle height and radius (output parameters o_1 to o_8) while keeping the other process parameters at their center values. As one would expected, the needle radius increases with increasing mask size. The solid lines shown in Figure 4 (a-b) are obtained from fits to Equation 5, where all the input parameters i_i, excluding mask diameter i_1, are kept at their standard process values.

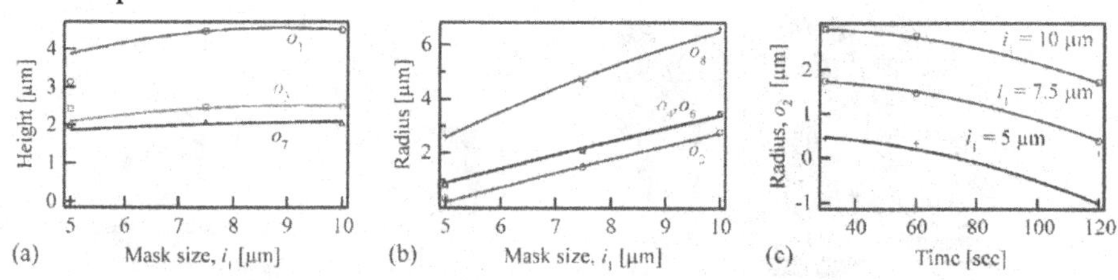

Figure 4. (a) Microneedle height and (b) radius o_i measured at different heights as a function of the mask diameter i_1. (c) Influence of the duration of the 1st isotropic etching step on the radius o_2 of the microstructures.

Table 1. Overview of process parameters and their main influence on the microneedle profile. This main influence represents the mean coefficient, which is obtained from the cubic solution. (increase ↑, decrease ↓ and no influence 0, the size of the arrows quantifies the influence).

Parameters	Process Step	Input	Values	o_1	o_2
Mask size [μm]	All	i_1	5 / 7.5 / 10	·	↑
Etching time [sec]	1st isotropic	i_2	30 / 60 / 120	↑	↓
Platen power [W]	1st isotropic	i_3	0.2 / 10 / 20	↑	↓
SF_6 flow [sccm]	1st isotropic	i_4	40 / 50 / 60	0	0
Etching time [sec]	Anisotropic	i_5	5 / 10	↓	↓
Passivation time [sec]	Anisotropic	i_6	5 / 10	↓	↓
Platen power [W]	Anisotropic	i_7	6 / 12 / 24	↑	↑
SF_6 flow [sccm]	Anisotropic	i_8	40 / 50 / 60	0	0
C_4F_8 flow [sccm]	Anisotropic	i_9	20 / 30 / 40	0	0
Etching time [sec]	2nd isotropic	i_{10}	60 / 90 / 120	·	·
Platen power [W]	2nd isotropic	i_{11}	0.2 / 10 / 20	↓	↓
SF_6 flow [sccm]	2nd isotropic	i_{12}	40 / 50	↑	↑
Throttle valve angle [°]	All	i_{13}	0.1 / 10 / 20	↓	↓

Figure 4 (b) illustrate the influence of 1st isotropic etch time on the radius o_2 at the top of the needle for different mask sizes i_1. Again, the needle radius decrease with the etching time as expected. The needle is overetched when $o_2 = 0$ is achieved, in case of a mask size of 5 μm the needles been overetched at an etch time of 70 sec.

Using Equations 3, 5 and 6 to extract coefficients b it is possible to estimate the shape of the microneedles. Figure 5 shows the example of a needle profile calculated using the linear, quadratic and cubic solution of the DoE analysis. Aside from the calculated needle profile, Figure 5 shows the respective difference Δf between simulated and real needle shape obtained from a FIB cut. It is obvious from Figure 5 (c) that the cubic solution results in the best fit. The fitting points match the measured data, since from a fundamental point of view a higher order polynomial will be more flexible to pass by most of the measured data points. Thus, the shapes of the microneedle may be estimated in the ranges of the process variation given in Table 1. Figure 6 shows the distributions of the residues, i.e. the differences between the measured and fitted values, of output o_4 obtained from data from 48 different needles and fits using the three different equations. The figure shows the respective residues of all needles used to calculate the fit coefficients. The linear solution [Figure 6 (a)] shows the worst fitting while again the best fit results are achieved with the cubic solution as shown in Figure 6 (c). Nevertheless, the three distributions resemble functions with mean value around zero, ruling out any systematic errors

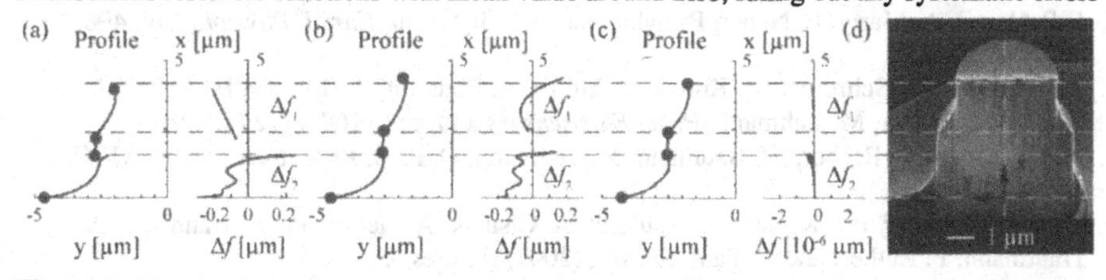

Figure 5. Simulated needle geometry with respective difference Δf_i, between simulated and real profile (i = 1, 2). (a) linear function, (b) quadratic, (c) cubic functions and (d) SEM micrograph obtained from FIB cross section from the needle.

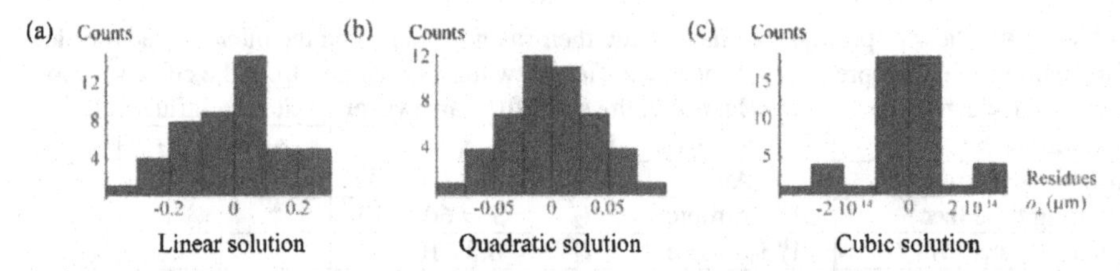

Figure 6. Residues distribution of output o_4 obtained from (a) linear, (b) quadratic and (c) cubic solution.

that could have been occurred.

The influence of all 13 parameters on the radius o_1 and height o_2 in terms of main effects is summarized in Table 1. These influences are assessed from the main coefficient $b_{i,n}$ of the cubic solution. The largest contribution on the geometry of the needles arises from the oxide mask size, as expected.

CONCLUSIONS

We report on the systematic characterization of the fabrication of microneedles used to realize microelectrodes for intracellular recordings. The fabrication process is described by 13 parameters which have been varied systematically in this study. The needle geometries are extracted from SEM micrographs obtained from FIB cross sections. The analysis of the parameters is based on the DoE methodology. Three interpolation procedures of different order have been proposed to predict the needle geometry with processing conditions. As expected, the highest order interpolation solution, i.e. cubic solution, shows the best result with the measured geometry. The characterization of the etching of the microneedles has been demonstrated.

ACKNOWLEDGMENTS

The authors gratefully acknowledge A. Baur, F. Dieterle and M. Reichel (IMTEK cleanroom service center) for useful discussions and help in needle fabrication. The authors also thank BMBF/VDE/VDI for funding the project MIBA – Mikrostrukturen und Methoden für die intrazelluläre Bioanalytik – under the project number 16SV2337.

REFERENCES

1. O.P. Hamil, A. Marty, E. Neher, B. Sakmann, F. J. Sigworth, *Eur. J. Physiol.*, Vol. 406, pp. 73-82 (1981).
2. W. Baumann, E. Schreiber, G. Krause, S. Stüwe, A. Podssun, S. Homma, H. Anlauf, I. Freund, R. Rosner, M. Lehmann, *Proc. Eurosensors XVI*, pp. 1169-1172 (2002).
3. A. Trautmann, P. Ruther, W. Baumann, M. Lehmann, O. Paul, *Proc. Eurosensors XVIII*, pp. 140-141 (2004).
4. J. Held, J. Gaspar, P.J. Koester, C. Tautorat, A. Cismak, A. Heilmann, W. Baumann, A. Trautmann, P. Ruther, and O. Paul, *MEMS*, (2007) (in press).
5. See, for example, D.C. Montgomery, *Design and Analysis of Experiments*, (Hoboken (NJ): Wiley, 2005).

Mater. Res. Soc. Symp. Proc. Vol. 1052 © 2008 Materials Research Society 1052-DD07-11

Novel Fabrication Process for the Integration of MEMS Devices with Thick Amorphous Soft Magnetic Field Concentrators

Simon Brugger[1], Wilhelm Pfleging[2], and Oliver Paul[1]
[1]Department of Microsystems Engineering (IMTEK), University of Freiburg, Freiburg, 79110, Germany
[2]Institute for Materials Research I, Forschungszentrum Karlsruhe GmbH, Karlsruhe, 76021, Germany

ABSTRACT

This paper reports a novel fabrication process enabling the integration of mechanical MEMS devices with thick amorphous soft magnetic field concentrators. The integration process combines silicon on insulator technology for the MEMS device fabrication and epoxy-resin-based attachment of 18-µm-thick amorphous soft magnetic ribbons followed by a wet chemical structuring process. The fabrication process is reported on the basis of a field-concentrator-based resonant magnetic sensor combining an electrostatically driven micromechanical resonator and a planar magnetic field concentrator with two narrow gaps. For realization of the concentrator gaps, the integration process is extended by micro-patterning of the soft magnetic ribbons via UV-laser ablation using an excimer laser system. The characterization of the fabricated resonant magnetic sensor using a stroboscopic video microscope for in-plane motion measurement shows a high sensitivity of 390 kHz/T at a magnetic flux density of 158 µT.

INTRODUCTION

Over the past years, several CMOS magnetic sensors were combined with planar magnetic concentrators to improve their magnetic sensing properties. For example, it has been shown that the use of magnetic concentrators considerably improves the resolution of CMOS magnetotransistors [1] and CMOS Hall devices [2,3]. Further, the integration of planer magnetic concentrators enables standard Hall sensors to measure in-plane magnetic fields and therefore facilitates the design and fabrication of three-axis magnetic sensor systems [4]. Since most concentrator-based magnetic sensing applications require magnetic concentrators with a thickness of 15 µm and more, high demands are placed on the wafer-level deposition process. Vacuum deposition processes for soft-magnetic materials, e.g. sputtering, enable soft-magnetic layers with a thickness of only a few microns. Thus, the electroplating of permalloy (NiFe) [1,5] and the attachment of macroscopically fabricated amorphous soft-magnetic ribbons [3,6] were previously explored for the realization of thick magnetic concentrators. To enable magnetic sensors with high resolution, the coercivity of the used soft magnetic material is required to be as low as possible. In contrast to crystalline soft magnetic materials, amorphous soft magnetic materials have a much lower coercivity and are particularly qualified for realizing planar magnetic concentrators. Since the coercivity of macroscopically fabricated cobalt-based amorphous ribbons is at least 10 times lower than that of thick electroplated amorphous layers [7], the attachment of amorphous ribbons is the most widely used technology for the integration

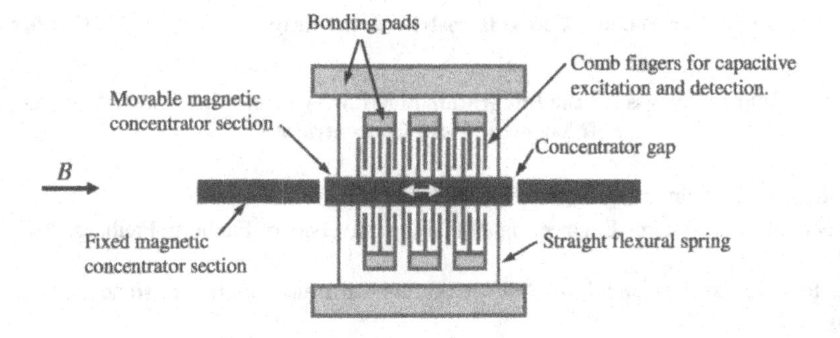

Figure 1. MEMS design of the resonant magnetic sensor combining an electrostatically driven micromechanical resonator and a geometrically optimized planar magnetic field concentrator with two gaps.

of magnetic concentrators. However to date, amorphous ribbon material has been used only for non-mechanical sensor systems.

This paper reports a novel fabrication process enabling the integration of mechanical MEMS devices with thick amorphous soft magnetic field concentrators. The integration process combines silicon on insulator (SOI) technology for the MEMS device fabrication and epoxy-resin-based attachment of amorphous soft magnetic ribbons followed by a wet chemical structuring process. The fabrication process is reported on the basis of a miniaturized field-concentrator-based resonant magnetic sensor based on a previously published sensing principle first demonstrated with a macroscopic proof-of-concept structure [8]. The sensor principle is shown in Figure 1. The structure combines an electrostatically driven micromechanical resonator and an optimized planar magnetic field concentrator with two narrow gaps experiencing a resonance frequency shift in the presence of a magnetic field [9]. For realizing the narrow concentrator gaps, the integration process is extended by micro-patterning of the soft magnetic ribbons via UV-laser ablation using an excimer laser system [10].

FABRICATION PROCESS

The integration process combines two major technologies: (i) silicon on insulator technology for the MEMS device fabrication and (ii) epoxy-resin-based attachment of an annealed ribbon of Metglas® 2714A for the magnetic concentrator. Metglas® 2714A provided by Metglas®, Inc. (Conway, SC, USA), is an 18-μm-thick ribbon of an amorphous cobalt-based alloy with a coercivity below 0.8 A/m. To achieve a hard axis with low and linear permeability along the magnetic concentrators, the ribbons were annealed in the presence of a magnetic field of 200 Oe. The used SOI wafer consists of a 25-μm-thick device layer with a resistivity of 0.01 Ωcm, a 1-μm-thick buried oxide and a 390-μm-thick handle wafer.

The main steps of the fabrication process are shown in Figure 2: (i) bonding pad metallization using physical vapor deposition (PVD) of 0.5 μm of aluminum followed by annealing at $T = 450°C$, (ii) deep reactive ion etching (DRIE) to structure the mechanical

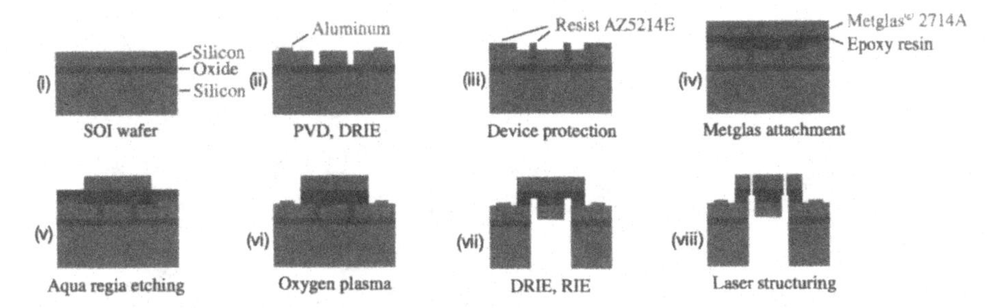

Figure 2: Schematic view of the fabrication process integrating a SOI-technology-based micromechanical resonator with an 18-μm-thick ribbon of Metglas® 2714A.

resonator, the comb fingers and the mechanical springs using a 900-nm-thick positive photoresist, (iii) protection of the resonator structure, the comb fingers and the bonding pad metal against epoxy resin using 2.5 μm hardbaked negative resist AZ5214E, (iv) spin coating of epoxy resin followed by the attachment of the Metglas® 2714A ribbon, (v) structuring of the magnetic concentrator using a 7-μm-thick photoresist and a 1:1:1 (v/v/v) aqua regia mixture of HNO_3 (69%), HCl (37 %) and H_2O at $T = 45°C$ and (vi) high energy oxygen plasma for removal of the remaining epoxy resin and protecting resist, (vii) rear DRIE and oxide removal to release the resonator from the substrate, and (viii) realization of the narrow concentrator gaps using laser-assisted patterning performed via UV-laser ablation.

LASER-ASSISTED PATTERNING

Micro-patterning of thick ribbons of Metglas® 2714A was performed via UV-laser ablation. The aim was to generate defect-free grooves with a width of 5 μm and a depth of about 20 μm in order to realize narrow concentrator gaps. Nevertheless, laser ablation of thick metallic films is still a challenge because debris and melt formation during laser ablation could reduce the structure accuracy and the reproducibility of the patterning process. In order to meet these high requirements for device manufacturing a process based on short pulse UV-laser radiation was selected. For this purpose an excimer laser system (Promaster, Optec s.a.) was used which operates with an ATLEX-500-SI (ATL Lasertechnik GmbH) at a wavelength of 248 nm. Figure 3 schematically shows the used excimer laser micro-machining system. The laser has a pulse length of 4-6 ns and it was expected that these short laser pulses significantly reduce thermal contributions to the ablation process [10]. Furthermore, the used excimer laser source directly generates a high beam homogeneity with intensity fluctuation better than 5%. This "flat top" profile is absolutely necessary to successfully provide laser-assisted ablation via mask imaging techniques. For patterning of grooves, a motorized aperture mask was used. The aperture consists of four independent shields which can be positioned with μm accuracy. This mask technique is a precise and flexible processing technique for the fabrication of the desired groove structures. The ablation field can easily be adapted to the desired device geometry. For the patterning of a single groove with a sharp edged geometry, 10^4 laser pulses with a laser fluence of 3.2 J/cm^2 and a laser repetition rate of 400 Hz were applied. For a further reduction of

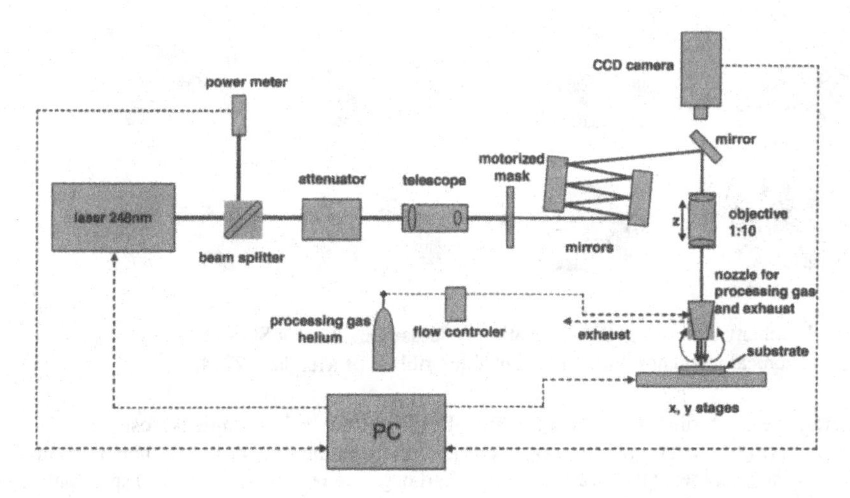

Figure 3. Schematic view of the excimer laser micro-machining system used for realization of the narrow concentrator gaps.

debris formation Helium was introduced via a special designed nozzle into the ablation zone (see Figure 3). After laser ablation a laser cleaning processing-step was established in order to remove re-deposited material. For the cleaning process 10 laser pulses with a fluence of 0.3 J/cm^2 and a repetition rate of 300 Hz were used.

RESULTS AND DISCUSSION

Figure 4 shows SEM pictures of the realized resonant magnetic microsensor with a concentrator gap distance d_g of about 5 µm. The pictures show the sensor after fabrication step (ii), (vi) and (viii). To characterize the fabricated device, the resonator was capacitively excited using the interdigitated comb fingers. Its oscillation was optically measured using the Micro System Analyzer (MSA) 400 of Polytec, a stroboscopic video microscope for in-plane motion measurement. Further, a magnetic flux density B was applied in the direction parallel to the magnetic concentrator using a Helmholtz coil.

Figure 5 shows the measured oscillation magnitude of the resonator as a function of the applied excitation frequency for several magnetic flux densities. The measurement reveals different resonance peaks for different values of the applied magnetic flux density. Further, the figure shows the extracted resonance frequency f_{res} of the resonant magnetic sensor as a function of the applied magnetic flux density B. The resonance frequency shift depends quadratically on the applied magnetic flux density and at $B = 0$ T shows a natural resonance frequency of 2652.5 Hz. Further, the device shows a quality factor $Q = 85$ in atmospheric pressure. A high sensitivity $S = df_{res}/dB$ of 390 kHz/T was extracted at a magnetic flux density of 158 µT. Because of the earth magnetic field component parallel to the magnetic concentrator, the parabolic

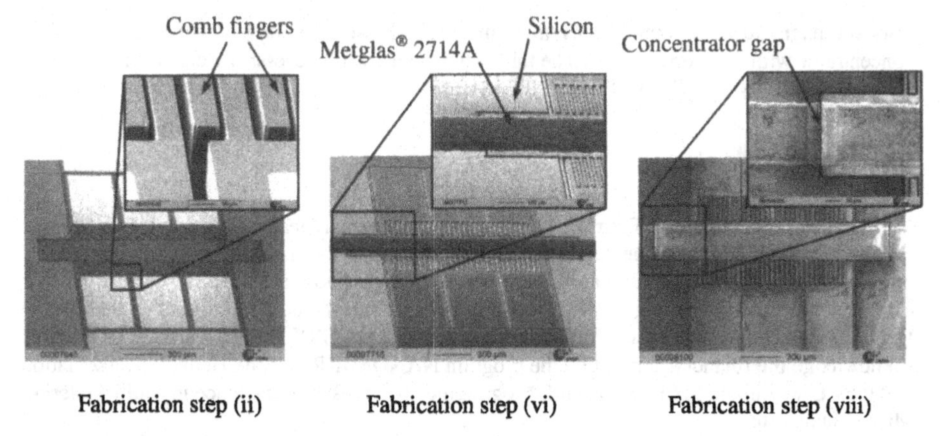

Figure 4. SEM pictures of the resonant magnetic microsensor after fabrication step (ii), (vi) and (viii).

response curve is slightly shifted horizontally. Thus, the applied magnetic flux density $\pm B$ result in different resonance frequencies.

CONCLUSIONS

A novel fabrication process enabling the integration of MEMS devices with thick amorphous magnetic field concentrators was reported. The process combines silicon on insulator technology and epoxy-resin-based attachment soft-magnetic ribbons followed by wet chemical structuring and by laser-assisted patterning. Advantage was taken of the process in fabricating a

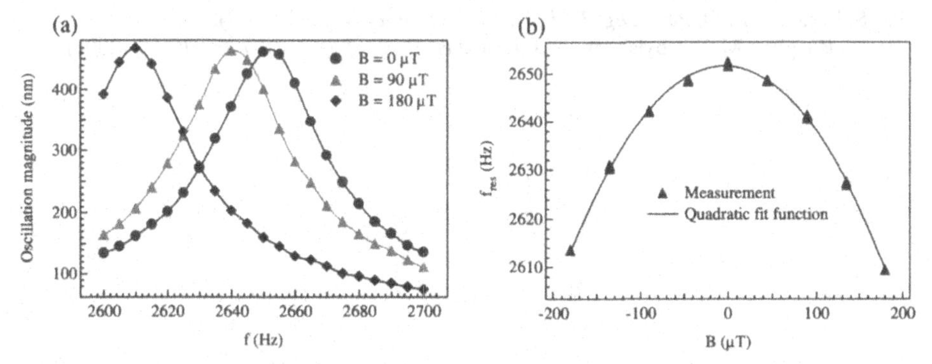

Figure 5. (a) Measured oscillation magnitude of the resonator as a function of the applied excitation frequency for several magnetic flux densities. (b) Measured resonance frequency f_{res} as a function of the applied magnetic flux density B.

resonant magnetic sensor combining a micromechanical resonator and a planar magnetic field concentrator with two narrow gaps. The fabricated device was successfully characterized and shows a sensitivity of 390 kHz/T at 158 μT.

ACKNOWLEDGMENTS

The authors thank A. Baur, M. Reichel and F. Dieterle (Clean Room Service Center of IMTEK) for their help with the device fabrication, H. Besser for his support in laser material processing, P. Ruther and J. Gaspar for helpful technical discussions and R. Hasegawa (Metglas®, Inc.) for providing us with the Metglas® 2714A ribbons. This work is funded by Deutsche Forschungsgemeinschaft (DFG) under contract no. GRK 1103/1. Further, we gratefully acknowledge the financial support by the program NANOMIKRO of the Helmholtz association and the EU within the Sixth Framework Programme "Network of Excellence in Multi-Material Micro Manufacture".

REFERENCES

1. M. Schneider, R. Castagnetti, M.G. Allen and H. Baltes, *Proc. of IEEE MEMS Conf.*, 151 (1995).
2. P.M. Drljača, F. Vincent, P.-A. Besse and R.S. Popović, *Sens. Actuators A* **97-98**, 10 (2002).
3. C. Schott, R. Racz and S. Huber, *Proc. of IEEE Sensors Conf.*, 959 (2005).
4. C. Schott, R. Racz and S. Huber, *Proc. of IEEE Sensors Conf.*, 977 (2004).
5. W.P. Taylor, M. Schneider and H. Baltes, *Dig. Tech. Papers Transducers* **2**, 1445 (1997).
6. R.S. Popović, P.M. Drljača and P. Kejik, *Sens. Actuators A* **129**, 94 (2006).
7. L. Perez, C. Aroca, P. Sánchez, E. López and M.C. Sánchez, *Sens. Actuators A* **109**, 208 (2004).
8. S. Brugger, P. Simon and O. Paul, *Proc. of IEEE Sensors Conf.*, 1016 (2006).
9. S. Brugger and O. Paul, *Dig. Tech. Papers Transducers*, 2377 (2007).
10. W. Pfleging, M. Przybylski and H.J. Brückner, *Proc. of SPIE* **6107**, 61070G-1 (2006).

MEMS Materials and Processes II

Mater. Res. Soc. Symp. Proc. Vol. 1052 © 2008 Materials Research Society 1052-DD08-01

Analysis and Measurement of Forces in an Electrowetting-Driven Oscillator

Nathan Brad Crane[1], Alex A Volinsky[1], Vivek Ramadoss[1], Michael Nellis[1], Pradeep Mishra[1], and Xiaolu Pang[1,2]
[1]Department of Mechanical Engineering, University of South Florida, 4202 E. Fowler Ave ENB 118, Tampa, FL, 33620
[2]Department of Materials Physics and Chemistry, University of Science and Technology Beijing, Beijing, 100083, China, People's Republic of

ABSTRACT

Electrowetting is a promising method for manipulating small volumes of liquid on a solid surface. This complex phenomenon couples electrical and fluid properties and offers many potential surprises. The complex electrical and capillary interactions in electrowetting are illustrated by an analysis of an electrowetting configuration that produces an oscillating droplet motion from a steady DC voltage input. The paper presents an analysis of the electrowetting forces to explain the oscillation and presents a new method for measuring electrowetting forces using a Hysitron Triboindenter. Initial results are compared with predictions from numerical models and simplified analytical solutions.

INTRODUCTION

Electrowetting is the change of apparent surface energy in an applied electric field [1]. By varying the electric field applied to liquid drops, they can be moved, split, merged, and mixed to enable lab on a chip microfluidics [2]. Many applications of electrowetting have been proposed in recent years, and many different configurations have been studied. Cooney *et.al.* have considered the relative advantages of three different electrowetting configurations for control of droplet motion: two plate designs and variations of single plate designs in which all electrical connections are made from the bottom [3]. The electrowetting response is determined by interactions between the electrical and surface energies of the systems that can combine to produce a large variety of behaviors. A system in which oscillatory droplet motion is produced from a steady DC input along with an explanation for this behavior are presented. This phenomenon could be useful in improving mixing in digital microfluidic systems.

Oscillation observations

This work considers the case in which both electrodes are located in the substrate and are isolated from the droplet by a dielectric layer. Test structures were fabricated by patterning aluminum electrodes on an oxidized silicon substrate and then spin coating an 800 nm layer of diluted Cytop 809M to act as both dielectric and hydrophobic layer. Uncovered 5 μl droplets of 1 M NaCl water solution were positioned asymmetrically over the aluminum pads. When a DC potential is applied, liquid drops frequently exhibit an oscillatory response at around 35 V. The motion frequency varied from 2 to 7 Hz with an oscillation amplitude of approximately 100-150 μm. The oscillation was seen less often with thicker Cytop coatings. The center of the oscillation is offset from the electrode boundary.

This oscillation is caused by variations in the electrowetting force with the droplet position. Electrowetting is commonly characterized by the change in contact angle on a substrate as a function of applied voltage. In the basic model, the contact angles (θ) will depend on the voltage (V) between the pad and the droplet and the liquid wetting angle (θ_0) without the applied voltage. These can then be related to the thickness (δ) and dielectric strength ($\varepsilon_0, \varepsilon_R$) of the Cytop layer by the Young Lippman equation [4]:

$$\cos\theta_1 = \cos\theta_o + \frac{\varepsilon_o \varepsilon_r V^2}{2\gamma_{lv}\delta} \quad (1).$$

The system energy is found by modeling the fluid using the contact angle data and the Young Lippman equation. The forces acting on the droplet are calculated by differentiating the energy with respect to the displacement of interest. A closed form solution is developed for the electrowetting force on a drop confined between an electrically insulating cover positioned over two substrate electrodes as illustrated in Figure 1(a). These predictions are compared with numerical modeling of surface forces using Surface Evolver [5, 6] and experimental measurements performed with a nanoindenter.

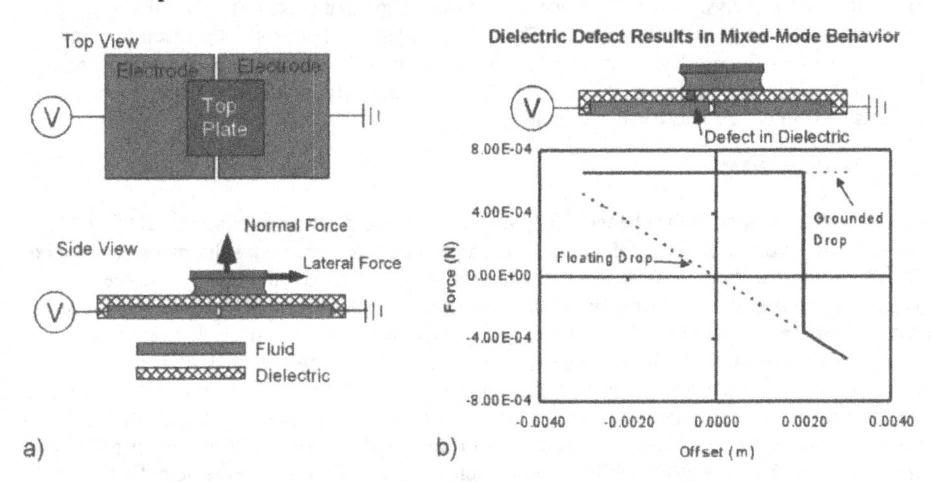

Figure 1. a) Schematic of the floating drop electrowetting configuration with square plate over the drop. b) Force displacement behavior of a mixed-mode drop with a transition at 0.002 m due to a defect in the dielectric layer. Dashed lines indicate the response of ideal grounded and floating droplets.

Analytical droplet model

In the case of interest, the droplet is positioned over two electrodes with a voltage applied across the electrodes. The electrodes are covered with a dielectric material. The droplet can be considered conductive so that the system is modeled as two capacitors in series. The voltage between each pad and the droplet will vary with the droplet position due to changes in the capacitor area and thus their capacitance. Neglecting the droplet resistance, the arrangement can be modeled as simple series capacitor circuit composed of two parallel plate capacitors. The voltages across the left and the right capacitors (V_L, V_R) are given as

$$V_L = \frac{A_R}{A_L + A_R} V_{tot},\ V_R = \frac{A_L}{A_L + A_R} V_{tot} \tag{2}$$

The change in contact angle is driven by the reduction in effective surface energy by the energy stored in the capacitors. Using these relationships, the total capacitive energy is given by

$$E = \tfrac{1}{2}\left(C_l V_L^2 + C_R V_R^2\right) \tag{3}$$

The area of the droplet over each electrode (A_L, A_R) is a function of the droplet position and the form of this function depends on the droplet shape. In general, this shape will vary with the offset from the equilibrium position. However, if the droplet is sandwiched between the substrate and an electrically insulating cover plate that is wet by the liquid, the droplet contact area and its contact shape with the substrate will approach the area of the cover plate as the droplet thickness decreases relative to the cover plate dimensions. With this simplification, the area values of the droplet beneath a square cover plate with edge length s can be related to the displacement (x) of the cover plate from the equilibrium position:

$$A_L = s\left(\frac{s}{2} - x\right),\ A_R = s\left(\frac{s}{2} + x\right),\ -\frac{s}{2} < x < \frac{s}{2} \tag{4}$$

The energy and force as a function of position are then

$$E = \frac{V_{tot}^2 \varepsilon_0 \varepsilon_R}{8\delta}\left(s^2 - 4x^2\right)$$
$$F_x = \frac{dE}{dx} = -\frac{\varepsilon_0 \varepsilon_R}{\delta} V_{tot}^2 x \tag{5}$$

where the force as a function of cover plate position is found by differentiating the energy with respect to the displacement. The static equilibrium position is at the center of the two electrodes with equal area on each electrode. In cases of limited viscous and electrical energy loss, the drop could oscillate around the equilibrium when disturbed. However, the oscillation would be centered on the gap and should not depend on the thickness of the dielectric coating. In contrast, the center of oscillations was offset from the electrode gap. While an offset could be caused by asymmetric electrowetting [7], contact angle measurements on the Cytop substrate show modest polarity dependence at low voltages.

This behavior could be explained by the presence of a hole in the dielectric layer. A hole would short the capacitor on one side so that no voltage will be applied across the region of the droplet over the shorted electrode. The full voltage drop will occur across the other side. In this case, the electrowetting force will be constant with displacement. The force is given by

$$F_x = -\frac{\varepsilon_0 \varepsilon_R}{2\delta} s V_{tot}^2 \tag{6}$$

If there is a local defect over one of the electrodes, hybrid behavior could be observed in which there is a transition from one mode to another as illustrated in Figure 1b). Oscillation is likely in this situation because there is no point at which the force is zero as required for equilibrium.

Electrowetting force measurement

Typically, electrowetting is characterized by the change in contact angle. If information is desired about the resulting energy or forces, this is estimated by modeling. This work proposes for the first time to directly measure these effects. This is particularly useful in characterizing complicated phenomena and detecting local phenomena that might be difficult to detect on the basis of a contact angle change.

Electrowetting forces were measured using a custom tip in a Hysitron Triboindenter. The probe was assembled by aligning a 9x9 mm^2 glass cover slip with the axes of the Hysitron Triboindenter optical system. A blunt tip is then placed into the Hysitron transducer and this is brought into contact with the center of the coverslip. A drop of cyanoacrylate adhesive is applied to bond the coverslip to the blunt tip and the assembly was allowed to cure overnight at room temperature. Forces are measured by applying a droplet of known volume over the electrodes and then positioning the indenter over the droplet. The indenter is brought down towards the substrate until it is completely wet by the liquid. Typical heights were 400 μm for a 42 μl drop and 250 μm for a 30 μl drop. After each test, the droplet was replaced with a new droplet to reduce the impact of evaporation and contact angle hysteresis on the test results.

The test configuration was modeled using Surface Evolver to verify that the area of the liquid contact was adequately approximated by the assumption of a square profile. This was accomplished by relating the surface energy of the faces to the actual area over each electrode as calculated in the Surface Evolver model. After adjusting the analytical model for the effects of a finite gap between the electrodes, the areas agreed to within 1.5%.

RESULTS AND DISCUSSION

Initially data was collected at zero voltage to test the stability of the measurement system. Figure 2a) shows a typical plot of the lateral and normal forces. A significant drift in the normal force is observed due to the liquid evaporation. However, the lateral force is relatively stable. The average drift rates measured for the normal and lateral forces were –0.835 and –0.085 μN/s, respectively. This work focuses on the measurement of the lateral forces.

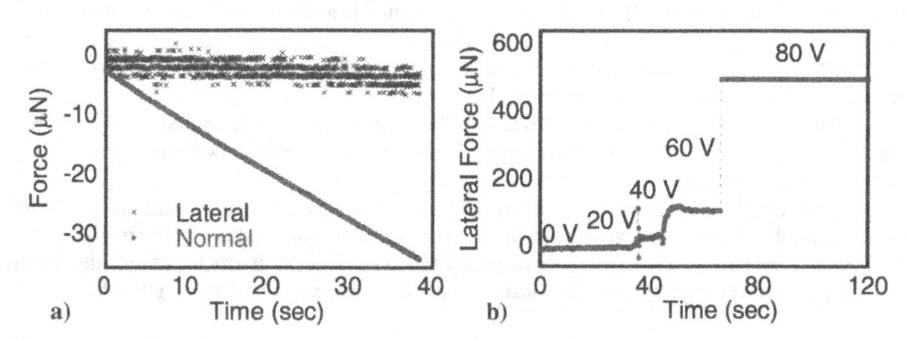

Figure 2. Electrowetting force data. a) Normal and lateral force measurements with zero applied voltage. b) Lateral force measured with steps of 0, 20, 40, 60, and 80 V applied.

Figure 2b) shows the data from a 0.1M NaCl drop with voltage steps of 0, 20, 40, 60, 80 V. The last three were clearly measured. These values were compared to theoretical predictions and good agreement was observed at 40 and 60 V when modeled as a floating drop. The Young-Lippman equation was fit to the contact angle measurements for 1 M NaCl solution for a voltage range of 0-50 V to find an effective thickness of 1 μm when assuming a relative dielectric constant of 2.1. The 80 V measurement is matched well by the grounded droplet model. This suggests that the coating on one side failed at the higher voltage. These results are summarized in Table I.

Table I. Comparison of measured electrowetting forces with analytical predictions.

Applied Voltage (V)	Measured Force (μN)	Predicted Force (μN)	Prediction Method
20	6	11	Floating Drop
40	41	44	Floating Drop
60	113	98	Floating Drop
80	505	535	Grounded Drop

A series of droplets were also measured with a constant 100 V applied but with different offsets from the electrode gap. These tests were conducted with DI water to reduce potential degradation of the coating. A typical dataset from these tests is illustrated in Figure 3a). From this data the peak and steady forces were extracted and summarized in Figure 3b) as a function of the central position offset. This data shows a possible transition from floating to grounded drops as indicated above.

These early results suggest that this system can be effective in measuring electrowetting forces and can provide insight into experimental observations of droplet behavior. These early measurements have detected force transitions indicative of coating defects. With further refinement, these methods could be used to provide quick characterization of electrowetting materials and characterize the ability of electrowetting drops to apply forces to other interacting elements. This method can also provide a means to directly measure other parameters of interest in electrowetting such as dielectric decay and the influence of liquid conductivity on system response.

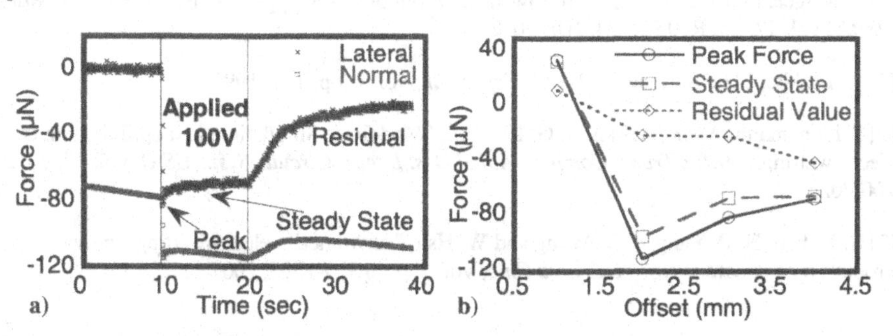

Figure 3. Electrowetting force measurements. a) Representative data sample with 100 V applied. b) Force measurements variation with position for DI water drops. Sign reversal is consistent with local shorting or dielectric breakdown.

CONCLUSIONS

When electrowetting is done with a series capacitance circuit, the resulting force is shown to depend on the area of contact with the electrodes. The voltage across each capacitor is a function of the area of the electrode covered by the liquid drop. This can be used to create an oscillating drop motion from a steady voltage input when there is an abrupt transition between the circuit characteristics.

Understanding of the force variations will enable improvements in electrowetting droplet control and simplified droplet control schemes that reduce the number of independent electrodes required for simple droplet motions. The novel method for electrowetting force measurement presented here will also enable better understanding of anomalous behavior observed in many electrowetting systems.

ACKNOWLEDGEMENTS

Nathan Crane would like to acknowledge the support of the University of South Florida Research Education Initiative Program under Grant Number FMMD04. Alex Volinsky would like to acknowledge the support from NSF under CMMI-0631526 and CMMI-0600231 grants.

REFERENCES

[1] F. Mugele, J. C. Baret and D. Steinhauser, "Microfluidic mixing through electrowetting-induced droplet oscillations," *Appl. Phys. Lett.,* vol. 88, pp. 204106, MAY 15. 2006.

[2] S. K. Cho, H. J. Moon and C. J. Kim, "Creating, transporting, cutting, and merging liquid droplets by electrowetting-based actuation for digital microfluidic circuits," *J Microelectromech Syst,* vol. 12, pp. 70-80, FEB. 2003.

[3] C. G. Cooney, C. Y. Chen, M. R. Emerling, A. Nadim and J. D. Sterling, "Electrowetting droplet microfluidics on a single planar surface," *Microfluid. Nanofluid.,* vol. 2, pp. 435-446, SEP. 2006.

[4] F. Mugele and J. C. Baret, "Electrowetting: From basics to applications," *J. Phys. -Condes. Matter,* vol. 17, pp. R705-R774, JUL 20. 2005.

[5] K. Brakke, "The Surface Evolver," *Exp. Math,* vol. 1, pp. 141, 1992.

[6] J. Lienemann, A. Greiner and J. G. Korvink, "Modeling, Simulation, and optimization of electrowetting," *IEEE Trans. Comput. -Aided Des. Integr. Circuits Syst. (USA),* vol. 25, pp. 234, 2006/02/.

[7] S. K. Fan, H. P. Yang, T. T. Wang and W. Hsu, "Asymmetric electrowetting - moving droplets by a square wave," *Lab on a Chip,* vol. 7, pp. 1330-1335, 2007.

Mater. Res. Soc. Symp. Proc. Vol. 1052 © 2008 Materials Research Society 1052-DD08-02

Bottom-Up Fabrication of Individual SnO_2 Nanowires-Based Gas Sensors on Suspended Micromembranes

Albert Romano-Rodriguez[1], Francisco Hernandez-Ramirez[1,2], Joan Daniel Prades[1], Albert Tarancon[1], Olga Casals[1], Roman Jimenez-Diaz[1], Miguel Angel Juli[1,2], Juan Ramon Morante[1], Sven Barth[3,4], Sanjay Mathur[3,4], Andreas Helwig[5], Jan Spannhake[5], and Gerhard Mueller[5]
[1]Electronics, University of Barcelona, Martí i Franquès 1, Barcelona, E-08028, Spain
[2]NTEC106, S.L., Mare de Deu dels Desemparats 12 baixos, L'Hospitalet de Llobregat, E-08903, Spain
[3]Chemistry, University Würzburg, Würzburg, D-97070, Germany
[4]Nanocrystalline Materials and Thin Film Systems, Leibniz Institut of New Materials, Saarbruecken, D-66123, Germany
[5]IW-SI Sensors, Electronics & System Integration, EADS Deutschland GmbH, Muenchen, D-81663, Germany

ABSTRACT

Bottom – up techniques were used to obtain gas sensors based on individual SnO_2 nanowires placed over microhotplates with integrated heaters. These nanowires were electrically connected to pre-patterned microelectrodes by means of Focused Ion Beam (FIB) nanofabrication methodologies. The performance of these sensors, which exhibit reproducible and stable responses, was evaluated as a function of different gas atmospheres and power dissipated by the heater, demonstrating that this technological approach could be used to develop functional devices based on nanomaterials.

INTRODUCTION

The new properties of nanomaterials with respect to bulk materials have attracted great research interest because of their potential applications in functional devices [1]. Although great advances have been realized in the synthesis and characterization of nanomaterial's fundamental properties, the fabrication of reliable and reproducible devices based on these nanomaterials is still rare due to the difficulties in studying them [2].

In this work, bottom – up techniques were used to fabricate sensors based on individual SnO2 nanowires placed over microhotplates with heaters, which allows a fast modulation of their working temperature and their sensing characteristics. This approach is based on previous work in which similar nanowires have been connectged to standard photolighographically defined microelectodees on oxidized silicon wafers [3].

The stability of some of these devices was evaluated as a function of the operating time and the applied current, showing good performance for weeks and with currents below $I = 100$ nA. At higher currents, self heating related problems, such as degradation of the electrical contacts and irreversible damage of the nanowires, were observed.

EXPERIMENT

Monocrystalline SnO_2 nanowires were synthesized by chemical vapor deposition following a process explained previously [4]. These nanowires were grown as single-crystalls with dislocation-free bodies. Their main growth direction was [100] with interplanar spacing according to the rutile structure of SnO_2. The length and radii of these nanowires varied between 2 and 7 microns, and between 30 and 70 nm respectively. The grown nanowires were dispersed over the microhotplate with an integrated heater and lithographically pre-patterned platinum interdigitated electrodes (figure 1.a.). Similar structures have been reported previously [5].

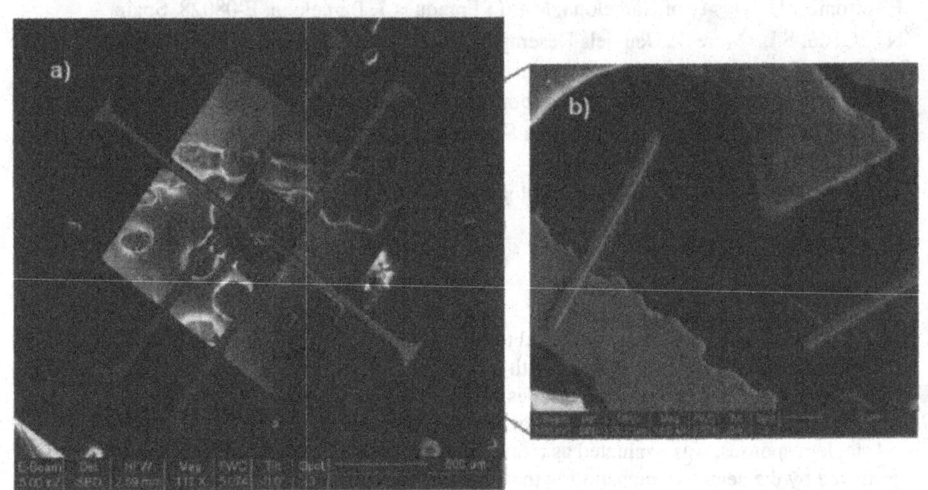

Figure 1. (a) Image of one microhotplate with integrated heater and (b) detail of a single nanowire connected with Pt strips to the interdigitated electrodes.

Contacts between the electrodes and nanowires were fabricated with a FEI Dual-Beam Strata 235 FIB instrument with a trimethyl - methylcyclopentadienyl – platinum ($(CH_3)_3CH_3C_5H_4Pt$) injector to deposit platinum [6].

To guarantee that the nanofabrication process did not alter the properties of the nanowires, deposition of platinum strips over these structures was performed by means of an electron beam induced process while the rest of the contacts, up to the pre – patterned microelectrodes, was produced with ion – beam induced depositions (figure 1.b). This step reduces the required fabrication time, which takes approximately two hours for each device, limiting this technique in large scale processes [3, 5]. Electrical measurements were performed with the help of a circuit designed to guarantee low currents and avoid any undesired current fluctuation [7, 8].

All the experiments were performed in a home-made chamber, where the gas flow was maintained between 50 and 200 ml min^{-1}. Accurate gas mixtures were prepared by combining different gases passing through mass flow controllers.

DISCUSSION

The electrical stability as a function of operating time and the applied current was measured for up to ten devices. No significant degradation was observed after several weeks operating with currents below I = 100 nA. On the contrary, the devices were quickly destroyed when currents exceeded this value, because of uncontrolled self – heating effects produced during the measurement [9]. In this case, two degradation processes were clearly identified: breaking of the nanowires and melting of the contacts. This last effect is related to the evaporation of carbon present in high concentrations (around 70 %) in the FIB – assisted depositions. Similar modifications on these interconnections have been reported elsewhere [10]

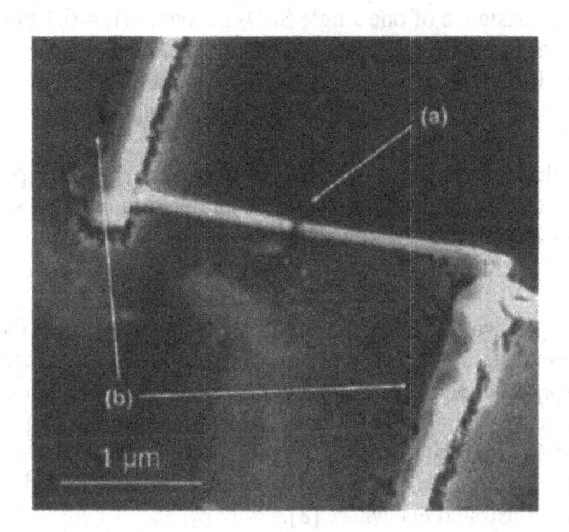

Figure 2. Sensor destroyed after applying a high current of I = 100 µA. (a) The rupture of the nanowire, (b) and the degradation of the contacts due to the evaporation of carbon from the Pt deposition can be clearly observed.

Depending on the diameter and length of the nanowires, electrical resistances were estimated ranging from a few kiloohms to a few megaohms. The results were in concordance with the previously reported resistivity values found on similar nanowires [3].

The effect of the temperature on the resistance of the device was also studied. Figure 3a shows the variation of the resistance with time when different pulses of electrical power to the heater were applied. Increasing power, directly related to temperature (Figure 3b), leads to a decrease in the nanowire's resistance, demonstrating the semiconductor characteristics of the nanowire. The modulation of the resistance produced by heater pulses disappeared after a few seconds of switching it off. If applied power was above P = 45 mW (T ≈ 400 K), the recovery time increased by a few minutes suggesting that modifications of the adsorbed species at the nanowire surface are produced: changes are seen to oxygen and water molecules [11, 12].

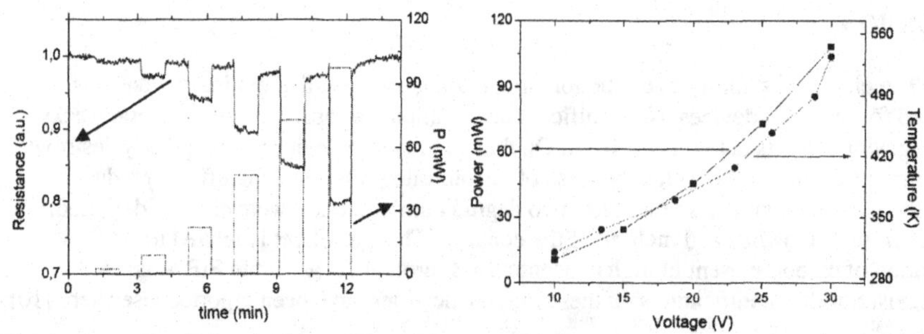

Figure 3. (a) Evolution in time of the resistance of one single SnO_2 nanowire (L = 6.4 μm, R = 50 ± 5 nm) to series of increasing pulses of electrical power to the heater. The experiment was performed in synthetic air. (b) Calibration of the dissipated power by the heater P and the effective temperature of the microhotplate T [13] as function of applied voltage V.

The operating temperature of these devices can be varied by applying controlled power to the heater. This process is fast enough to reach a complete thermal stabilization in a few seconds (Figure 3a), demonstrating that the optimal working conditions were easily modulated in a fast and controlled way.

Gas sensing experiments were performed with different atmospheres to evaluate the capabilities of these devices as functional elements. It is well known that the electrical resistance of SnO2 is modulated by oxygen adsorption and desorption, which are thermally activated processes [11, 12]. This point is demonstrated in figure 4 where one of these sensors responds to pulses of synthethic air (SA) (20 % of O_2) alternated with nitrogen (N_2) pulses with a dissipated power of the heater of P = 23 mW (T ≈ 340 K). On the contrary, no response is observed when the heater is switched off, as it was demonstrated elsewhere [8].

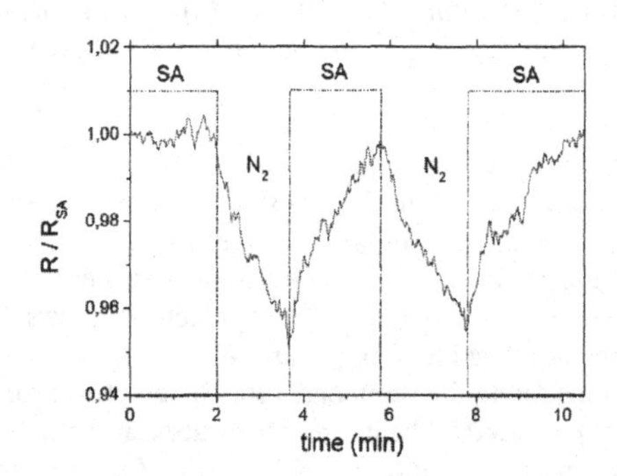

Figure 4. Response of the device to nitrogen and synthetic air pulses with a dissipated power in the heater of P = 23 mW (T ≈ 340 K). The resistance drops when nitrogen is passed through the chamber demonstrating oxygen desorption. The dashed lines represent the synthetic air pulses.

In this work, we also report the possibility of using these devices to detect carbon monoxide (CO). CO is a reducing gas which leads to a reversible reduction of the SnO_2 resistance. Applying power to the heater of P =56 mW (T ≈ 393 K), the device is able to detect and discriminate CO concentrations of 50, 100 and 200 ppm (figure 5).

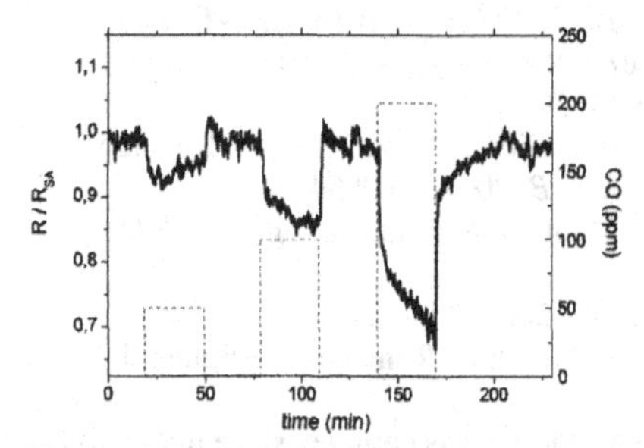

Figure 5. Response of the device to three CO pulses of 50, 100 and 200 ppm with a dissipated power in the heater of P = 56 mW (T ≈ 393 K). The dashed lines indicate CO pulses.

CONCLUSIONS

A bottom-up fabrication process for gas microsensors based on single nanowires is shown to be effective and reliable. Individual SnO_2 nanowires were connected to pre-patterned microelectrodes placed on microhotplates. These devices show good stability as a function of operating time when currents below 100 nA are applied. Additionally, the integration of a heater in these devices leads to the possibility of enhancing the adsorption / desorption of gas molecules on the surface of the nanowires, improving their responses as gas sensors. The possibility of using this technique to obtain functional devices was demonstrated.

ACKNOWLEDGMENTS

This work has been partially supported by the EU through the project NANOS4 of the 6th FMP. The support of the Spanish Ministry of Education and Science (MEC) is also acknowledged through the projects MAGASENS and CROMINA, through the FPU grants of several authors (F-H.-R., O.C. and J.D.P) and through the Torres Quevedo PTQ05 –02 – 0301 program (F.H.-R.). Thanks are due to the German Science Foundation (DFG) for supporting this work in the frame of the priority program on nanomaterials – *Sonderforschungsbereich 277* – at the Saarland University, Saarbruecken, Germany. The authors would like to acknowledge the valuable suggestions of Dr. O. Ruiz during the development of the electronic circuit.

REFERENCES

1. Xing-Jiu Huang and Yang-Kyu Choi *Sensors Actuators B* 2007 **122** 659.
2. Y. Chen, C. Zhu, M. Cao, T. Wang, *Nanotenology* 2007, **18**, 285502.
3. Francisco Hernández-Ramírez, Albert Tarancón, Olga Casals, Jordi Rodríguez, Albert Romano-Rodríguez, Joan R Morante, Sven Barth, Sanjay Mathur, Tae Y Choi, Dimos Poulikakos, Victor Callegari and Philipp M Nellen, *Nanotechnology* 2006 **17** 5577

4. S. Mathur, S. Barth, H. Shen, J. C. Pyun, U. Werner, *Small* 2005, **1**, 713.
5. J. Spannhake, A. Helwig, G. Müller, G. Faglia, G. Sberveglieri, T. Doll, T. Wassner, M. Eickhoff. *Sensors and Actuators B* 2007, **124**, 421 – 428.
6. Strata DB 235. Product data sheet, FEI Company.
7. E. Pescio, A. Ridi, A. Gliozzi, *Review of Scientific Instruments*. 2000, **71**, 1740.
8. F Hernandez-Ramirez, J D Prades, A Tarancon, S Barth, O Casals, R Jiménez–Diaz, E Pellicer, J Rodriguez, M A Juli, A Romano-Rodriguez, J R Morante, S Mathur, A Helwig, J Spannhake and G Mueller, *Nanotechnology* 2007 **18** 495501
9. F. Hernández – Ramírez, A. Tarancón, O. Casals, E. Pellicer, J. Rodríguez, A. Romano – Rodríguez, J. R. Morante, S. Barth, S. Mathur. *Phys. Rev. B*. 2007, **76** 085429
10. A. Botman, J. J. L. Mulders, R. Weemaes and S. Mentink, *Nanotechnology* **17**, 15 (14 August 2006), 3779-3785
11. A. Kolmakov, M. Moskovits, *Annu. Rev. Mater. Res.* 2004, **34**, 83.
12. F. Hernandez–Ramirez, A. Tarancon, O. Casals, J. Arbiol, A. Romano–Rodriguez, J. R. Morante, *Sensors and Actuators B* 2006, **121**, 13.
13. Accurate estimation of the real temperature of the nanowires is complex since uncontrolled self-heating effects on these nanostructures should be considered. For this reason, in this work only dissipated power and nominal temperatures at the centre of the membrane are mentioned.

Mater. Res. Soc. Symp. Proc. Vol. 1052 © 2008 Materials Research Society 1052-DD08-06

Nature-Inspired Microfluidic Manipulation Using Magnetic Actuators

S. N. Khaderi[1], D. Ioan[2], J. M. J. den Toonder[3], and P. R. Onck[1]
[1]Physics and Applied Physics, University of Groningen, Micromechanics of Materials, Nijenborgh 4, Groningen, 9747 AG, Netherlands
[2]Universitatea Politehnica din Bucuresti, Bucharest, Romania
[3]Philips Research, Eindhoven, Netherlands

ABSTRACT

Magnetically actuated micro-actuators are proposed to propel and manipulate fluid in micro-channels. As the fluid flows at low Reynolds number in such systems, the actuator should move in an asymmetric manner. The proposed actuators are polymer films with embedded magnetic particles, which are actuated by an external magnetic field. Based on the nature of the particles, the films can be either ferromagnetic or super-paramagnetic. We have identified four configurations in which the actuator exhibits an asymmetric motion.

INTRODUCTION

A rapidly growing field in biotechnology is the fabrication of micro-fluidic devices for biomedical applications such as biosensors. Biosensors are micro-fabricated laboratories-on-a-chip, aimed at analyzing biological samples like bio-fluids (e.g. blood, saliva, urine). A biosensor typically consists of a system of microscopic channels, connecting micro-chambers (the labs) where dedicated tests are carried out. Classical means for fluid-propulsion do no longer suffice at these small length scales, which has led to a search for new methods. In this work we propose a new micron-scale propulsion mechanism, inspired by nature, based on artificial cilia.

Fluid dynamics in our daily experience of the physical world is dominated by gravity and inertia. However, at the micrometer scale gravity does not play a role and fluid dynamics is usually dominated by viscosity rather than inertia [1]. This has important consequences for fluid propulsion mechanisms. In particular, the actuators will be effective in propelling fluids only if their motion is cyclic, but asymmetric in shape change. Nature has solved this problem by means of hair-like structures, called cilia, whose beating pattern is non-reciprocating and consists of an effective and a recovery stroke, as shown in Fig. 1.

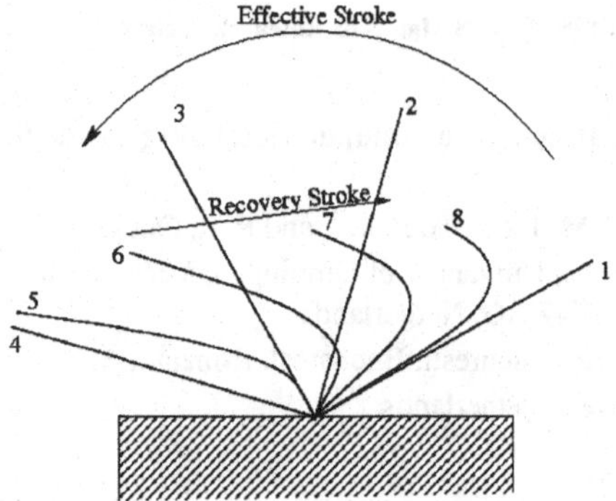

Figure 1. Asymmetric motion of a cilium, showing a separate effective and recovery stroke.

In this work we design artificial cilia that can be actuated by an external magnetic field. The artificial cilia are thin films consisting of a polymer matrix filled with magnetic nano-particles. Depending on the nature of the particles, the film can be either super-paramagnetic (SPM) or ferromagnetic with a remnant magnetization. The applied magnetic field is uniform but its magnitude and direction can be manipulated in time to get the desired asymmetric motion.

METHOD

The problem under consideration involves magnetostatics, solid dynamics and fluid dynamics, whose governing equations should be solved in a fully-coupled manner. However, in this paper the fluid is not explicitly accounted for, but is incorporated through a drag-force on the film. The drag-force is proportional to the velocity of the film, with coefficients of proportionality C_1 and C_2, in the tangential and normal directions, respectively. We solve the equations corresponding to magnetostatics and solid mechanics, which are coupled through the magnetic couple generated by the fields. Maxwell's equations are solved for the magnetization in an implicit manner [2], yielding the magnetization and local magnetic fields. The magnetic couple is then calculated which acts as an external forcing term, giving the dynamic motion of the film. The film is assumed to be an assemblage of beam elements. The motion of the film is calculated by taking the inertia and geometric nonlinearity into consideration.

RESULTS

As mentioned before, the actuating member should move in an asymmetric manner to effectively propel fluid. In this section we discuss several configurations that are able to do so. By tuning the applied field, the initial geometry of the film and its magnetic nature (permanently magnetic or super-paramagnetic), we have identified four configurations that mimic ciliary motion. For the results presented, the thickness of the film is taken to be 2 μm and the elastic modulus to be 1 MPa.

A partly magnetic film with cracks

The natural cilium is found to have a varying stiffness in the effective and recovery stroke [3]. To use this concept we need to have the film to possess a large bending stiffness in the effective stroke while pushing the fluid and to possess a low stiffness during the recovery stroke. This can be achieved by introducing cracks in one side of the film, while only a part of the film is magnetic. The film, which is initially straight, is attached at the left end and has cracks of size 0.75 μm at the bottom. By magnetizing only a part of the film, it is expected to behave like a flexible oar (as also mentioned by Purcell [1]). Only 20% of the film, the end near to the fixed part, is magnetic. The assumed remnant magnetization is 15 kA/m, with the magnetization vector pointing from the fixed end to the free end. The drag coefficients used are $C_1 = 30$ Ns/m^3 and $C_2 = 60$ Ns/m^3. The applied magnetic field is increased linearly to 145 mT in the y direction in 0.6 ms, then rotated by 90^0 in the next 1.2 ms and finally reduced to zero in the next 0.2 ms. The movement of the film under the action of the applied magnetic field is shown in Fig. 2. When the external magnetic field is applied, the magnetic couple act on the magnetized portion of the film in a counter-clockwise manner, thus rotating the film about the fixed end. Now the drag forces are acting on the top part of the film, which close the cracks, making the film stiff. When the applied field is switched off (Fig. 2(e)), the film will recover elastically and the drag forces act on the bottom part of the film, which open the cracks making the film compliant. Such an interaction of magnetic couple, elastic forces and drag forces results in an asymmetric motion.

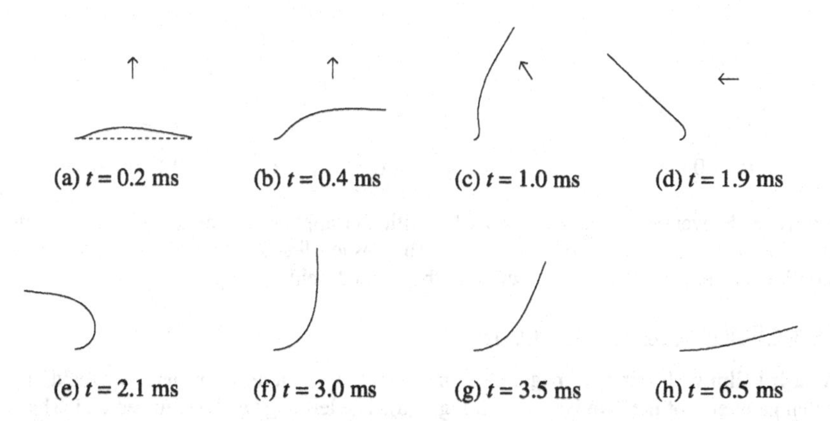

Figure 2. Film with cracks with only a part (20%) magnetized. The dashed line shows the initial position of the film. The arrow shows the direction of the applied field.

Buckling of a straight magnetic film

A straight horizontal magnetic film with a slight perturbation is used to get the desired asymmetric motion. The film is assumed to have a uniform magnetization with the magnetization vector pointing along the film length, from the fixed end at the left to the free end at the right. The remnant magnetization of the film is taken to be 15 kA/m. The length of the film is 100 μm. The drag coefficients used are C_1 = 30 Ns/m^3 and C_2 = 60 Ns/m^3. An external magnetic field of 30 mT is applied in the negative x direction from t = 0 to t = 1 ms and the field is reduced to zero in the next 0.2 ms. Initially, the magnetization and the applied field are parallel, but with opposite sign, so that the magnetic couple is zero. However, with any perturbation of the film, the equilibrium state becomes unstable and the film will buckle away from the straight configuration. By assuming a uniform magnetization in the film and neglecting drag forces, the critical field can be calculated. When it buckles, the film tries to curl such that the magnetization in the film is aligned with the applied field (Fig. 2(a)–2(d)). When the applied field is removed, the film returns back to its initial position by elastic recovery. For this configuration, the effective fluid propulsion will take place when the film recovers elastically. Whether the film will buckle up or down depends critically on the sign of the initial imperfection.

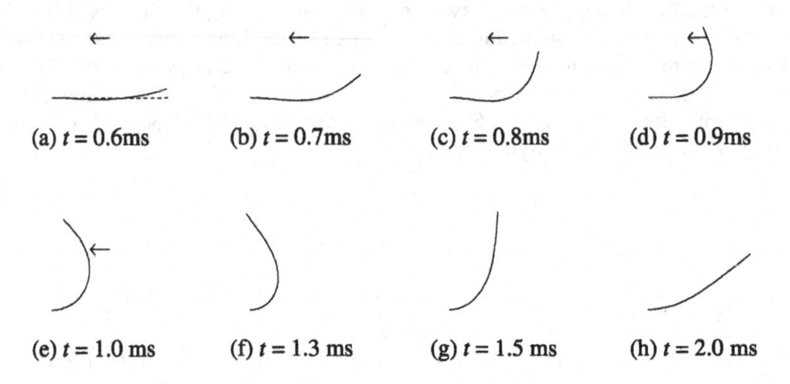

Figure 3. Movement of a perturbed film with the applied magnetic field in opposite direction to the magnetization of the film. The dashed line shows the initial position of the film. The arrow shows the direction of the applied field.

Curled film in uniform magnetic field

A curled film with remnant magnetization is subjected to a uniform magnetic field. The initial geometry of the film is shown in Fig. 4(a). The left edge is the clamped end. The

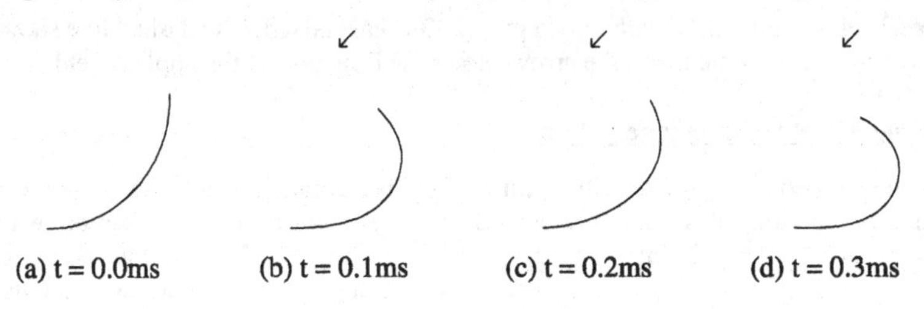

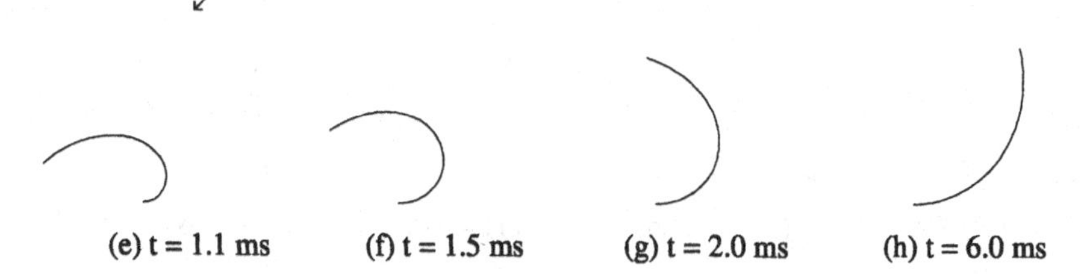

Figure 4. Movement of a permanently magnetic film. The arrow shows the direction of applied field.

direction of the magnetization is along the film with the magnetization vector pointing from the clamped end to the free end. The remnant magnetization of the film is taken to be 15 kA/m. An external field of magnitude 9 mT is applied at 225^0 to the x axis from t = 0 ms to t = 1 ms and then linearly reduced to zero in the next 0.2 ms. The drag coefficients used are C_1 = 5 Ns/m^3 and C_2 = 60 Ns/m^3. The radius of curvature of the film is 100 μm. The physical mechanism of asymmetry is akin to the previous case, except that the film is made to buckle in a predetermined way by a curled geometry, which turns out to exhibit a large asymmetry in motion. Again, the propulsive action in the effective stroke takes place during elastic recovery. In the portion near the fixed end, the magnetic couple acts in a clockwise sense and in the portion near the free end, in a counter-clockwise sense. As a result, these couple tend to bend the film, such that the curvature of the film increases, bringing the ends of the film closer together (this kind of behavior is common to a natural cilium). Here, the asymmetry is larger than the previous configuration.

Tapered super-paramagnetic film

A super-paramagnetic film which is anisotropic in its magnetic susceptibility and which is tapered along its length is subjected to a rotating magnetic field. The assumed susceptibilities are 4.6 and 0.8 in the tangential and normal directions, respectively. The thickness of the film varies linearly along its length, being 2 μm at the left (attached) end and decreasing to 1μm at the right end. A magnetic field of 30 mT is rotated from 0^0 to 180^0 in t = 10 ms and then kept constant for the rest of the time. The drag coefficients used are C_1 = 30 Ns/m^3 and C_2 = 60 Ns/m^3. When the rotating field is applied, the free end portion of the beam is rotated through 135^0 and a U bend is formed near the fixed end (Fig. 5(e) and 5(f)). It is to be noted that the portion of the beam near the free end is nearly straight. This is because the magnetization of the film in this region is aligned to the applied field (which is evident from the magnetization vectors shown on the film in

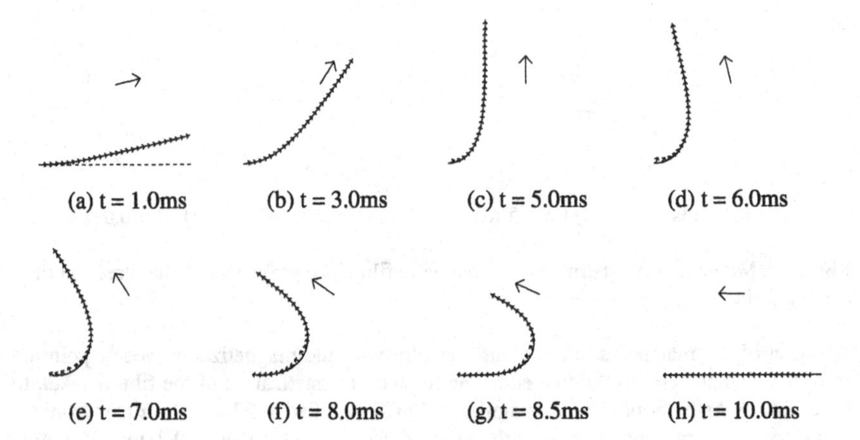

Figure 5. Super-paramagnetic film in a rotating magnetic field. The big arrow shows the direction of the applied magnetic field. The small arrows show the magnetization.

Fig. 5). As a result, the normal component of the field vanishes, and hence the moment is zero as well. In this situation it is in the part of the film near the U bend, where the magnetic couple distribution (which tends the film to bend) balances the elastic forces (which tends the film to become straight again), hence "freezing-in" the bent shape. When compared with other parts of the film, at the U bend portion of film the magnetic couple are large. On the portion of the film between the free end and the U bend, anticlockwise moments are acting and in the portion between the fixed end and the U bend, clockwise moments are acting (Figs. 5(e), 5(f), 5(g)). Under the influence of such a system of moments, the film becomes more curved, the end to end distance decreases and the film recovers. As the beam recovers elastically the U bend propagates to the free end of the beam. It is to be noted that the film recovers in the presence of magnetic forces, i.e. the recovery is not an elastic one, but controlled by magnetic forces, keeping the film low. This phenomenon can be exploited to provide a large asymmetry in motion of the film during the forward and return stroke. A larger field will result in larger curvature of the U bend, forcing the film to stay lower, enhancing the efficiency of the return stroke.

The four proposed configurations are summarized in Fig. 6. Of these four configurations, the curled magnetic film and the SPM film exhibit the largest asymmetry in motion.

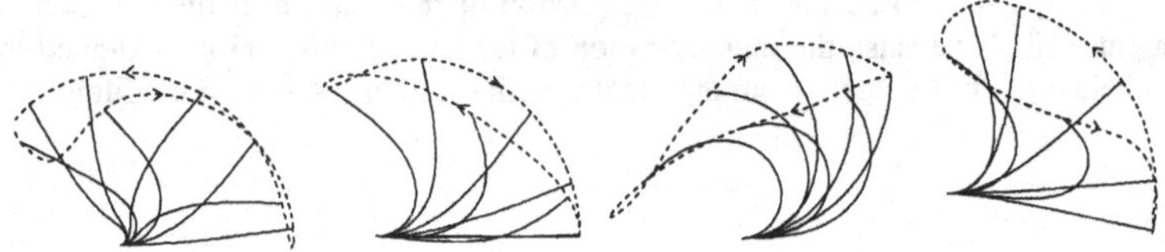

Figure 6. Asymmetric motion generated by the four configurations in Figs. 2, 3, 4 and 5, respectively.

ACKNOWLEDGMENTS

This work is a part of the 6th Framework European project 'Artic', under contract STRP 033274.

REFERENCES

1. E. M. Purcell, Life at low Reynolds number, *American Journal of Physics* **45**, 3-11 (1977).
2. J. D. Jackson, *Classical Electrodynamics*, 1974, John Wiley.
3. J. Gray, The Mechanism of Ciliary Movement, *Proceedings of The Royal Society of London B* **93**, 104-121 (1922).

Mater. Res. Soc. Symp. Proc. Vol. 1052 © 2008 Materials Research Society 1052-DD08-07

MEMS-Based MHz Silicon Ultrasonic Nozzles for Production of Monodisperse Drops

Y. L. Song[1,2], Chih H. Cheng[3], Ning Wang[1], Shirley C. Tsai[4], Yuan F. Chou[3], Ching T. Lee[2], and Chen S. Tsai[1]
[1]Electrical Engineering and Computer Science, University of California, Irvine, CA, 92697
[2]Electrical Engineering, National Cheng-Kung University, Tainan, Taiwan
[3]Mechanical Engineering, National Taiwan University, Taipei, Taiwan
[4]Chemical Engineering, California State University, Long Beach, CA, 90840

ABSTRACT

This paper reports production of 4.5 μm-diameter monodisperse water drops using a micro electro-mechanical system (MEMS)-based 1 MHz 3-Fourier horn ultrasonic nozzle. The required electrical drive voltage for atomization was 6.5 V at 964±1 kHz that is in good agreements with the values obtained by impedance measurement and by the three-dimensional (3-D) simulation using a commercial finite element analysis program. Such small diameter drops with geometrical standard deviation (GSD) as small as 1.2 and 90% inhale-able fine particle fraction (<5.8 μm-diameter) were achieved in ultrasonic atomization for the first time. Therefore, the MEMS-based MHz ultrasonic nozzles should have potential application to targeted delivery of reproducible doses of medicine to the respiratory system.

Keywords: Ultrasonic atomization, MEMS, Monodisperse drops, Silicon nozzles, Spray

INTRODUCTION

Atomization (spray) refers to the breakup of a volume of liquid into drops (or droplets). Monodisperse drops <10 μm in diameter are highly desirable in nanoparticles synthesis from heat sensitive precursor solutions because they can be processed at a conveniently low temperature and atmospheric pressure. Also, monodisperse particles 1 - 6 μm in diameter have multiple biomedical applications including pulmonary drug delivery and drug preparation for inhalation [1,2]. Although current ultrasonic nebulizers [3] that utilize micro-size mesh filters are capable of producing micron-size drops, the resulting drop-size distributions are rather broad and the electrical drive voltage and power required are relatively high [4]. In order to overcome these shortcomings, the MHz ultrasonic nozzles presented in this paper employ a novel design of multiple Fourier horns in resonance [5], which activates pure capillary wave atomization mechanism, resulting in monodisperse drops at greatly reduced electrical drive power requirement.

Silicon-based ultrasonic nozzles have a number of advantages over conventional metal-based bulk-type ultrasonic nozzles. Silicon possesses a relatively large electro-mechanical coupling coefficient and a high acoustic velocity and, thus, offers the possibility for realization of MHz ultrasonic nozzles. In addition, mass production of any resonator profile can be accomplished by MEMS-based fabrication technology. The first MEMS-based silicon ultrasonic nozzle with a single-horn resonator operating at 72 kHz demonstrated its ability to produce 20-35 μm-diameter drops and its potential application to fuel injection [6]. Some applications such as pulmonary drug delivery require production of much smaller monodisperse drops (i.e. 5 μm or smaller in diameter). Production of such drops demands a much higher ultrasonic frequency because the drop diameter is inversely proportional to the ultrasonic frequency to the 2/3 power [7]. In this paper, the experimental results of atomization using the MHz 3-Fourier horn nozzles are reported. The simulation results on the input impedance of such nozzles are compared to the experimental data. Very good agreements between the simulation predictions and the experimental results have been obtained.

DESIGN AND THREE-DIMENSIONAL (3-D) SIMULATION

As shown in Figure 1, the MEMS-based MHz ultrasonic nozzle is made of a piezoelectric drive section and a 3-Fourier horn silicon-resonator; it has a central channel for liquid flow. The dimensions of the nozzle are also shown in the figure. 3-D simulation was carried out using a commercial finite element method (FEM) (ANSYS Inc., Canonsburg, PA) for vibration mode shape analysis and harmonic analysis. Specific simulation procedures for vibration mode shape analysis were reported previously [5]. The material properties of silicon and PZT used in the simulation were taken from the literature [8,9]. Similar procedure was used in the harmonic analysis except that a voltage was applied. The harmonic analysis yields input impedance values within a specified frequency range of interest. The nozzle resonant frequencies are then identified from the resulting impedance plots. The resonant frequencies determined from the vibration mode shape analysis and that by the harmonic analysis are generally in very good agreement. Possible effect of acoustic losses was neglected in the simulation, however.

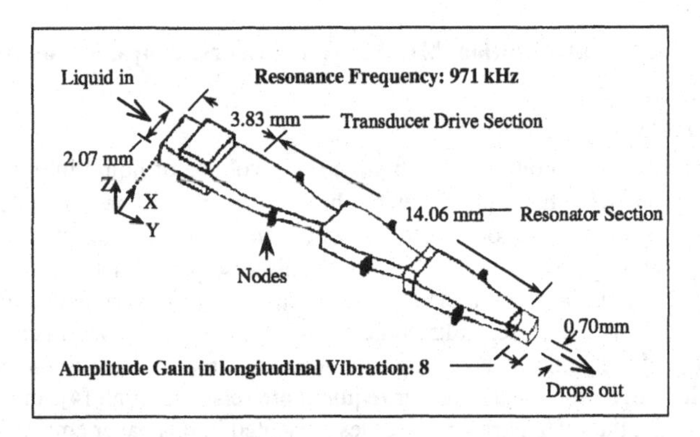

Figure 1 Dimensions and simulation results on the amplitude gain at the nozzle tip of a Si-based 3-Fourier horn 1.0 MHz ultrasonic nozzle with a central channel cross section of 200 x 200 μm^2.

The results of 3-D simulation of vibration mode shape for a 3-Fourier horn nozzle with a design frequency of 1.0 MHz and vibration amplitude magnification of two for each horn [10] are shown in Figure 1. No silver paste layers were included in the simulation. This figure shows that at the resonant frequency of 971 kHz, the three horns in cascade vibrate longitudinally with neither flexural nor lateral motion. Furthermore, the maximum magnitude of longitudinal vibration at each succeeding horn tip increases progressively, resulting in an overall amplitude gain of 8 (theoretical design $2^3 = 8$) at the tip of the 3-horn nozzle [5].

Figure 2 shows the 3-D simulation results for a 1.0 MHz 3-Fourier horn nozzle with and without thin layers of silver paste for bonding of the pair of PZT plates to the two silicon plates in the drive section. This figure shows that the resonance frequency shifted from 971 to 964 kHz when the two 5 μm-thick silver paste layers were included in the simulation with the following material properties for the silver paste: Young's modulus of 5.0 GPa, Poisson's ratio of 0.32, and density of 2710 kg/m^3.

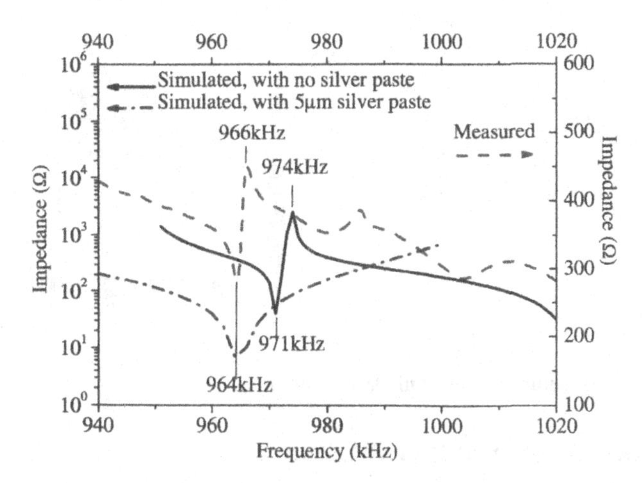

Figure 2 Comparison of impedance measurement (dashed curve) to simulation results for 1.0 MHz 3-horn nozzle: solid curve (no silver paste), dotted-dashed curve (5 μm-thick silver paste).

FABRICATION AND ATOMIZATION SETUP

The nozzles were fabricated using MEMS technology [5]. The central channel of the nozzle for liquid flow (X-axis) was in the direction of the primary flat of the silicon wafer, <110>, and measured 200 μm x 200 μm. The most critical fabrication step was the profiling of the silicon resonator halves (including the base sections where PZT plates are to be bonded) using inductive coupled plasma (ICP) etching. Subsequently, two resonator halves were glued together to form a central rectangular channel (200μm x 200μm) for liquid flow. Two PZT plates, one on each side, were then bonded to the resonator at the base section using silver paste; the central line of the PZT plates was aligned with the nodes of the resonator base section. The PZT plates were connected electrically to a radio frequency signal generator using coated copper wires 50μm in diameter. The 1 MHz three-Fourier horn ultrasonic nozzles measured 1.80 x 0.21 x 0.11 cm^3 and their input impedances were measured using Agilent Precision Impedance Analyzer Model #4294A.

A schematic diagram of the experimental setup for atomization is shown in Figure 3. Major components of the atomization setup are: (1) a PZT drive system to provide an alternating current (AC) electrical signal to the Si-based ultrasonic nozzle, (2) a Syringe Pump (<u>*kd*</u> Scientific Model #101) to provide a constant flow rate of liquid, (3) a CCD camera to take pictures or movies of the spray produced, and (4) a Malvern Particle Sizer Model 2600C (a laser light diffraction-based technique) for analysis of drop size and drop-size distribution [11]. As shown in the figure, the pair of PZT transducers in the nozzle was driven by the amplified AC voltage (by Amplifiers Research Model #75A250) from the Function Generator (Agilent Model 33120A), and the CCD camera provided the flexibility of Progressive Scan and Zoom.

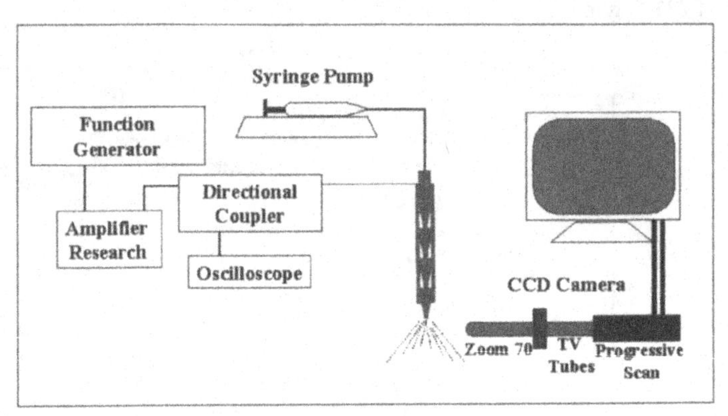

Figure 3 Experimental setup for ultrasonic atomization.

EXPERIMENTAL RESULTS AND DISCUSSION

<u>Impedance analysis</u>

Figure 2 also shows the measured input impedance plot of the microfabricated 1.0 MHz 3-Fourier horn ultrasonic nozzle used in the atomization experiment. Clearly, the resonance frequency of 964 kHz measured by the Agilent Precision Impedance Analyzer is in excellent agreement with that obtained by impedance simulation for the 1.0 MHz e 3-Fourier horn nozzle with 5 μm-thick silver paste layers. This resonance frequency is also in very good agreement with the drive ultrasonic frequency of 964±1 kHz for stable atomization as will be shown in the following subsection.

<u>Atomization</u>

Like the atomization using 0.5 MHz 3-Fourier horn ultrasonic nozzle [12], as water was pumped into the central channel and exits at the nozzle tip that vibrates longitudinally (along X-axis shown in Figure 1), a thin film of liquid formed at the tip of a 1.0 MHz 3-Fourier horn nozzle and a spray of monodisperse drops was produced simultaneously. The novel design of multiple Fourier horns in resonance facilitates pure capillary wave mechanism atomization [12] and, thus, yields monodisperse drops.

Specifically, Figure 4 shows that monodisperse drops 4.5 μm in mass median diameter (MMD) were produced when a drive voltage of 6.5 V at 964±1 kHz was applied to a 3-Fourier horn 1.0 MHz nozzle. The 4.5 μm-MMD measured is in good agreement with the drop diameter of 4.3 μm predicted from the capillary wavelength multiplied by the factor 0.34 for water with surface tension of 72 dyne/cm [7]. The factor of 0.34 has been used widely [4,6] since introduced by Lang in 1962. This figure also shows that 90% of the drops measured by laser diffraction technique (Malvern Particle Sizer 2600c) are smaller than 5.8 μm in diameter and the geometrical standard deviation (GSD) is as small as 1.2. GSD is defined as $\sqrt{\left(D^{84}/D^{50}\right)\left(D^{50}/D^{15}\right)}$, where D^{84}, D^{50}, and D^{15} are the drop diameters at cumulative undersize percentages of 84.13, 50.00, and 15.78, respectively (see Figure 4). It should be noted that the fraction of all particles smaller than 5.8 μm in diameter represents the inhale-able fine particle fraction [13] and aerosol with GSD up to 1.3 is commonly accepted as monodisperse in dust sampler evaluation [14]. Therefore, the MEMS-based MHz ultrasonic nozzle has potential application to targeted delivery of reproducible doses of medicine to the respiratory system.

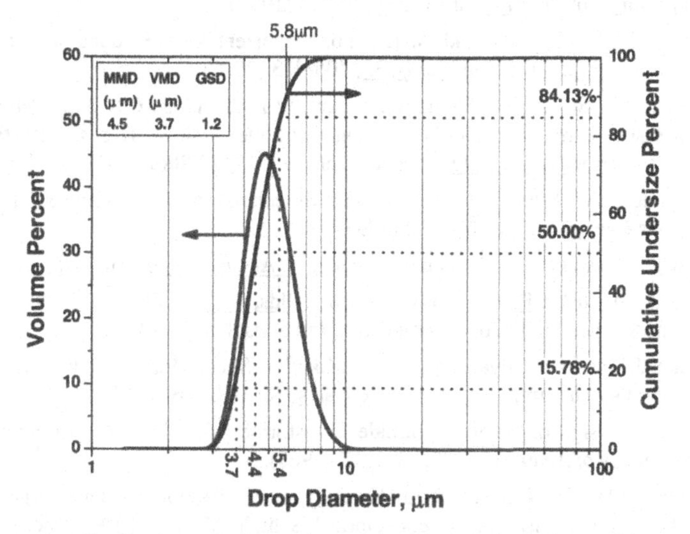

Figure 4 Drop size distributions of ultrasonic atomization of water when a drive voltage of 6.5 V at 964±1 kHz was applied to a 3-Fourier horn 1 MHz nozzle.

CONCLUSIONS

Monodisperse water drops 4.5 μm in diameter have been produced via ultrasonic atomization using the miniaturized 1 MHz 3-Fourier horn silicon ultrasonic nozzles. The 3-D simulations on vibration mode shape and input impedance of the 3-Fourier horn nozzles yield the resonant frequencies of pure longitudinal vibration in good agreement with the experimental values. The novel design of multiple Fourier horns in resonance facilitates pure capillary wave mechanism atomization, resulting in monodisperse drops at greatly reduced electrical drive voltage requirement.

Potential applications of such MHz ultrasonic nozzles include nanoparticles synthesis, 3-D spray coating for micro/nano electronics, pulmonary drug delivery, and drug preparation for inhalation.

ACKNOWLEDGEMENTS

Supports by National Institute of Health (NIH/NIBIB), USA and the National Science Council, Taiwan are gratefully acknowledged.

REFERENCES

[1] Clark, A.R., "Medical Aerosol Inhalers: Past, Present, and Future," Aerosol Science and Technology, 23, 374-391 (1995).

[2] Usmani, O.S., M.F. Biddiscombe, and P.J. Barnes, "Regional Lung Deposition and Bronchodilator Response as a Function of $_2$-Agonist Particle Size," Ame. J. of Respiratory and Critical Care Medicine, 172, 1497-1504 (2005).

[3] Taylor, K.M.G. and O.N.M. McCallion, "Ultrasonic nebulizers for pulmonary drug delivery," Int. J. of Pharmaceutics, 153, 93-104 (1997).

[4] Barreras, F., H. Amaveda, and A. Lozano, "Transient High-Frequency Ultrasonic Water Atomization," Experiments in Fluids, 33, 405-413 (2002).

[5] Tsai, S.C., Y.L. Song, T.K. Tseng, Y.F. Chou, W.J. Chen, and C.S. Tsai, "High Frequency Silicon-Based Ultrasonic Nozzles Using Multiple Fourier Horns," IEEE Trans. on Ultrasonics/Ferroelectrics and Frequency Control, 51, 277-286, 2004.

[6] Lal, A. and R.M. White, "Micromachined Silicon Ultrasonic Atomizer," Proc. of IEEE Ultrasonics Symposium, 1, 339-342 (1996).

[7] Lang, R., "Ultrasonic Atomization of Liquids," J. Acous. Soc. of America, 34, 6-8 (1962).

[8] Auld, B.A. Acoustic Fields and Waves in Solids, Vol. 1, Chapter 8, "Piezoelectricity", Wiley-Interscience Publication, John Wiley and Sons, NY (1973).

[9] Wortman, J.J. and R.A. Evans, "Young's Modulus, Shear Modulus, and Poisson's Ratio in Silicon and Germanium," J. of Applied Physics, 36, 153-156 (1965).

[10] Eisner, E., "Design of Sonic Amplitude Transformers for High Magnification," *J. of the Acoust. Society of America*, 35, 1367-1377 (1963).

[11] Hirleman, E.D., V. Oechsle, and N.A. Chigier, "Response Characteristics of Laser Diffraction Particle Size Analyzers: Optical Sample Volume Extent and Lens Effects," Optical Engineering, 23, 610-619 (1984).

[12] Tsai, S.C., Song, Y.L., Tsai C.S., Chou Y.F., and Cheng C.H., "Ultrasonic Atomization Using MHz Silicon-Based Multiple-Fourier Horn Nozzles," Appl. Phys. Lett., 88, 014102, Jan. 2, 2006 (also Virtual Journal of Nanoscale Science and Technology, January 16, 2006).

[13] Dalby, R., M. Spallek, T. Voshaar, "A review of the development of Respimat Soft Mist™ Inhaler," Int. J. of Pharmaceutics, 283, 1-9 (2004).

[14] Oeseburg, F. and F.M. Benschop, "Aerosol Generator for Sampling Efficiency Determinations under Field Conditions," J. Aerosol Science, 22, 159-180 (1991).

Select Paper from Symposium N

Mater. Res. Soc. Symp. Proc. Vol. 1037 © 2008 Materials Research Society 1037-N02-04

Enabling the Desktop NanoFab with DPN® Pen and Ink Delivery Systems

Joseph S. Fragala[1], R. Roger Shile[1], and Jason Haaheim[2]
[1]MEMS, NanoInk, Inc., 215 E. Hacienda Ave., Campbell, CA, 95008
[2]DPN Applications, NanoInk, Inc., 8025 Lamon Ave, Skokie, IL, 60077

ABSTRACT

Depositing a wide range of materials as nanoscale features onto diverse surfaces with nanometer registration and resolution are challenging requirements for any nanoscale processing system. Dip Pen Nanolithography® (DPN®), a high resolution, scanning probe-based direct-write technology, has emerged as a promising solution for these requirements [1,2]. Many different materials can be deposited directly using DPN, including alkane thiols, metal salts and nanoparticles, metal oxides, polymers, DNA, and proteins. Indirect deposition allows the creation of many interesting nanostructures For instance, mercaptohexadecanoic Acid (MHA) may be deposited via DPN and then used as a template to create arrays of antibodies, which then bond specifically to antigens on the surface of viruses or cells, to create cell or virus arrays. The DPN system is designed to allow registration to existing features on a writing substrate via optical alignment or nanoscale alignment using the core AFM platform. This allows, for instance, the nanoscale deposition of sensor materials directly onto monolithic electronic chips with both sensing and circuit features.

To enable the DPN process, novel pen and ink delivery systems have been designed and fabricated using MEMS technology [3,4]. These MEMS devices bridge the gap between the macro world (instrument) and the nano world (nanoscale patterns). The initial MEMS devices were simple and robust both in design and fabrication to get products into the marketplace quickly. The first MEMS-based DPN device was a passive pen array based on silicon nitride AFM probe technology from Cal Quate's group at Stanford [5,6]. The next two devices (an inkwell chip and a thermal bimorph active pen) were more complicated and took considerable effort to commercialize. In this work, some of the difficulties in bringing brand new MEMS devices from the prototype stage into production will be shared. The subsequent MEMS products have become even more complicated both in design and fabrication, but the development process has improved as well. For example, the 2D nanoPrintArray has 55,000 pens in one square centimeter for high throughput writing over large areas [7]. The 2D arrays enable templated self assembly of nanostructures giving researchers the ability to control the placement of self assembled features rather than allowing the self assembly to occur randomly.

Applications of DPN technology vary from deposition of DNA or proteins in nanoarrays for disease detection or drug discovery, to deposition of Sol-gel metal oxides for gas sensors, and to additive repair of advanced phase-shifting photomasks.

INTRODUCTION

Most observers consider the development of MEMS devices to be straightforward: conceive a product, design it, fabricate it, test it, and it is done. Since most papers on MEMS devices cover few of the problems discovered along the way, this perception is reinforced. But, it is wrong. Many MEMS products take years to develop, and the MEMS devices for enabling DPN are not different. Key elements to reducing the time and number of iterations to get to a final product include the use of evolutionary (rather than novel) designs, using existing subsystems and processes, and employing experienced design, process development and wafer fabrication staff.

DPN can deposit materials with dimensions as fine as 15 nm, but is limited by writing speed throughput. To increase throughput, two paths are available: increasing the rate at which ink comes off the tip; or, increasing the number of tips. Increasing the diffusion rate of the inks is possible but usually difficult, so we have focused on increasing the number of tips. The first 1D pen arrays had up to 26 pens. 1D pen arrays with up to 250 pens have been used by the Mirkin group [8]. The first example of a working 2D pen array had 25,000 pens and the most recent development has 55,000 pens in a one centimeter squared format [9]. The path from 1 to 55,000 working pens has included a few design and process issues needing innovation.

EXPERIMENT

In order to illustrate the development of production-worthy devices for DPN arrays, we introduce first the general microfabrication process which served as the starting point for this development. In particular, this process incorporated the use of a Pyrex handle wafer.

The fabrication process for the 2D array with a Pyrex handle chip begins with a layer of silicon dioxide grown thermally on a silicon wafer and patterned lithographically with 10um square openings. The oxide patterns are then used as an etch mask for an anisotropic silicon etch forming pyramidal tip molds in the silicon wafer. The oxide etch mask is removed in hydrofluoric acid and an LPCVD low stress nitride is deposited onto the wafer. The nitride is patterned lithographically to form the arrays of cantilevers. Separately, a Pyrex wafer is coated with chromium and patterned lithographically to form rectangular openings. The Pyrex is etched in a reactive ion etcher forming recesses to allow the cantilevers free movement, and then the Cr is removed. The nitrided silicon wafer and the Pyrex wafer are then bonded electrostatically and the back of the Pyrex wafer is scribed, 300 um deep, into 1 cm squares by diamond saw. The silicon wafer is etched away in a KOH/water solution leaving the nitride cantilever arrays attached to the Pyrex. Finally, each chip is broken apart from the wafer along the diced scribe lines.

The fabrication process for the 2D array with a silicon handle chip begins with the same silicon nitride deposition on the etched silicon mold wafer. One micron each of lift-off resist and photoresist are then spun onto the mold wafer and patterned. Chromium, platinum and gold (30 nm/60 nm/400 nm thick) are evaporated onto the wafer. The metal on the resist is lifted away using hot resist stripper in an ultrasonic bath. The mold wafer is now ready for gold thermocompression bonding. The handle wafer is patterned lithographically and etched in a reactive ion etcher forming the recesses for the cantilever movement. The handle wafer is oxidized with 1000 nm of thermal silicon oxide. The backside of the handle wafer is patterned with rectangular openings in the oxide, which form blind etched recesses for separating the chips from the wafer after etching. Also, six rectangular openings in the oxide are etched which form through-wafer viewports for allowing tilt and tip leveling of the array. The front side of the handle wafer is spun with 5 microns of lift-off resist and then 1 micron of photoresist. These layers are patterned to form openings on top of the ridges between the etched recesses. Cr/Pt/Au is evaporated and lifted off using the same process as for the nitride wafer. The handle and nitride wafers are then aligned and bonded using Suss MA/BA-6 aligner and SB-6 bonder. The wafers are then etched in a tetramethyl ammonium hydroxide/water solution for 16 to 18 hours until the mold wafer is etched away and the viewports are etched through. The chips are then snapped off the wafer along the etched scribe lines.

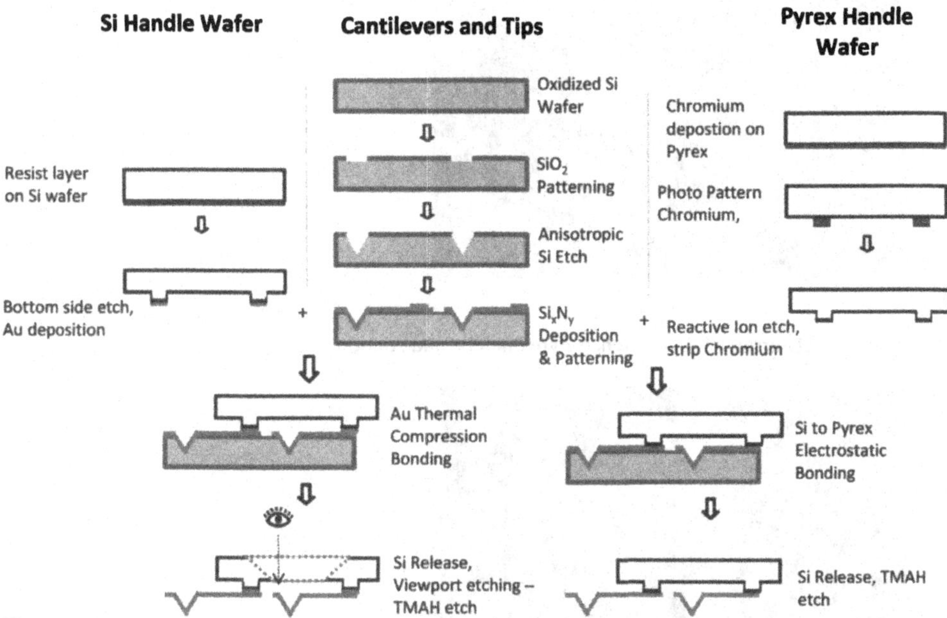

Figure 1. Fabrication processes for 2D array with silicon and Pyrex handle wafers

DISCUSSION

The first attempt at fabricating a 2D array used the same hydrofluoric acid wet etchant as in the single pen process. This gave the Pyrex recess a pleasing, smooth-appearance surface. However, it led to the nitride cantilevers sticking down to the recess surface after etching, due to a phenomenon known in surface micromachining circles as stiction. As a consequence, a new Pyrex dry etch process was developed using trifluoromethane in a Technics Micro-RIE table top tool. The dry etched surface was rough enough to stop the stiction problem from occurring again, with the advantage of better lateral dimension control of recessed patterns.

The next intermittent problem was peeling strips of nitride cantilevers from the Pyrex handle after KOH wet etch to remove the mold wafer. By observing the wafers before the KOH etch was completed, we discovered the Pyrex material, on which the nitride was bonded, was modified by the bonding process, making it etch faster in KOH than the rest of the Pyrex. This problem was solved by replacing the silicon etchant KOH with TMAH, which has higher etch selectivity to silicon oxide. Many 2D arrays were produced with this process, with a high yield and great DPN results.

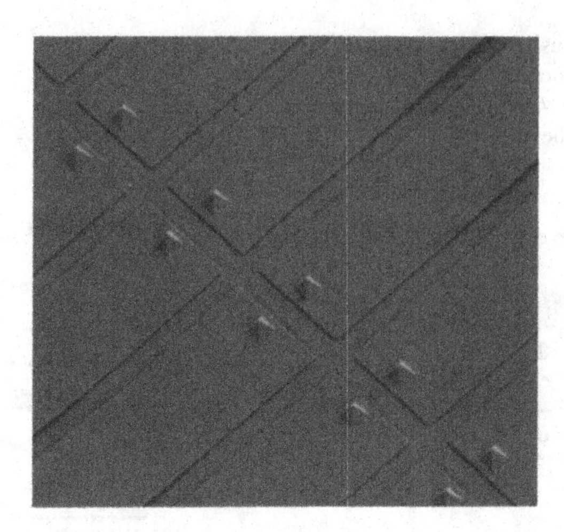

Figure 2. 2D array with wet etched Pyrex handle and stiction

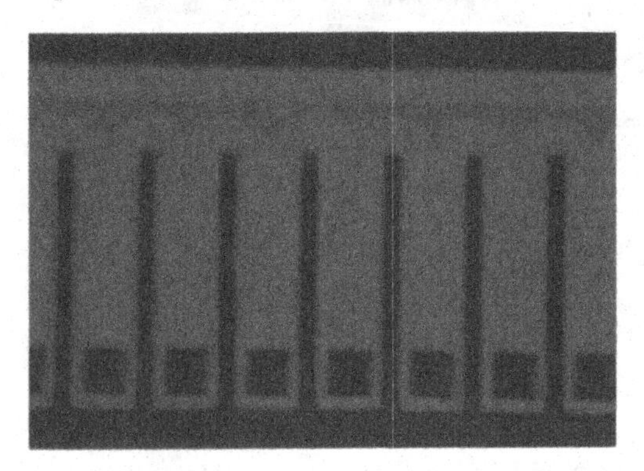

Figure 3. Undercut Pyrex in bonded area under nitride during KOH etching

These early 2D array chips were mounted and leveled on the NSCRIPTOR® a rather crude process. First the chip was coated with ink, then it was placed on a dummy writing surface; next, a drop of epoxy was placed on the back of the Pyrex handle chip, a standard DPN pen chip (with the cantilever snapped off) was mounted and then lowered into the epoxy, and finally the epoxy was allowed to cure. This procedure gave adequate two-axis tilt and tip leveling, but was far from a convenient process for most customers. A plastic mount, called a wedge, was designed with high strength magnets for direct attachment to the NSCRIPTOR. This approach necessitated viewports drilled through the Pyrex handle for viewing the cantilevers during leveling. As the tips touch the surface the cantilever bends and the reflection of ambient light changes, allowing the operator to level the array interactively.

Unfortunately, the laser-drilled holes were too narrow to allow viewing the cantilevers from the back with the existing optics. The handle wafer was then redesigned to be made from silicon so that anisotropically-etched holes could be used for the viewports. Anisotropic etching of silicon along the <111> plane gives holes with a 54.74° sidewall angle. This gives wide viewports at the top (allowing easy viewing of the cantilever on a lower focal plane) with a narrow opening at the bottom (allowing close regular packing of the cantilevers).

Changing to a silicon handle, however, meant replacing anodic bonding with a different wafer bonding process. We chose gold thermocompression bonding after having previously developed a process for DPN Active Pens. Thermocompression bonding had numerous challenges to solve before we had a working process. Ti, a common adhesion layer in metal stacks, could not be used because it would etch away in the HF solution used to remove oxide at the end of the etching process. Cr is the other common adhesion layer, but diffuses quickly through gold grain boundaries and poisons the bonding surface. Platinum was therefore added as a diffusion barrier between the Cr and the Au. Non-uniform bonding was fixed by resurfacing and aligning the bonding plates in the wafer bonding machine and using 0.5 mm thick graphite paper between the wafer and the top bonding plate.

The RIE process for etching the recesses in the silicon handle briefly caused a recurrence of stiction until the etch depth of the recess was increased. To allow a deeper recess, a thick (> 5 microns) lift-off layer process was developed to allow spinning of resist over the recess topology.

The next problem was cantilever breakage towards the end of the etching process. This problem was traced to etching from both sides of the mold wafer partially freeing the cantilevers and then tearing them off the handle wafer as the thinned mold wafer flexed from intrinsic stresses. This problem was solved by oxidizing the mold wafer after the nitride was patterned but before the metals were deposited.

Figure 4. 2D array cantilever breakage

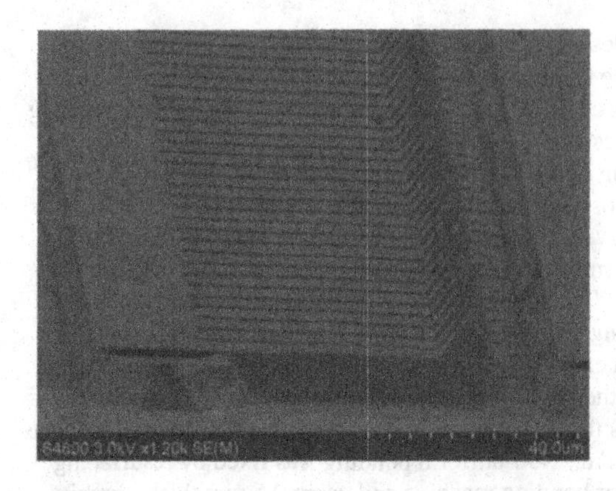

Figure 5. High yield 2D array scanning electron micrograph

After fabrication, the 2D array chips were mounted on customized tooling and attached to the NSCRIPTOR for test. The test was performed by writing with octadecane thiol (ODT) onto a silicon substrate coated with a thin film of gold. The ODT bonds to the gold and was used as a resist to wet etch the gold giving a high contrast image.

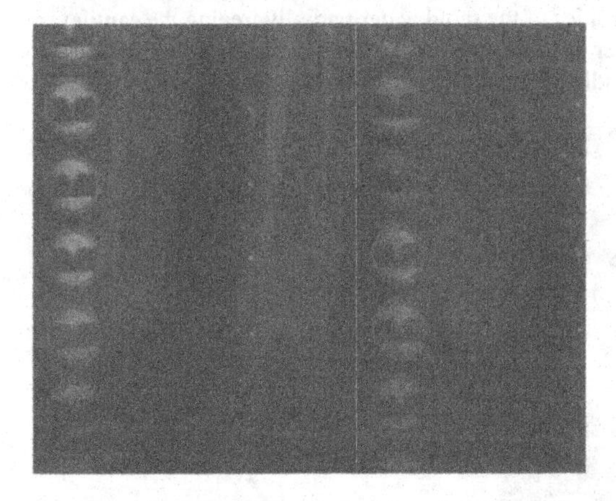

Figure 6. DPN of the NanoInk logo with 80 nm pixel size

CONCLUSIONS

Iterative design and process optimization were performed for a 2D array of 55,000 DPN pens. High yield arrays were mounted on the NSCRIPTOR instrument and used to write complex patterns with nanometer resolution. Potential uses for these arrays include templating for indirect assembly of biological materials such as proteins and DNA, fabrication of custom substrates for SERS (Surface Enhanced Raman Spectroscopy), and templates for control and orientation of other nanoscale objects such as carbon nanotubes.

ACKNOWLEDGMENTS

The authors thank the DPN Applications, Instrumentation and MEMS teams for their help, especially Pam Simao for her superb MEMS fabrication work. The authors wish to thank Chad Mirkin for his vision and support. The authors wish to thank Al Henning for his comments and discussion. This work was supported by the DARPA Applications of Molecular Electronics, BAA number 05-19. The authors wish to thank Program Managers Morley Stone and Cindy Daniell for their support.

REFERENCES

1. R.D. Piner, J. Xu, F. Xu, S. Hong, C.A. Mirkin, "Dip-Pen Nanolithography", *Science* **283**, 661-663 (1999).
2. C.A. Mirkin, "Dip-Pen Nanolithography: Automated Fabrication of Custom Multi-component, Sub-100-Nanometer Surface Architectures," *MRS Bull.*, , **26**, 535-538 (2001).
3. M. Zhang, D. Bullen, S-W. Chung, S. Hong, K. Ryu, Z. Fan, C. Mirkin, C. Liu, "A MEMS nanoplotter with high-density parallel dip-pen nanolithography probe arrays," *Nanotechnology*, **13**, 212-217, (2002).
4. D. Bullen, S-W. Chung, X. Wang, J. Zou, C. Mirkin, C. Liu, "Parallel dip pen nanolithography with arrays of individually addressable cantilevers," *Applied Physics Letters*, **84**, No. 5, 789-791, (2004).
5. R.J. Grow, S.C. Minne, S.R. Manalis, C.F. Quate, "Silicon Nitride Cantilevers with Oxidation Sharpened Tips for Atomic Force Microscopy," *J. MEMS*, **11**, No. 4, 317-321 (2002).
6. T.R. Albrecht, S. Akamine, T.E. Carver, C.F. Quate, "Microfabrication of Cantilever Styli for the Atomic Force Microscope," *J. Vac. Sci. A*, (1990)
7. K. S. Salaita, Y. Wang, J. Fragala, C. Liu, C.A. Mirkin, "Massively Parallel Dip-Pen Nanolithography with 55,000-Pen Two-Dimensional Arrays," *Angew. Chem. Int. Ed.*, **45**, 7220-7223 (2006).
8. K. Salaita, S.W. Lee, L. Huang, C. A. Mirkin, "Sub-100 nm, Centimeter-Scale, Parallel Dip-Pen Nanolithography," *Small*, 1, No. 10, 940-945 (2005).
9. K. Salaita, Y. Wang, C.A. Mirkin, "Applications of dip-pen nanolithography," *Nature Nanotechnology*, **2**, 145-155 (2007).

AUTHOR INDEX

SUBJECT INDEX

Printed in the United States
By Bookmasters